Fundamentals of Enzyme Kinetics

To Marilú

Fundamentals of Enzyme Kinetics

Third edition

by Athel Cornish-Bowden

PORTLAND PRESS

Published by Portland Press Ltd.

Portland Press Ltd.
59 Portland Place
London W1B 1QW
U.K.
e-mail: editorial@portlandpress.com
www.portlandpress.com

British Library Cataloguing-in-Publication Data
A catalogue record for this book is available from the British Library

ISBN 1 85578 158 1

Typeset by SR Nova Private Limited, Bangalore, India
Printed by Information Press Ltd, Oxford, UK

Contents

3   Practical Aspects of Kinetic Studies

6 Tight-binding and Irreversible Inhibitors

7 Reactions of More than One Substrate

8   Use of Isotopes for Studying Enzyme Mechanisms

9   Effect of pH on Enzyme Activity

12   Kinetics of Multi-Enzyme Systems

13　Fast Reactions

14　Estimation of Kinetic Constants

Preface to the Third Edition

It was said of the great statistician R. A. Fisher (and doubtless of other scientists of similar stature) that whenever he introduced a result with the words "it can easily be shown that..." one could be sure that two or three hours of hard work would be in store for anyone wishing to verify it. As a student I thought that many authors used this formula as a way to avoid explaining things that they could not explain. I hasten to add that in Fisher's case I am sure there was no lack of ability, though there may have been a lack of appreciation of the difficulties that his readers had. When I was writing the earliest version of this book, therefore, I made a resolution never to claim that anything was easy unless I was quite sure that it was. In the more than 25 years that have passed since then I believe I have kept this resolution, though sometimes I have had to revise my views about what was simple enough to be left unexplained. Each revision has thus involved making some steps in logical or algebraic arguments more explicit, and the present revision is no exception. Above all I have striven for clarity, being guided by a slogan from Keith Laidler (1998): "Correctness, cogency, clarity: these three, but the greatest of these is clarity". Errors can be corrected, weak arguments can be strengthened, but lack of clarity leaves a fog that may take years to dispel.

The emphasis throughout is on understanding enzyme kinetics, not on covering every aspect of the subject in an encyclopaedic style. So I have preferred to describe the principles that will allow readers to proceed as far as they want in any direction. In the words of Kuan-tzu (as quoted by Parzen, 1980): "If you give a man a fish, he will have a single meal; if you teach him how to fish, he will eat all his life".

I make no apology for continuing to illustrate concepts with abundant graphs, including the straight-line graphs that biochemists have used for three-quarters of a century. A recently published book on enzyme kinetics bases its claim to offer a radically new treatment on the complete avoidance of these until late in the book (mentioning them then only to disparage them); more generally, many scientists argue that the universal availability of computers has made graphical methods obsolete. It is more difficult to find professional statisticians with a real knowledge of data analysis who would agree, because they know that fooling a computer is far easier than fooling the human eye, and that fooling the eye with a table of numbers is easier than fooling it with a graph. Given that there is little to "see" in a biochemical experiment and that almost all our information comes at second hand from instruments, it is essential to convert it into something visible by means of well-chosen graphs. At the same time judicious use of the computer is equally necessary — not just graphs, not just computation, but both, in partnership — and in this spirit I have not only retained but have expanded the final chapter of the book, which has been a well received feature of the earlier editions.

This edition has benefitted greatly from the comments of many people who have read it in all or in part: Robert Alberty, Keith Brocklehurst, Marilú Cárdenas, Robert Eisenthal, David Fell, Herbert Friedmann, Marc Jamin, Jannie Hofmeyr, Keith Laidler, Gösta Pettersson, Ana Ponces, Valdur Saks, Marius Schmidt and Keith Tipton. I have not followed all of their suggestions, and they are anyway not responsible for any faults that remain, but I have followed most of them, and I am extremely grateful for all of their comments. It is a pleasure also to acknowledge Dr. Mireille Bruschi, the Director of the Laboratory of *Ingénierie et Bioénergétique des Protéines* of the CNRS, for her general support and for her success in creating a congenial working environment.

It would be nice to think that there were no typographical or other errors in this book. Nice, yes, but if past experience is any guide, not very realistic, so a list of corrections will be maintained at http://bip.cnrs-mrs.fr/bip10/fek.htm. That there are as few errors as there are is due in no small measure to the efforts of Marilú Cárdenas, who spent more than forty hours with me in checking the proofs.

Athel Cornish-Bowden

Marseilles, April 2003

Chapter 1

Basic Principles of Chemical Kinetics

1.1 Abbreviations and symbols

This book follows as far as possible the recommendations of the International Union of Biochemistry and Molecular Biology (International Union of Biochemistry, 1982). However, as these allow some latitude and anyway do not cover all of the cases that we shall need it is useful to begin by noting some points that apply generally in the book. First of all, it is important to recognize that a chemical substance and its concentration are two different entities and need to be represented by different symbols. The recommendations allow square brackets around the chemical name to be used without definition for its concentration, so [glucose] is the concentration of glucose, [A] is the concentration of a substance A, and so on. In this book I shall use this convention for names that consist of more than a single letter, but it has the disadvantage that the profusion of square brackets can lead to forbiddingly complicated equations in enzyme kinetics (see some of the equations in Chapter 7, for example, and imagine how they would look with square brackets). Two simple alternatives are possible: one is just to put the name in italics, so the concentration of A is A, for example, and this accords well with the standard convention that chemical names are written in roman (upright) type and algebraic symbols are written in italics. However, experience shows that many readers barely notice whether a particular symbol is roman or italic, and so it discriminates less well between the two kinds of entity than one would hope. For this reason I shall use the lower-case italic letter that corresponds to the symbol for the chemical entity, so a is the concentration of A, for example. If the chemical symbol has any subscripts these apply unchanged to the concentration symbol, so a_0 is the concentration of A_0, for example. Both of these systems (and others) are permitted by the recommendations as long as each symbol is defined when first used. This provision is satisfied in this book, and it is good to follow it in general, because almost nothing that authors consider obvious is perceived as obvious by all their readers. In the problems at the ends of the chapters, incidentally, the symbols may not be the same as those used in the corresponding chapters: this is intentional, because in the real world one cannot always expect the questions that one has to answer to be presented in familiar terms.

As we shall see, an enzyme-catalysed reaction virtually always consists of two or more steps, and as we shall need symbols to refer to the different steps it is necessary to have some convenient indexing system to show which symbol refers to which step. The recommendations do not impose any particular system, but, most important, they do require the system in use to be stated. Because of the different ways in which, for example, the symbol k_2 has been used in the biochemical literature one should never assume in the absence of a clear definition what is intended. The system preferred by the recommendations is used in this book: for a reaction of n steps, these are numbered $1, 2 \ldots n$; lower-case italic k with a positive subscript refers to the kinetic properties of the forward step corresponding to the subscript, for example, k_2 refers to the forward direction of the second step; the same with a negative subscript refers to the corresponding reverse reaction, for example, k_{-2} for the second step; a capital italic K with a subscript refers to the thermodynamic (equilibrium) properties of the whole step and is typically the ratio of the two kinetic constants, for example, $K_2 = k_2/k_{-2}$.

Finally a word on abbreviations. Much of the modern literature is rendered virtually unintelligible to non-specialist readers by a profusion of unnecessary abbreviations. These save little space, and little work (because with modern word-processing equipment it takes no more than a few seconds to expand all of the abbreviations that one may have found it convenient to use during preparation), but the barrier to comprehension that they represent is formidable. Abbreviations have accordingly been completely eliminated from this book, apart from a few survivors (like ATP) that are more easily understood by most biochemists than the words they stand for.

1.2   Order of a reaction

1.2.1   Order and molecularity

Chemical kinetics as a science began in the middle of the nineteenth century, when Wilhelmy (1850) was apparently the first to recognize that the rate at which a chemical reaction proceeds follows definite laws, but his work attracted little attention until it was taken up by Ostwald towards the end of the century (see Laidler, 1993). Wilhelmy realized that chemical rates depended on the concentrations of the reactants, but before considering some examples we need to examine how chemical reactions can be classified.

One way is according to the *molecularity*, which defines the number of molecules that are altered in a reaction: a reaction A → P is *unimolecular* (sometimes called *monomolecular*), and a reaction A + B → P is *bimolecular*. One-step reactions of higher molecularity are extremely rare, if they occur at all, but a reaction A + B + C → P would be *trimolecular* (or *termolecular*). Alternatively one can classify a reaction according to its *order*, a description of its kinetics that defines how many concentration terms must be multiplied together to get an expression for the rate of reaction. Hence, in a *first-order reaction* the rate is proportional to one concentration; in a *second-order reaction* it is proportional to the product of two concentrations or to the square of one concentration; and so on.

For a simple reaction that consists of a single step, or for each step in a complex reaction, the order is usually the same as the molecularity (though this may not be apparent if one concentration, for example that of the solvent if it is also a reactant, is so large that it is effectively constant). However, many reactions consist of sequences of unimolecular and bimolecular steps, and the molecularity of the complete reaction need not be the same as its order. Indeed, a complex reaction often has no meaningful order, as the rate often cannot be expressed as a product of concentration terms. As we shall see in later chapters, this is almost universal in enzyme kinetics, where not even the simplest enzyme-catalysed reactions have simple orders. Nonetheless, the individual steps in enzyme-catalysed reactions nearly always do have simple orders, usually first or second order, and the concept of order is important for understanding enzyme kinetics. The binding of a substrate molecule to an enzyme molecule is a typical example of a second-order bimolecular reaction in enzyme kinetics, whereas conversion of an enzyme–substrate complex into products or into another intermediate is a typical example of a first-order unimolecular reaction.

The rate v of a first-order reaction A → P can be expressed as

$$v = \frac{dp}{dt} = -\frac{da}{dt} = ka = k(a_0 - p) \tag{1.1}$$

in which a and p are the concentrations of A and P respectively at any time t, k is a *first-order rate constant*[*] and a_0 is a constant. The first two equality signs

[*]Some authors, especially those with a strong background in physics, object to the term "rate constant" (preferring "rate coefficient") for quantities like k in equation 1.1 and for many similar quantities that will occur in this book, on the perfectly valid grounds that they are not constant, because they vary with temperature and with many other conditions. However, the use of the word "constant" to refer to quantities that are constant only under highly restricted conditions is virtually universal in biochemical kinetics (and far from unknown in chemical kinetics), and it is hardly practical to abandon this usage in this book. See also the discussion at the end of Section 9.4.3.

in this equation represent alternative definitions of the rate v: because every molecule of A that is consumed becomes a molecule of P, it makes no difference to the mathematics whether the rate is defined in terms of the appearance of product or disappearance of reactant. It may make a difference experimentally, however, because experiments are not done with perfect accuracy, and the relative changes in p are much larger than those in a in the early stages of a reaction. For this reason it will usually be more accurate to measure increases in p than decreases in a.

The third equality sign in the equation is the one that specifies that this is a first-order reaction, because it states that the rate is proportional to the concentration of reactant A. Finally, if the time zero is defined in such a way that $a = a_0$ and $p = 0$ when $t = 0$, the stoicheiometry allows the values of a and p at any time to be related according to the equation $a + p = a_0$, thereby allowing the last equality in the equation.

Equation 1.1 can readily be integrated by separating the two variables p and t, bringing all terms in p to the left-hand side and all terms in t to the right-hand side:

$$\int \frac{dp}{a_0 - p} = \int k dt \tag{1.2}$$

therefore

$$-\ln(a_0 - p) = kt + \alpha \tag{1.3}$$

in which α, the constant of integration, can be evaluated by noting that there is no product at the start of the reaction, so $p = 0$ when $t = 0$. Then $\alpha = -\ln(a_0)$, and so

$$\ln\left(\frac{a_0 - p}{a_0}\right) = -kt \tag{1.4}$$

Taking exponentials of both sides we have

$$(a_0 - p)/a_0 = e^{-kt} \tag{1.5}$$

which can be rearranged to give

$$p = a_0(1 - e^{-kt}) \tag{1.6}$$

Notice that the constant of integration α was included in this derivation, evaluated and found to be non-zero. Constants of integration must always be included and evaluated when integrating kinetic equations; they are rarely found to be zero.

The commonest type of bimolecular reaction is one of the form A + B → P + Q, in which two different kinds of molecule A and B react to give products. In this example the rate is likely to be given by a second-order expression of the form

$$v = \frac{dp}{dt} = kab = k(a_0 - p)(b_0 - p) \tag{1.7}$$

in which k is now a *second-order rate constant*. (Conventional symbolism does not, unfortunately, indicate the order of a rate constant.) Again, integration is readily achieved by separating the two variables p and t:

$$\int \frac{dp}{(a_0 - p)(b_0 - p)} = \int k dt \tag{1.8}$$

For readers with limited mathematical experience, the simplest and most reliable method for integrating the left-hand side of this equation is to look it up in a standard table of integrals. It may also be done by multiplying both sides of the equation by $(b_0 - a_0)$ and separating the left-hand side into two simple integrals:

$$\int \frac{dp}{a_0 - p} - \int \frac{dp}{b_0 - p} = \int (b_0 - a_0) k dt \tag{1.9}$$

Hence

$$-\ln(a_0 - p) + \ln(b_0 - p) = (b_0 - a_0)kt + \alpha \tag{1.10}$$

Putting $p = 0$ when $t = 0$ we find $\alpha = \ln(b_0/a_0)$, and so

$$\ln\left[\frac{a_0(b_0 - p)}{b_0(a_0 - p)}\right] = (b_0 - a_0)kt \tag{1.11}$$

or

$$\frac{a_0(b_0 - p)}{b_0(a_0 - p)} = e^{(b_0 - a_0)kt} \tag{1.12}$$

The following special case of this result is of interest: if a_0 is negligible compared with b_0, then p must also be negligible compared with b_0 at all times, because p can never exceed a_0 on account of the stoicheiometry of the reaction. So $(b_0 - a_0)$ and $(b_0 - p)$ can both be written with good accuracy as b_0 and equation 1.12 simplifies to

$$p = a_0(1 - e^{-kb_0 t}) \tag{1.13}$$

This is of exactly the same form as equation 1.6, the equation for a first-order reaction. This type of reaction is known as a *pseudo-first-order reaction*, and kb_0 is a *pseudo-first-order rate constant*. Pseudo-first-order conditions occur naturally when one of the reactants is the solvent, as in most hydrolysis reactions, but it is also advantageous to create them deliberately, to simplify evaluation of the rate constant (see Section 1.5).

A trimolecular reaction, such as $A + B + C \rightarrow P + \ldots$, does not normally consist of a single trimolecular step involving a three-body collision, which would be inherently unlikely; consequently it is not usually third-order. Instead it is likely to consist of two or more *elementary steps*, such as $A + B \rightarrow X$ followed by $X + C \rightarrow P$. In some reactions the kinetic behaviour as a whole is largely determined by the rate constant of the step with the smaller rate constant, accordingly known as the *rate-limiting step* (or, more objectionably, as the *rate-determining step*). However, although such terms are widespread in chemistry they involve some conceptual confusion (discussed in Section 13.1.3) and as far as possible are best avoided. When there is no clearly defined rate-limiting step the rate equation is typically complex, with no integral order. Some trimolecular reactions do display third-order kinetics, however, with $v = kabc$, where k is now a *third-order rate constant*, but it is *not* necessary to assume a three-body collision to account for third-order kinetics. Instead, we can assume a two-step mechanism, as above but with the first step rapidly reversible, so that the concentration of X is given by $x = Kab$, where K is the equilibrium constant for binding of A to B, the association constant of X. The rate of reaction is then the rate of the slow second step:

$$v = k'xc = k'Kabc \tag{1.14}$$

where k' is the second-order rate constant for the second step. Hence the observed third-order rate constant is actually the product of a second-order rate constant and an equilibrium constant.

Some reactions are observed to be of *zero order*, with a constant rate, independent of the concentration of reactant. If a reaction is zero order with respect to only one reactant, this may simply mean that the reactant enters the reaction after the rate-limiting step. However, some reactions are zero order overall, which means that they are independent of all reactant concentrations. These are invariably catalysed reactions and occur if every reactant is present in such large excess that the full potential of the catalyst is realized. Enzyme-catalysed reactions commonly approach zero-order kinetics as the limit at very high reactant concentrations.

1.2.2   Determination of the order of a reaction

The simplest means of determining the order of a reaction is to measure the rate v at different concentrations a of the reactants. A plot of $\ln v$ against $\ln a$ is then a straight line with slope equal to the order. If all the reactant concentrations are altered in a constant ratio, the slope of the line is the overall order. It is useful to know the order with respect to each reactant, however, and this can be found by altering the concentration of each reactant separately, keeping the other concentrations constant. The slope of the line is then equal to the order with respect to the variable reactant. For example, if the reaction is second-order in A and first-order in B,

$$v = ka^2b \tag{1.15}$$

then

$$\ln v = \ln k + 2\ln a + \ln b \tag{1.16}$$

Hence a plot of $\ln v$ against $\ln a$ (with b held constant) has a slope of 2, and a plot of $\ln v$ against $\ln b$ (with a held constant) has a slope of 1. These plots are illustrated in Figure 1.1. If the plots are drawn with the slopes measured from the progress curve (as plots of concentration against time), the concentrations of *all* the reactants change with time. Therefore, if valid results are to be obtained, either the initial concentrations of the reactants must be in stoicheiometric ratio, in which event the overall order is found, or (more usually) the "constant" reactants must be in large excess at the start of the reaction, so that the changes in their concentrations are insignificant. If neither of these alternatives is possible or convenient, the rates must be obtained from a set of measurements of the slope at zero time, that is to say measurements of initial rates. This method is usually preferable for kinetic measurements of enzyme-catalysed reactions, because the progress curves of enzyme-catalysed reactions often do not rigorously obey simple rate equations for extended periods of time. The progress curve of an enzyme-catalysed

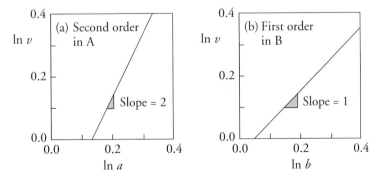

Figure 1.1. Determination of the order of reaction. The lines are drawn for a reaction that is second-order in A and first-order in B, so the slopes of the plots are 2 and 1 respectively. The appearance of the plots (though not the numerical values) would be the same if logarithms to base 10 or any other base were used instead of natural logarithms, provided that the same changes were made in both coordinates.

reaction (see Section 2.9) often requires a more complicated equation than the integrated form of the rate equation derived for the initial rate, because of progressive loss of enzyme activity, inhibition by accumulating products and other effects.

1.3   Dimensions of rate constants

Dimensional analysis provides a simple and versatile technique for detecting algebraic mistakes and checking results. It depends on the existence of a few simple rules governing the permissible ways of combining quantities of different dimensions, and on the frequency with which algebraic errors result in dimensionally inconsistent expressions. Concentrations can be expressed in M (or $mol\,l^{-1}$), and reaction rates in $M\,s^{-1}$. In an equation that expresses a rate v in terms of a concentration a as $v = ka$, therefore, the rate constant k must be expressed in s^{-1} if the left- and right-hand sides of the equation are to have the same dimensions. All first-order rate constants have the dimensions of $time^{-1}$, and by a similar argument second-order rate constants have the dimensions of $concentration^{-1} \times time^{-1}$, third-order rate constants have the dimensions of $concentration^{-2} \times time^{-1}$, and zero-order rate constants have the dimensions of $concentration \times time^{-1}$.

Knowledge of the dimensions of rate constants allows the correctness of derived equations to be checked easily: the left- and right-hand sides of any equation (or inequality) must have the same dimensions, and all terms in a summation must have the same dimensions. For example, if $(1 + t)$ occurs in an equation, where t has the dimensions of time, then the equation is incorrect, even if the "1" is intended to represent a time that happens to have the numerical value of 1. Rather than mixing dimensioned constants and variables in an expression in this way it is better to write the unit after the number, $(1\,s + t)$ for example, or to give the constant a symbol, $(t_0 + t)$ for example, with a note in the text defining t_0 as $1\,s$. Although both alternatives appear more clumsy than just writing $(1 + t)$ they avoid confusion. Section 8.6.1 contains an example, equation 8.17, where clarity requires inclusion of units inside an equation.

Quantities of different dimensions can be multiplied or divided, but must not be added or subtracted. Thus, if k_1 is a first-order rate constant and k_2 is a second-order rate constant, a statement such as $k_1 \gg k_2$ is meaningless, just as $5\,g \gg 25\,°C$ is meaningless. However, a pseudo-first-order rate constant such as k_2a has the dimensions of $concentration^{-1} \times time^{-1} \times concentration$, which simplifies to $time^{-1}$; it therefore has the dimensions of a first-order rate constant, and *can* be compared with other first-order rate constants.

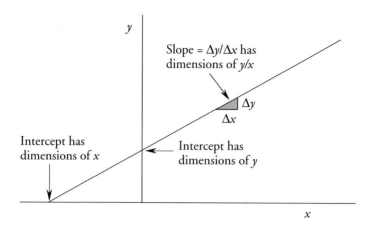

Figure 1.2. Application of dimensional analysis to graphs.

Another major principle of dimensional analysis is that one must not use a dimensioned quantity as an exponent or take its logarithm. For example, e^{-kt} is permissible, if k is a first-order rate constant, but e^{-t} is not. An apparent exception is that it is often convenient to take the logarithm of what appears to be a concentration, for example when pH is defined as $-\log[H^+]$. The explanation is that the definition is not strictly accurate and to be dimensionally correct one should define pH as $-\log\{[H^+]/[H^+]^0\}$, where $[H^+]^0$ is the value of $[H^+]$ in the standard state, corresponding to pH $= 0$. As $[H^+]^0$ has a numerical value of 1 it is usually omitted from the definition. Whenever one takes the logarithm of a dimensioned quantity in this way, a standard state is implied whether stated explicitly or not.

Dimensional analysis is particularly useful as an aid to remembering the slopes and intercepts of commonly used plots, and the rules are simple: any intercept must have the same dimensions as whatever variable is plotted along the corresponding axis, and a slope must have the dimensions of the ordinate (y) divided by those of the abscissa (x). These rules are illustrated in Figure 1.2.

1.4   Reversible reactions

All chemical reactions are reversible in principle, and for many the reverse reaction is readily observable and must be allowed for in the rate equation:

$$\underset{a_0 - p}{A} \underset{k_{-1}}{\overset{k_1}{\rightleftharpoons}} \underset{p}{P} \tag{1.17}$$

In this case,

$$v = \frac{dp}{dt} = k_1(a_0 - p) - k_{-1}p = k_1 a_0 - (k_1 + k_{-1})p \tag{1.18}$$

This differential equation is of exactly the same form as equation 1.1, and can be solved in the same way:

$$\int \frac{dp}{k_1 a_0 - (k_1 + k_{-1})p} = \int dt \tag{1.19}$$

Therefore

$$\frac{\ln[k_1 a_0 - (k_1 + k_{-1})p]}{-(k_1 + k_{-1})} = t + \alpha \tag{1.20}$$

Setting $p = 0$ when $t = 0$ gives $\alpha = -\ln(k_1 a_0)/(k_1 + k_{-1})$, and so

$$\ln\left[\frac{k_1 a_0 - (k_1 + k_{-1})p}{k_1 a_0}\right] = -(k_1 + k_{-1})t \tag{1.21}$$

Taking exponentials of both sides, we have

$$\frac{k_1 a_0 - (k_1 + k_{-1})p}{k_1 a_0} = e^{-(k_1 + k_{-1})t} \tag{1.22}$$

which can be rearranged to give

$$p = \frac{k_1 a_0 [1 - e^{-(k_1 + k_{-1})t}]}{k_1 + k_{-1}} = p_\infty [1 - e^{-(k_1 + k_{-1})t}] \tag{1.23}$$

where $p_\infty = k_1 a_0/(k_1 + k_{-1})$. This is the value of p after infinite time, because the exponential term approaches zero as t becomes large.

1.5   Determination of first-order rate constants

It is common for a reaction to be first-order in every reactant, and it is then often possible to carry it out under pseudo-first-order conditions overall by keeping every reactant except one in large excess. In many practical situations, therefore, the problem of determining a rate constant can be reduced to the problem of determining a first-order rate constant. We have seen in equation 1.6 that for a simple first-order reaction,

$$p = a_0(1 - e^{-kt}) \tag{1.24}$$

and in the more general case of a reversible reaction, equation 1.23:

$$p = p_\infty [1 - e^{-(k_1 + k_{-1})t}] \tag{1.25}$$

So

$$p_\infty - p = p_\infty e^{-(k_1 + k_{-1})t} \tag{1.26}$$

Therefore,

$$\ln(p_\infty - p) = \ln p_\infty - (k_1 + k_{-1})t \tag{1.27}$$

Thus a plot of $\ln(p_\infty - p)$ against t gives a straight line of slope $-(k_1 + k_{-1})$. Before pocket calculators became universally available this was usually expressed in terms of logarithms to base 10:

$$\log(p_\infty - p) = \log p_\infty - \frac{(k_1 + k_{-1})t}{2.303} \tag{1.28}$$

so that a plot of $\log(p_\infty - p)$ against t gives a straight line of slope $-(k_1 + k_{-1})/2.303$. However, it is nowadays just as convenient to retain the form in terms of natural logarithms[*].

Guggenheim (1926) pointed out a major objection to this plot: it depends heavily on an accurate value of p_∞. In the general case of a reversible reaction with p_∞ different from a_0 an accurate value of p_∞ is difficult to obtain, and even in the special case of an irreversible reaction with p_∞ identical to a_0 the instantaneous concentration of A at zero time may be difficult to measure accurately. Guggenheim suggested measuring two sets of values p_i and p_i' at times t_i and t_i', such that every $t_i' = t_i + \tau$, where τ is a constant. Then, from equation 1.26,

$$p_\infty - p_i = p_\infty e^{-(k_1 + k_{-1})t_i} \tag{1.29}$$

$$p_\infty - p_i' = p_\infty e^{-(k_1 + k_{-1})(t_i + \tau)} \tag{1.30}$$

By subtraction,

$$p_i' - p_i = p_\infty [1 - e^{-(k_1 + k_{-1})\tau}] e^{-(k_1 + k_{-1})t_i} \tag{1.31}$$

---

[*]An argument could be made for dispensing with common logarithms (to base 10) altogether in modern science, as they are now virtually never used as an aid to arithmetic. However, this will hardly be practical as long as the pH scale continues to be used, and in historical references, such as that in the legend of Figure 2.1, it would be incorrect to imply that natural logarithms were used if they were not. Finally, when graphs need to span several orders of magnitude (as in Figure 2.4) it is much easier for the reader to interpret a scale marked in decades than in powers of e. Otherwise, however, there usually is no good reason to use common logarithms, and then, as in Figure 1.1, they have been replaced in this edition with natural logarithms.

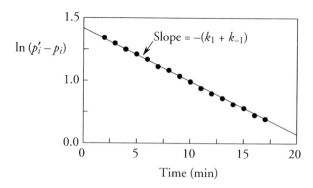

Figure 1.3. The Guggenheim plot. This plot allows a first-order rate constant to be determined without requiring an accurate value for the degree of reaction at equilibrium. Symbols are as follows: p, p', concentrations of product at times t and $t + \tau$ respectively, where τ is a constant.

Taking logarithms,

$$\ln(p_i' - p_i) = \ln p_\infty + \ln[1 - e^{-(k_1 + k_{-1})\tau}] - (k_1 + k_{-1})t_i \qquad (1.32)$$

This has the form

$$\ln(p_i' - p_i) = \text{constant} - (k_1 + k_{-1})t_i \qquad (1.33)$$

So a plot of $\ln(p_i' - p_i)$ against t_i gives a straight line of slope $-(k_1 + k_{-1})$, as illustrated in Figure 1.3. It is known as a *Guggenheim plot*, and does not require an estimate of p_∞. As k_1/k_{-1} is equal to the equilibrium constant, which can be estimated independently, the values of the individual rate constants k_1 and k_{-1} can be calculated from the two combinations.

The Guggenheim plot is insensitive to deviations from first-order kinetics: it can give an apparently good straight line even if first-order kinetics are not accurately obeyed. For this reason it should not be used to determine the order of reaction, which should be established independently. The same comment applies to the related Kézdy–Swinbourne plot, the subject of Problem 1.3 at the end of this chapter.

1.6 Catalysis

To this point we have discussed the dependence of reaction rates on concentrations as if the only concentrations that needed to be considered were those of the reactants, but this is obviously too simple: more than two centuries ago Fulhame (1794) noted that many reactions would not proceed at a detectable rate unless the mixture contained certain necessary non-reactant components (most notably water). In a major insight that did not become generally adopted

Elizabeth Fulhame

Almost all that is known of Elizabeth Fulhame is derived from her book *An Essay on Combustion*, which she published privately in 1794. She appears to have been the wife of Dr Thomas Fulhame, a physician who obtained his doctorate from the University of Edinburgh on the basis of a study of puerperal fever. The interest of her work for enzymology lies not only in her description of catalysis, a generation before Berzelius, but also in the emphasis that she placed on the role of water and in the fact that she was possibly the first to realize that a chemical reaction might require more than one step. She was a pioneer in the study of the effect of light on silver salts, and her discovery of photoreduction marks a first step in the development of photography.

in chemistry until many years afterwards, she realized that her observation was most easily interpreted by supposing that such components were consumed in the early stages of the reaction and regenerated at the end. Her work was largely forgotten by the time that Berzelius (1836) introduced the term *catalysis* for this sort of behaviour. He considered it to be an "only rarely observed force", unlike Fulhame, who had come to the opposite conclusion that water was necessary for virtually all reactions. Both points of view are extreme, of course, but at least in enzyme chemistry the overwhelming majority of known reactions do require water. To a considerable degree the study of enzyme catalysis is the study of catalysis in aqueous solution, and as the relevant terminology will be introduced later in the book when we need it, there is little to add here, beyond remarking that despite its age the classic book by Jencks (1969) remains an excellent source of general information on catalysis in chemistry and biochemistry, for readers who need more emphasis on chemical mechanisms than is found in the present book.

1.7   The influence of temperature and pressure on rate constants

1.7.1   The Arrhenius equation

From the earliest studies of reaction rates, it has been evident that they are profoundly influenced by temperature. The most elementary consequence of this is that the temperature must always be controlled if meaningful results are to be obtained from kinetic experiments. However, with care, one can use temperature much more positively and, by carrying out measurements at several temperatures, deduce important information about reaction mechanisms.

The studies of van't Hoff (1884) and Arrhenius (1889) form the starting point for all modern theories of the temperature dependence of rate constants. Harcourt (1867) had earlier noted that the rates of many reactions approximately doubled for each $10\,°C$ rise in temperature, but van't Hoff and Arrhenius attempted to find a more exact relationship by comparing kinetic observations with the known properties of equilibrium constants. Any equilibrium constant K varies with the absolute temperature T in accordance with the van't Hoff equation,

$$\frac{\mathrm{d}\ln K}{\mathrm{d}T} = \frac{\Delta H^0}{RT^2} \tag{1.34}$$

where R is the gas constant and ΔH^0 is the standard enthalpy change in the reaction. But K can be regarded as the ratio k_+/k_- of the rate constants k_+ and k_- for the forward and reverse reactions (because the net rate of any reaction is zero at equilibrium). So we can write

$$\frac{\mathrm{d}\ln(k_+/k_-)}{\mathrm{d}T} = \frac{\mathrm{d}\ln k_+}{\mathrm{d}T} - \frac{\mathrm{d}\ln k_-}{\mathrm{d}T} = \frac{\Delta H^0}{RT^2} \tag{1.35}$$

This equation can be partitioned as follows to give separate expressions for k_+ and k_-:

$$\frac{\mathrm{d}\ln k_+}{\mathrm{d}T} = \frac{\Delta H^0_+}{RT^2} + \lambda \tag{1.36}$$

$$\frac{\mathrm{d}\ln k_-}{\mathrm{d}T} = \frac{\Delta H^0_-}{RT^2} + \lambda \tag{1.37}$$

where λ is a quantity about which nothing can be said *a priori* except that it must be the same in both equations (because otherwise it would not vanish when one equation is subtracted from the other). Thus far this derivation follows from thermodynamic considerations and involves no assumptions. However, it proved difficult or impossible to show experimentally that the term λ in these equations was necessary. So Arrhenius postulated that its value was in fact zero, and that the temperature dependence of any rate constant k could be expressed by an equation of the form

$$\frac{\mathrm{d}\ln k}{\mathrm{d}T} = \frac{E_\mathrm{a}}{RT^2} \tag{1.38}$$

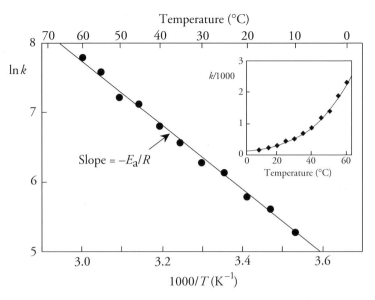

Figure 1.4. The Arrhenius plot. The activation energy E_a is calculated from the slope. The inset shows the same data in linear coordinates.

where E_a is the *activation energy* and corresponds to the standard enthalpy of reaction ΔH^0 in the van't Hoff equation. Integration with respect to T gives

$$\ln k = \ln A - \frac{E_a}{RT} \tag{1.39}$$

where $\ln A$ is a constant of integration. This form of the Arrhenius equation is the most convenient for graphical purposes, as it shows that a plot of $\ln k$ against $1/T$ is a straight line of slope $-E_a/R$, or, if $\log k$ is plotted against $1/T$, the slope is $-E_a/2.303R$. This plot, illustrated in Figure 1.4, is known as an *Arrhenius plot*, and provides a simple method of evaluating E_a.

1.7.2 Elementary collision theory

It is instructive to relate the rates of reactions in the gas phase with the frequencies of collisions between the reactant molecules. After taking exponentials, equation 1.39 may be rearranged to give

$$k = A e^{-E_a/RT} \tag{1.40}$$

According to the Maxwell–Boltzmann distribution of energies among molecules, the number of molecules in a mixture that have energies in excess of E_a is proportional to $e^{-E_a/RT}$. We can therefore interpret the Arrhenius equation to mean that molecules can take part in a reaction only if their energy exceeds some threshold value, the *activation energy*. In this interpretation, the constant A ought to be equal to the frequency of collisions, Z, at least for bimolecular reactions, and it certainly follows from equation 1.40 that A is the limit of the rate constant when $1/T = 0$, at infinite temperature. For some simple reactions in the gas phase, such as the decomposition of hydrogen iodide, A is indeed equal to Z, but in general it is necessary to introduce a factor P,

$$k = PZe^{-E_a/RT} \qquad (1.41)$$

and to assume that, in addition to colliding with sufficient energy, molecules must also be correctly oriented if they are to react. The factor P is then taken to be a measure of the probability that the correct orientation will be adopted spontaneously, so we modify the interpretation above to say that at infinite temperature every collision is productive if the orientation is correct.

With this interpretation of the factor P equation 1.41 accords reasonably well with modern theories of reaction rates in the gas phase. However, virtually all of the reactions that interest biochemists concern complicated molecules in the liquid phase, and collision frequencies have less relevance for these. Thus we need a theory that explains the experimental observations in a way that is as appropriate in aqueous solution as it is in the gas phase.

1.7.3   Transition-state theory

The *transition-state theory* (sometimes called the *theory of absolute reaction rates*) is derived largely from the work of Eyring (1935), and was fully developed in the book by Glasstone, Laidler and Eyring (1940). It is so called because it relates the rates of chemical reactions to the thermodynamic properties of a particular high-energy state of the reacting molecules, known as the *transition state*. (The term *activated complex* is also sometimes used, but it is best avoided in discussions of enzyme reactions, in which the word *complex* is often used with a different meaning.) As a reacting system passes along a notional "reaction coordinate", it must pass through a continuum of energy states, as illustrated in Figure 1.5, and at some stage it must pass through a state of maximum energy. This maximum energy state is the transition state, and should be clearly distinguished from an *intermediate*, which represents not a maximum but a metastable minimum on the reaction profile. No intermediates occur in the reaction profile shown in the main part of Figure 1.5, but a two-step example is shown in the inset, with one intermediate and two transition states. A bimolecular reaction can be represented as

$$A + B \xrightarrow{\ K^{\ddagger}\ } X^{\ddagger} \longrightarrow P + Q \qquad (1.42)$$

where $X^{\ddagger}$ is the transition state. It is assumed to be in *quasi-equilibrium* with A and B, an imaginary state in which the entire system (including products

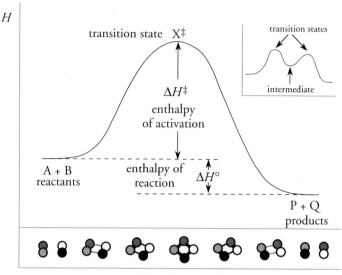

Figure 1.5. Reaction profile according to transition-state theory. The diagrams below the abscissa axis indicate the meaning of the reaction coordinate for a simple bimolecular reaction, but they should not be interpreted too exactly. The inset shows a more complicated example in which the reaction consists of more than one step, with multiple maxima and hence multiple transition states. In such examples states of minimum energy along the reaction profile are called *intermediates*.

P and Q) was at equilibrium before the products were abruptly swept away. For a fuller discussion of what this means see, for example, Laidler, Meiser and Sanctuary (2002); the important point is that the sudden absence of P and Q has no effect on the concentration of $X^{\ddagger}$, which is related to those of A and B by an ordinary equilibrium expression:

$$[X^{\ddagger}] = K^{\ddagger}[A][B] \tag{1.43}$$

where $K^{\ddagger}$ is given by

$$\Delta G^{\ddagger} = -RT \ln K^{\ddagger} = \Delta H^{\ddagger} - T\Delta S^{\ddagger} \tag{1.44}$$

and $\Delta G^{\ddagger}$, $\Delta H^{\ddagger}$ and $\Delta S^{\ddagger}$ are the Gibbs energy, enthalpy and entropy of formation, respectively, of the transition state from the reactants. The concentration of $X^{\ddagger}$ is therefore given by

$$[X^{\ddagger}] = [A][B]e^{\Delta S^{\ddagger}/R}e^{-\Delta H^{\ddagger}/RT} \tag{1.45}$$

Given the way the quasi-equilibrium was described, the transition-state species in equilibrium with A and B are ones that in the immediate past were molecules of A and B. Because of this the first step in equation 1.42 must be written with an irreversible arrow: it is a mistake, found in many accounts

Non. reversibility

of the theory, including those in earlier editions of this book, to represent this as a reversible reaction. The practical importance of this is that molecules that reach $X^{\ddagger}$ from the left in equation 1.42 are like bodies propelled up a slope towards a col: any that reach it are virtually certain to continue down the slope on the other side.

As written, equation 1.45 contains no information about time, like any true thermodynamic equation. We can introduce time by taking account of the natural vibrations that the transition state can undergo. For all but one of the vibration modes the transition state is in no way special: most chemical bonds vibrate in the same way as they would in an ordinary molecule. The exception is the bond that becomes broken in the reaction: its vibration frequency can be calculated from the same quantum-mechanical principles that underlie other vibrations, but it is assumed to have no restoring force, so once the bond starts to break it continues to break. We shall again consider molecular vibrations in Section 8.6.1, in which Figure 8.4 illustrates curves for the dependence of energy on bond length in a breaking C—H bond. Notice that for short bond lengths the curves are of similar shape, but for stretched bonds they are quite different: the curve for the ground state has a minimum and that for the transition state does not.

It follows from considerations of this kind (for more detail, see Laidler, Meiser and Sanctuary, 2002) that equation 1.45 allows calculation of the concentration of transition state, and the vibration frequency for the breaking bond allows the rate constant for the breakdown of $X^{\ddagger}$ to be calculated as RT/Nh, where N is the Avogadro constant and h is Planck's constant. (The numerical value of RT/Nh is about 6.25×10^{12} s^{-1} at 300 K.) The second-order rate constant for the complete reaction is therefore

$$k = \frac{RT}{Nh} e^{\Delta S^{\ddagger}/R} e^{-\Delta H^{\ddagger}/RT} \tag{1.46}$$

Taking logarithms, we obtain

$$\ln k = \ln\left(\frac{RT}{Nh}\right) + \frac{\Delta S^{\ddagger}}{R} - \frac{\Delta H^{\ddagger}}{RT} \tag{1.47}$$

and differentiating,

$$\frac{d \ln k}{dT} = \frac{\Delta H^{\ddagger} + RT}{RT^2} \tag{1.48}$$

Comparison of this with equation 1.38, the Arrhenius equation, shows that the activation energy E_a is not equal to $\Delta H^{\ddagger}$, but to $\Delta H^{\ddagger} + RT$. Moreover, E_a is not strictly independent of temperature, so the Arrhenius plot ought to be curved (not only because of the obvious variation of RT with temperature, but also because $\Delta H^{\ddagger}$ is not strictly temperature-independent). However, the expected curvature is so slight that one would not normally expect to detect it (and the curvature one does often detect in Arrhenius plots is usually attributable to other causes); the variation in k that results from the factor T in equation 1.46 is trivial in comparison with variation in the exponential term.

As both A and E_a in equation 1.39 can readily be determined in practice from an Arrhenius plot, both $\Delta H^{\ddagger}$ and $\Delta S^{\ddagger}$ can be calculated, from

$$\Delta H^{\ddagger} = E_a - RT \tag{1.49}$$

$$\Delta S^{\ddagger} = R \ln\left(\frac{ANh}{RT}\right) - R \tag{1.50}$$

The enthalpy and entropy of activation of a chemical reaction provide valuable information about the nature of the transition state, and hence about the reaction mechanism. A large enthalpy of activation indicates that a large amount of stretching, squeezing or even breaking of chemical bonds is necessary for the formation of the transition state.

The entropy of activation gives a measure of the inherent probability of the transition state, apart from energetic considerations. If $\Delta S^{\ddagger}$ is large and negative, the formation of the transition state requires the reacting molecules to adopt precise conformations and approach one another at a precise angle. As molecules vary widely in their conformational stability, that is to say in their rigidity, and in their complexity, one might expect the values of $\Delta S^{\ddagger}$ to vary widely between different reactions. They do, though establishing the variation with certainty is difficult for the sort of reactions that interest biochemists because of the restricted temperature range over which they can usually be studied (see Section 10.3). The molecules that are important in metabolic processes are mostly large and flexible, and so uncatalysed reactions between them are inherently unlikely, which means that $-\Delta S^{\ddagger}$ is usually large.

Equation 1.46 shows that a catalyst can increase the rate of a reaction either by increasing $\Delta S^{\ddagger}$ (in practice this usually means decreasing the positive quantity $-\Delta S^{\ddagger}$) or by decreasing $\Delta H^{\ddagger}$, or both. It is likely that both effects are important in enzyme catalysis, though definite evidence of this cannot usually be obtained because the uncatalysed reactions are too slow for their values of $\Delta S^{\ddagger}$ and $\Delta H^{\ddagger}$ to be measured.

In all of this it must not be forgotten that the solvent, normally water in enzyme-catalysed reactions, is a part of the system and that entropy effects in the solvent can contribute greatly to entropies of activation. It is an error, therefore, and possibly a serious one, to try to interpret their magnitudes entirely in terms of ordering or disordering of the reactants themselves. Solvent effects can be of major importance in reactions involving ionic or polar species.

1.7.4 Effects of hydrostatic pressure on rate constants

I shall not discuss pressure effects extensively in this book (for more detail, see Laidler and Bunting, 1973), but it is convenient to mention them briefly, both because their treatment has some similarities with that of temperature, and because they can provide valuable information about the mechanistic details of chemical reactions.

The major difference between temperature and pressure effects on reactions in liquid solution is that whereas it is easy to change the rate of a reaction by increasing the temperature, an increase of a few degrees being usually sufficient to produce an easily measurable change, large increases in pressure, typically much more than 100 bar, are necessary to produce comparable results. This difference results from the very low compressibility of water and other liquids: to produce a chemical effect the increase in pressure must alter the volume occupied by the reacting molecules — something easy to achieve for reactions in the gas phase, but much more difficult in the liquid phase.

Another difference is that although enthalpies of activation are always positive, so all rate constants increase with temperature, volumes of activation can be either positive or (more commonly) negative, and so rate constants may change in either direction with increasing pressure. Forming the transition state for any reaction often implies bringing the reacting molecules into closer proximity than they would be in a stable system, especially if the reacting groups are ions of the same sign, but it can also imply bringing them further apart, especially if they are oppositely charged. These possibilities can be distinguished experimentally by examining the effect of pressure: increasing the pressure favours formation of a transition state that occupies a smaller volume than the ground state, so the reaction should be accelerated by increased pressure and has a negative volume of activation; conversely, if the transition state occupies a larger volume than the ground state its formation will be retarded by increased pressure, and the reaction will show a positive volume of activation.

Effects on the solvent molecules can make the major contribution to the magnitudes of volumes of activation in practice, just as they can for those of entropies of activation, and in chemical reactions the values of the two parameters are often highly correlated (Laidler and Bunting, 1973). The same may well apply to enzyme reactions, though it is made more difficult to establish experimentally by the difficulty of studying an enzyme-catalysed reaction over a wide enough temperature range to allow accurate estimation of the entropy of activation.

Problems

1.1 The data in the following table were obtained for the rate of a reaction with stoicheiometry $A + B \rightarrow P$ at various concentrations of A and B. Determine the order with respect to A and B and suggest an explanation for the order with respect to A.

[A] (mM)	10	20	50	100	10	20	50	100
[B] (mM)	10	10	10	10	20	20	20	20
v (μmol l^{-1} s^{-1})	0.6	1.0	1.4	1.9	1.3	2.0	2.9	3.9
[A] (mM)	10	20	50	100	10	20	50	100
[B] (mM)	50	50	50	50	100	100	100	100
v (μmol l^{-1} s^{-1})	3.2	4.4	7.3	9.8	6.3	8.9	14.4	20.3

1.2 Check the following statements for dimensional consistency, assuming that t represents time (units s), v and V represent rates (units $M s^{-1}$ or $mol l^{-1} s^{-1}$), and a, p, s and K_m represent concentrations (units M):

(a) In a plot of v against v/s, the slope is $-1/K_m$ and the intercept on the v/s axis is K_m/V.

(b) In a bimolecular reaction $2A \rightarrow P$, with rate constant k, the concentration of P at time t is given by $p = a_0^2 kt/(1 + 2a_0 kt)$.

(c) A plot of $t/\ln(s_0/s)$ against $(s_0 - s)/\ln(s_0/s)$ for an enzyme-catalysed reaction gives a straight line of slope $1/V$ and ordinate intercept V/K_m.

1.3 Kézdy, Jaz and Bruylants (1958) and Swinbourne (1960) independently suggested an alternative to the Guggenheim plot derived from equations 1.29–30 by dividing one by the other. Show that the resulting expression for $(p_\infty - p_i)/(p_\infty - p_i')$ can be rearranged to show that a plot of p_i' against p_i gives a straight line. What is the slope of this line? If several plots of the same data are made with different values of τ, what are the coordinates of the point of intersection of the lines?

1.4 Many reactions display an approximate doubling of rate when the temperature is raised from 25 °C to 35 °C. What does this imply about their enthalpies of activation? ($R = 8.31 \, J \, mol^{-1} \, K^{-1}$, $0 °C = 273$ K, $\ln 2 = 0.693$.)

1.5 In the derivation of the Arrhenius equation (Section 1.7.1) a term λ was introduced and subsequently assumed to be zero. In the light of the transition-state theory (Section 1.7.3), and assuming (not strictly accurately) that the enthalpy of activation does not change with temperature, what would you expect the value of λ to be at 300 K (27 °C)?

1.6 Some simple reactions involving nitric oxide (NO) have two unusual kinetic features: they follow third-order kinetics, so that, for example, the reaction with molecular oxygen has a rate proportional to $[NO]^2[O_2]$, and their rates decrease with increasing temperature. Suggest a simple way to explain these observations without requiring a trimolecular step and without contradicting the generalization that all elementary rate constants increase with temperature.

Chapter 2
Introduction to Enzyme Kinetics

2.1   The idea of an enzyme–substrate complex

The rates of enzyme-catalysed reactions were first studied in the latter part of the nineteenth century. At that time, no enzyme was available in a pure form, methods of assay were primitive, and buffers were not used to control the pH. Moreover, it was customary to follow the course of the reaction over a period of time, in contrast to the more usual modern practice of measuring initial rates at various different substrate concentrations, which gives results that are easier to interpret.

Most of the early studies were concerned with enzymes from fermentation, particularly invertase[*], which catalyses the hydrolysis of sucrose:

$$\text{sucrose} + \text{water} \rightarrow \text{glucose} + \text{fructose} \qquad (2.1)$$

O'Sullivan and Tompson (1890) made a thorough study of this reaction. They found it to be highly dependent on the acidity of the mixture and that provided that "the acidity was in the most favourable proportion", the rate was proportional to the amount of enzyme. It decreased as the substrate was consumed, and seemed to be proportional to the sucrose concentration, though there were slight deviations from the expected curve. At low temperatures, invertase showed an approximate doubling of rate for an increase of temperature of 10 °C. However, unlike most ordinary chemical reactions, the reaction displayed an apparent *optimum temperature* (Section 10.2), above which the rate fell rapidly to zero. Invertase proved to be a true catalyst, as it was not destroyed or altered in the reaction (except at high temperatures), and a sample was still active after catalysing the hydrolysis of 100 000 times

---

[*]In present-day work invertase is usually known as *β*-fructofuranosidase. However, in this chapter we are mainly concerned with its historical role in the development of enzyme kinetics and it is appropriate to retain the name that was usual at the time.

Eduard Buchner (1860–1917)

The birth of modern biochemistry is accepted to be Eduard Buchner's demonstration that a cell-free extract of yeast could catalyse alcoholic fermentation, as it greatly weakened (and eventually destroyed) the vitalist attitudes that dominated much of biological thinking in the nineteenth century. Buchner was born and educated in Munich and developed his interest in fermentation under the supervision of his elder brother, the bacteriologist Hans Buchner. He was awarded the Nobel Prize in Chemistry in 1907, but his career was brought to a premature end by his death from wounds sustained in action at the Roumanian front.

its weight of sucrose. Finally, the thermal stability of the enzyme was greatly increased by the presence of its substrate: "Invertase when in the presence of cane sugar [sucrose] will stand a temperature fully 25 °C greater than in its absence. This is a very striking fact, and, as far as we can see, there is only one explanation of it, namely the invertase enters into combination with the sugar." Wurtz (1880) had reached a similar conclusion previously in relation to the papain-catalysed hydrolysis of fibrin, observing a precipitate that he suggested might be a papain–fibrin compound that acted as an intermediate in the hydrolysis.

Brown (1892) placed the idea of an *enzyme–substrate complex* in a purely kinetic context. In common with a number of other workers, he found that the rates of enzyme-catalysed reactions deviated from second-order kinetics. Initially, he showed that the rate of hydrolysis of sucrose in fermentation by live yeast appeared to be independent of the sucrose concentration. The conflict between his results and those of O'Sullivan and Tompson was not at first regarded as serious, because catalysis by isolated enzymes was regarded as fundamentally different from fermentation by living organisms, as physiological chemistry was still dominated by ideas of vitalism. Despite cogent opposition from Berthelot (1860), the support of Pasteur (1860) for vitalism meant that it was not finally overthrown until the end of the century, when Buchner (1897) showed that a cell-free (non-living) extract of yeast could catalyse alcoholic fermentation.

This work can be regarded as the creation of biochemistry as a distinct science, and it prompted Brown (1902) to re-examine his earlier results. After confirming that they were correct, he showed that purified invertase behaved in a similar way. He suggested that involvement of an enzyme–substrate complex in the mechanism placed a limit on the rate that could

Victor Henri (1872–1940)

Victor Henri was born in Marseilles of Russian aristocratic parents. He was educated in Paris and St Petersburg, and obtained his first doctorate (on tactile sensations) at Göttingen. His more famous doctorate was granted by the Sorbonne on the basis of his thesis on *Les lois générales de l'action des diastases*, in which he developed his ideas on enzyme catalysis and kinetics. He was highly active throughout his career, which included teaching posts in Paris, Moscow, Zürich and Liège, and published more than 500 papers. His work was mainly in physical chemistry, but he also made contributions to other fields and collaborated with Alfred Binet, the pioneer in intelligence testing, with whom he wrote a book on intellectual fatigue.

be achieved: provided that the complex existed for a brief instant of time before breaking down to products, then a maximum rate would be reached when the substrate concentration was high enough for all the enzyme to be present as enzyme–substrate complex. The rate at which complex is formed would become significant at lower concentrations of substrate, and the rate of hydrolysis would then depend on the substrate concentration.

More detail about the early history of enzymology, including a new English translation of Buchner's paper, may be found in a recent book (Cornish-Bowden, 1997).

2.2   The Michaelis–Menten equation

Henri (1902, 1903) criticized Brown's model of enzyme action on the grounds that it assumed a fixed lifetime for the enzyme–substrate complex between its abrupt creation and decay. He proposed instead a mechanism that was conceptually similar to Brown's but which was expressed in more precise mathematical and chemical terms, with an equilibrium between the free enzyme and the enzyme–substrate and enzyme–product complexes.

Now although Brown and Henri reached essentially correct conclusions, they did so on the basis of experiments that were open to serious objections. O'Sullivan and Tompson experienced great difficulty in obtaining coherent results until they realized the importance of acid concentration. Brown prepared the enzyme in a different way and found the addition of acid to be unnecessary (presumably his solutions were sufficiently buffered by the

Leonor Michaelis (1875–1949)

Leonor Michaelis was born in Berlin. After a year working as assistant to Paul Ehrlich he studied clinical medicine and developed an early interest in controlling the hydrogen-ion concentration. In the years preceding the First World War he used his mastery of this subject to become one of the leaders in studying enzyme-catalysed reactions. This was an impressively productive period for him, and his famous paper with Maud Menten is just one of 94 publications, including five books, in the five years from 1910 to 1914. Several of his papers from these years are still cited, and one of the books, *Die Wasserstoffionenkonzentration*, became the standard work on pH, buffers and related topics. In the 1920s he spent three years as professor of biochemistry in Nagoya, Japan, and subsequently moved to the USA; from 1926 to 1929 he was at the Johns Hopkins University, and from 1929 until his death he was at the Rockefeller Institute for Medical Research. Until the end of his life he remained active in research, concerned mainly with the study of free radicals.

natural components of the yeast), and Henri did not discuss the problem. Apart from O'Sullivan and Tompson, the early investigators of invertase made no allowance for the mutarotation of the glucose produced in the reaction, although this certainly affected their results because they used polarimetric methods for following the reaction.

With the introduction of the concept of hydrogen-ion concentration, expressed by the logarithmic scale of pH (Sørensen, 1909), Michaelis and Menten (1913) realized the necessity of carrying out definitive experiments with invertase. They controlled the pH of the reaction by the use of acetate buffers, they allowed for the mutarotation of the product and they measured *initial rates* of the reaction at different substrate concentrations. When initial rates are used, complicating factors such as the reverse reaction, product inhibition and inactivation of the enzyme can be avoided and much simpler rate equations can be used. In spite of these refinements Michaelis and Menten's results agreed well with Henri's, and they proposed a mechanism essentially the same as his:

$$E + A \rightleftarrows EA \rightarrow E + P \tag{2.2}$$

Like Henri, they assumed that the reversible first step was fast enough to be represented by an equilibrium constant for substrate dissociation, $K_s = ea/x$,

Maud Leonora Menten (1879–1960)

Maud Menten was born in Port Lambton, Ontario, and in 1911 she became one of the first Canadian women to receive a medical doctorate. Her work with Leonor Michaelis on invertase was an interlude in a career devoted mostly to pathology and the more medical aspects of biochemistry and physiology. She co-authored more than 70 publications, and was the first to use electrophoretic mobility to study human haemoglobins. She spent most of her working life at the University of Pittsburgh, but went back to Canada after her retirement, where her research continued at the Medical Institute of British Columbia until it was brought to an end by ill health. She then returned to Ontario to spend the remainder of her life not far from the place of her birth.

in which x is the concentration of the intermediate EA, so that $x = ea/K_s$. The instantaneous concentrations of free enzyme and substrate, e and a respectively, are not directly measurable, however, and so they must be expressed in terms of the initial, measured, concentrations e_0 and a_0, using the stoicheiometric relationships $e_0 = e + x$ and $a_0 = a + x$. From the first of these, x cannot be greater than e_0, and so, if a_0 is much larger than e_0 it must also be much larger than x. So $a = a_0$ with good accuracy, and the expression for x becomes $x = (e_0 - x)a/K_s$, which can be rearranged to give

$$x = \frac{e_0}{(K_s/a) + 1} \tag{2.3}$$

The second step in the reaction, EA $\rightarrow$ E + P, is a simple first-order reaction, with a rate constant that may be defined as k_2, so that

$$v = k_2 x = \frac{k_2 e_0}{(K_s/a) + 1} = \frac{k_2 e_0 a}{K_s + a} \tag{2.4}$$

Michaelis and Menten showed that this theory, and equation 2.4, could account accurately for their results with invertase. Because of the definitive nature of their experiments, which have served as a standard for most subsequent enzyme-kinetic measurements, Michaelis and Menten are regarded as the founders of modern enzymology, and equation 2.4 (in its modern forms, equations 2.14 and 2.15) is generally known as the *Michaelis–Menten equation*.

Similar equations had been derived earlier by Henri (1902, 1903), and some authors believe his contribution to have been undervalued. He did not reach Michaelis and Menten's essential insight of recognizing that analysis in terms of initial rates was much simpler than struggling with time courses, but he did write the following equation for the rate of reaction (Henri, 1902):

$$\frac{dx}{dt} = \frac{K\Phi(a-x)}{1 + m(a-x) + nx} \tag{2.5}$$

in which a was the total amount of sucrose, x was the amount of product at time t, Φ was the amount of enzyme, and K, m and n were constants. If concentrations are treated as proportional to amounts, then putting the initial-rate condition $x = 0$ into this equation and making appropriate changes to the symbols makes it identical to equation 2.4. In his paper Henri stopped at equation 2.5, but in his doctoral thesis (1903, pages 90–91) he introduced the initial-rate condition and commented that the hyperbolic form of the dependence of initial rate on sucrose concentration agreed well with the experimental results. He also considered the possibility of inhibition by product present at time zero, and wrote an equation for it that is equivalent to equation 2.55 below (Section 2.8).

At about the same time as Michaelis and Menten were working, Van Slyke and Cullen (1914) obtained similar results with the enzyme urease. They assumed a similar mechanism, with the important difference that the first step was assumed to be irreversible:

$$\underset{e_0-x}{E} + \underset{a}{A} \xrightarrow{k_1} \underset{x}{EA} \xrightarrow{k_2} \underset{p}{E+P} \tag{2.6}$$

Here there are no reversible reactions, and there can be no question of representing x by an equilibrium constant; instead,

$$\frac{dx}{dt} = k_1(e_0 - x)a - k_2 x \tag{2.7}$$

Van Slyke and Cullen implicitly assumed that the intermediate concentration was constant, so $dx/dt = 0$, and hence $k_1(e_0 - x)a - k_2 x = 0$, which may be rearranged to give $x = k_1 e_0 a/(k_2 + k_1 a)$; substituting this into the rate equation $v = k_2 x$ gives

$$v = k_2 x = \frac{k_1 k_2 e_0 a}{k_2 + k_1 a} = \frac{k_2 e_0 a}{\dfrac{k_2}{k_1} + a} \tag{2.8}$$

This equation is of the same form as equation 2.4, with k_2/k_1 instead of K_s (which would be k_{-1}/k_1 if the rate constants in equation 2.2 were numbered) and is empirically indistinguishable from it.

At about the same time as these developments were taking place in the understanding of enzyme catalysis, Langmuir (1916, 1918) was reaching similar ideas about the adsorption of gases on solids. His treatment was much more general, but the case he referred to as *simple adsorption* corresponds closely to the type of binding assumed by Henri and by Michaelis and Menten. Langmuir recognized the similarity between solid surfaces and enzymes, although he imagined the whole surface of an enzyme to be "active", rather than limited areas or *active sites*. Hitchcock (1926) pointed out the similarity between the equations for the binding of ligands to solid surfaces and to proteins, and the logical process was completed when Lineweaver and Burk (1934) extended Hitchcock's ideas to catalysis.

2.3 The steady state of an enzyme-catalysed reaction

2.3.1 The Briggs–Haldane treatment

Whether we treat the first step of enzyme catalysis as an equilibrium or as an irreversible reaction, we make unwarranted and unnecessary assumptions about the magnitudes of the rate constants. As we have seen, both formulations lead to the same form of the rate equation, and Briggs and Haldane (1925) examined a more general mechanism that includes both as special cases:

$$\underset{e_0 - x}{\text{E}} + \underset{a}{\text{A}} \underset{k_{-1}}{\overset{k_1}{\rightleftharpoons}} \underset{x}{\text{EA}} \overset{k_2}{\longrightarrow} \text{E} + \underset{p}{\text{P}} \qquad (2.9)$$

This leads to the following rate equation:

$$\frac{dx}{dt} = k_1(e_0 - x)a - k_{-1}x - k_2x \qquad (2.10)$$

Briggs and Haldane argued that a *steady state* would be reached in which the concentration of intermediate was constant, with $dx/dt = 0$; then

$$k_1(e_0 - x)a - k_{-1}x - k_2x = 0 \qquad (2.11)$$

Collecting terms in x and rearranging leads to the following expression for the steady-state value of x:

$$x = \frac{k_1 e_0 a}{k_{-1} + k_2 + k_1 a} \qquad (2.12)$$

As before, the rate is given by $v = k_2x$:

$$v = \frac{k_1k_2e_0a}{k_{-1} + k_2 + k_1a} = \frac{k_2e_0a}{\dfrac{k_{-1} + k_2}{k_1} + a} \tag{2.13}$$

If the Briggs–Haldane treatment is presented in this way it is easy to be seduced into believing that it is more general than that of Michaelis and Menten, and innumerable textbooks (including the first edition of this one) have indeed suggested that it is the "right" approach that has superseded the more "naive" one of Michaelis and Menten. In reality, however, there are two serious objections to equation 2.9 that leave it as only a slight improvement on equations 2.2 and 2.6: first, it assumes that the reaction is irreversible, whereas all real enzyme-catalysed reactions are reversible; second, it shows only one intermediate complex, with substrate bound to enzyme, which means that the mechanism treats substrate and product in a way that is conceptually unsymmetrical. I shall return to these points in Section 2.7.1. Meanwhile, the proper lesson to draw from this section is not that equation 2.13 shows the "right" way to write the kinetic equation for a one-substrate reaction, but that unless there is good evidence for a pre-equilibrium one should analyse any mechanism in terms of the steady-state assumption.

2.3.2 The Michaelis–Menten equation

Equation 2.13 can be written in the following more general form

$$v = \frac{k_{cat}e_0a}{K_m + a} \tag{2.14}$$

in which k_2 has been written as k_{cat} (for reasons that will be considered shortly) and $(k_{-1} + k_2)/k_1$ as K_m, the *Michaelis constant*. This is the *Michaelis–Menten equation* (called the *Henri–Michaelis–Menten equation* by some authors), the fundamental equation of enzyme kinetics. The name is nowadays applied to the equation derived using the steady-state assumption, *not* to the form obtained by Michaelis and Menten using the equilibrium assumption. Equation 2.14 is more general than equation 2.13, which was derived for a particular mechanism, equation 2.9, and it applies to many mechanisms more complex than the simplest two-step Michaelis–Menten mechanism. That is why k_2 of equation 2.13 was replaced by k_{cat}; in general one cannot assume that k_{cat} is equivalent to the rate constant for the second step of the reaction or that K_m is equivalent to $(k_{-1} + k_2)/k_1$.

Although k_{cat} may not refer to a single step of a mechanism, it does have the properties of a first-order rate constant, defining the capacity of the enzyme–substrate complex, once formed, to form product. It is known as the *catalytic constant*, symbolized by k_{cat} (or sometimes as k_0). It is also sometimes called the *turnover number*, because it is a reciprocal time and defines the number of catalytic cycles (or "turnovers") the enzyme can undergo in unit time, or the number of molecules of substrate that one molecule of enzyme can convert into products in one unit of time.

At least in the early stages of investigating an enzyme, the true enzyme molarity e_0 is usually unknown, which complicates use of the Michaelis–Menten equation in the form shown as equation 2.14. The difficulty is commonly avoided by combining k_{cat} and e_0 into a single constant $V = k_{cat}e_0$, the *limiting rate*, which because of its dependence on the enzyme concentration is *not a fundamental property of the enzyme*. Thus it is often convenient to write the Michaelis–Menten equation as follows:

$$v = \frac{Va}{K_m + a} \tag{2.15}$$

To avoid confusion with v, V is usually spoken aloud as "vee-max", and is sometimes printed as V_{max} or V_m. These terms and symbols derive from the old name (still often used) of *maximum velocity* for V, now discouraged by the International Union of Biochemistry and Molecular Biology (International Union of Biochemistry, 1982) because it does not define a maximum in the mathematical sense but a limit. The symbol V_m is especially to be avoided because it misleadingly suggests that the subscript m corresponds to the one in K_m. (I have occasionally seen K_{max} in answers to examination questions!) In fact, the m in K_m stands for Michaelis, and it was the former and more logical custom to write it as K_M.

2.3.3   Units of enzyme activity

For enzymes whose molar concentration cannot be measured, either because the enzyme has not been purified or because its molecular mass is unknown, it is often convenient to define a unit of catalytic activity. The traditional "unit" of enzymologists (often symbolized as IU, for "international unit") is the amount of enzyme that can catalyse the transformation of 1 μmol of substrate into products in 1 min under standard conditions. This unit is still in common use, because the corresponding unit in the International System[*] of units, the *katal*, symbolized as kat, has not been widely adopted. A perceived problem is

---

[*]This was adopted by the 11th General Conference on Weights and Measures in 1960 (see International Union of Pure and Applied Chemistry, 1993), and is nearly always referred to as SI, from the French term *Système International*.

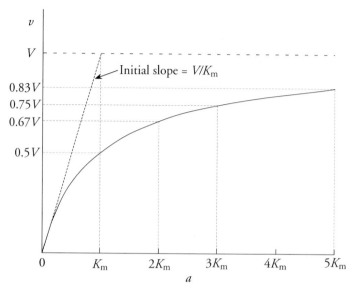

Figure 2.1. Dependence of initial rate v on the substrate concentration a for a reaction obeying the Michaelis–Menten equation. As a plot of this kind was *not* used or advocated by Michaelis and Menten it is inappropriate to refer to it as a "Michaelis–Menten plot". They actually used a plot of v against $\log a$ (Figure 2.4), and used its maximum slope of $0.577V$ ($= 0.25V \ln 10$) to estimate V and hence to estimate K_m.

that 1 kat, the amount of activity sufficient to catalyse the transformation of 1 mol of substrate into products in 1 s under standard conditions, is regarded as being excessively large: a typical enzyme activity in a cell extract may be about 1 unit/ml, or about 20 μkat/l. Nonetheless, more severe objections to the farad as the unit of capacitance have not prevented its general acceptance by electrical engineers, though they more often use its submultiples. Similarly, there are submultiples of the katal, such as 1 μkat $= 60$ units or 1 nkat $= 0.06$ unit, that are convenient in magnitude for laboratory use.

2.3.4 The curve defined by the Michaelis–Menten equation

The curve defined by equation 2.15 is shown in Figure 2.1. It is a rectangular hyperbola through the origin, with asymptotes $a = -K_m$ and $v = V$. (The relationship between this description and the way rectangular hyperbolas are typically defined and exemplified in elementary mathematics courses is explored in Problem 2.3 at the end of this chapter.) At very small values of a the denominator of the right-hand side of equation 2.15 is dominated by K_m, so a is negligible compared with K_m and v is directly proportional to a:

$$v \approx \frac{k_{cat}e_0 a}{K_m} = \frac{Va}{K_m} \qquad (2.16)$$

and the reaction is approximately second-order overall, first-order in a. As k_{cat}/K_m has this fundamental meaning as the second-order rate constant for the reaction $E + A \rightarrow E + P$ at low substrate concentrations, it should not be regarded just as the result of dividing k_{cat} by K_m. It is given the name *specificity constant*, for reasons that will be explained shortly (Section 2.4), and may be symbolized as k_A, where the subscript, A in this example, indicates which substrate it refers to. The reciprocal of V/K_m, or K_m/V, has dimensions of time and may be called the *specificity time*. It is the time that would be required to consume all of the substrate if the enzyme were acting under first-order conditions and maintained the same initial rate indefinitely (Cornish-Bowden, 1987). Although this is rather an abstract definition for ordinary assay conditions, it makes more sense in relation to the cell, where many enzymes operate under first-order conditions and maintain the same rate for long periods (because their substrates are replenished): in these circumstances the specificity time is the time required to replace all of the substrate.

When a is equal to K_m, equation 2.15 simplifies to the following:

$$v = Va/2a = 0.5V \qquad (2.17)$$

Thus the rate has half of its limiting value in these conditions, and K_m can be *defined* as the concentration at which $v = 0.5V$. This is an operational definition that applies regardless of the particular mechanism to which the Michaelis–Menten equation is applied.

At very large values of a the denominator of the right-hand side of equation 2.14 is dominated by a, so K_m is negligible in comparison with a and the equation simplifies to

$$v \approx k_{cat}e_0 = V \qquad (2.18)$$

The reaction is now approximately zero order in a, all of the active sites of the enzyme are occupied by the substrate and it is said to be *saturated*. This is the reason for the name *limiting rate* for V.

In many textbooks of biochemistry, and even in some specialist books on enzyme kinetics, the plot of v against a is drawn badly enough to be highly misleading. Students are easily given a quite wrong impression of the shape of the curve, suggesting that V can be estimated from such a plot of experimental observations by finding the point at which v "reaches" its limiting value. The main fault lies in drawing a curve that flattens out too abruptly and then drawing an asymptote too close to the curve. In fact v never reaches V at a finite value of a, as ought to be clear from examination of equation 2.15, and even when $a = 10K_m$ (a higher concentration than is used in many experiments) the value of v is still almost 10% less than V. This point may perhaps be

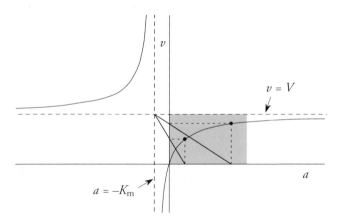

Figure 2.2. Dependence of initial rate v on the substrate concentration a for a reaction obeying the Michaelis–Menten equation. The part of the curve from $a = 0$ to $5K_m$, the shaded part of the Figure, is the same as in Figure 2.1, but the range of values shown is much wider and includes physically impossible values, in order to illustrate the relationship of the curve to the two asymptotes, which intersect at the point $(-K_m, V)$.

grasped more clearly by examining a much greater proportion of the curve defined by equation 2.15 than is given in Figure 2.1: such a view may be seen in Figure 2.2, which extends in both directions far beyond the usual a-range from zero to a few times K_m. The inclusion of physically impossible negative values explains the relationship of the curve to the usual two-limb hyperbolas found in mathematics textbooks, and it also shows that when one estimates K_m and V from a typical set of observations one is in effect trying to locate the point of intersection of the two asymptotes of an infinite curve from observations along a short arc. It is for this reason that estimation of K_m and V is not a trivial problem but one that requires considerable care: I shall return to it in Section 2.6 and again in Chapter 14.

2.3.5 Mutual depletion kinetics

In deriving equation 2.13 we assumed that the substrate was in sufficient excess over the enzyme that one could treat the free and total substrate concentrations as identical. Kinetic experiments are usually carried out in conditions where this assumption is valid, but in experiments at unusually high enzyme concentrations or low substrate concentrations it may be necessary to distinguish between the free and total substrate concentrations as a and a_0 respectively. One must then replace a in equation 2.10 by $a_0 - x$, because the substrate concentration as well as the enzyme concentration is depleted by formation of enzyme–substrate complex. This leads, instead of equation 2.11, to an equation that contains x^2 when multiplied out, a quadratic and not a linear equation for x.

As this type of system has greater practical importance for tight-binding inhibitors than for substrates, we shall defer more detailed examination until Chapter 6. For the moment it is sufficient to note that when substrate depletion is taken into account the (total) substrate concentration at which $v = 0.5V$ is not K_m but $K_m + 0.5e_0$ (compare with equation 6.9).

2.3.6 Ways of writing the Michaelis–Menten equation

As we have seen, there are three parameters that can be considered as parameters of the form of the Michaelis–Menten equation that applies when the enzyme concentration e_0 is known: the catalytic constant k_{cat}, the specificity constant k_A and the Michaelis constant K_m. As these are related by the identity $k_A \equiv k_{cat}/K_m$ it is evident that as alternatives to equation 2.14 the Michaelis–Menten equation can be expressed in terms of k_{cat} and k_A:

$$v = \frac{k_{cat}k_A e_0 a}{k_{cat} + k_A a} \tag{2.19}$$

or in terms of k_A and K_m:

$$v = \frac{k_A e_0 a}{1 + a/K_m} \tag{2.20}$$

The usual formulation in terms of k_{cat} and K_m owes more to history than anything else, and is in no way more fundamental than equations 2.19–20. The study of enzymes might well have been more convenient if the subject had developed differently and equation 2.19 had become the most familiar form of the Michaelis–Menten equation: this would simplify discussion of numerous aspects of the subject, such as the different kinds of inhibition (Chapters 5 and 6), reactions with multiple substrates (Chapter 7), effects of pH (Chapter 9) and the estimation of parameters by graphical or statistical means (Chapter 14). Nonetheless, in this book I follow the usual practice of writing it as equation 2.14 or 2.15, because these are by far the commonest forms in the research literature.

Regardless of how the equation itself is written, the essential point of this discussion is that the fundamental parameters of the Michaelis–Menten equation are k_{cat} and k_A, and many aspects of enzyme behaviour are most easily understood as effects on one or other of them. Rather than thinking of k_A as a derived quantity, therefore, it is better to think of it as fundamental, so that K_m is best thought of as the ratio k_{cat}/k_A.

2.4   Specificity

It is natural to be impressed by the capacity of enzymes to act as highly efficient catalysts: they permit reactions that for practical purposes do not occur at all under ordinary conditions, such as the decarboxylation of orotidine 5′-phosphate, with an estimated first-order rate constant of 3×10^{-16} s^{-1} for spontaneous decomposition in neutral aqueous solution at 25 °C (Radzicka and Wolfenden, 1995); they also accelerate reactions that are already fast without a catalyst, such as the dehydration of bicarbonate, for which the corresponding rate constant is 0.13 s^{-1}. Both reactions are catalysed by specific enzymes, the first by orotidine 5′-phosphate decarboxylase and the second by carbonic anhydrase. Radzicka and Wolfenden (1995) introduced the idea of *catalytic proficiency* as a measure of the capacity of an enzyme to accelerate a reaction beyond its uncatalysed rate, and by their criteria orotidine 5′-phosphate decarboxylase is a highly proficient enzyme. However, it is not evident that accelerating a slow process is necessarily more difficult than accelerating a fast one, and experience in human affairs would suggest the opposite: the arrangement of large rocks into monuments was achieved around 4000 years ago, even though the process does not occur naturally at a perceptible rate, whereas the routine transport of people and goods at more than twice the speed of a galloping horse was achieved much more recently. In any case, virtually any reaction can be accelerated, essentially without limit, by carrying it out at high temperature under extreme conditions, a property that was, of course, exploited by Radzicka and Wolfenden (1995) when they estimated the uncatalysed rate of decarboxylation of orotidine 5′-phosphate. What should impress us, therefore, about enzymes is not that they are excellent catalysts for certain reactions but that they are extremely poor catalysts for the overwhelming majority of other reactions: orotidine 5′-phosphate decarboxylase, for example, is a useless catalyst for dehydration of bicarbonate, even though that is also a decarboxylation reaction. In other words, what should impress us about enzymes is their *specificity*. To study this, however, we need to define the term precisely.

Experimental investigations of enzymes are usually done with only one substrate present in the reaction mixture at a time, without any alternative substrates able to undergo the same reaction. This is not at all the same as saying that the enzyme *requires* two or more substrates for the reaction to be complete. For example, hexokinase catalyses a reaction between glucose and ATP, in which glucose and ATP are not alternatives to one another but are both separately required for the reaction to be possible, so this requirement as such has nothing to do with competition between substrates. On the other hand hexokinase will accept fructose and other hexoses as alternatives to glucose, and if both glucose and fructose are simultaneously present they are *competing substrates*.

The reason for studying one substrate at a time is that competing substrates tend to complicate the analysis, usually without providing more information than would be obtained by studying the substrates separately*. However, this implies an important difference between experimental practice and the physiological conditions in which enzymes usually exist: most enzymes are not perfectly specific for a single substrate and must often select between several that are available simultaneously. To be physiologically meaningful, therefore, enzyme specificity must be defined in terms of how well the enzyme can discriminate between substrates present in the same reaction mixture.

This interpretation of what specificity is follows from a thorough discussion by Fersht (1977), who was concerned with the biologically important question of how the enzymes involved in protein synthesis can distinguish between structurally similar substrates (such as isoleucine and valine) and avoid having an intolerable frequency of errors. It is now widely accepted, but in a recent article Koshland (2002) contests it, arguing that a parameter known as a specificity constant ought to "provide a means of contrasting the specificities of different enzymes towards their substrates", to compare, for example, the specificity of a kinase for different carbohydrate substrates with that of a proteinase for different peptide substrates. In this book I shall follow Fersht's interpretation, which appears more appropriate for discussing the physiological problems that natural selection has had to solve. None of this means that the specificity of an enzyme cannot be determined by studying the different substrates separately, but it does mean that the parameters for the individual substrates need to be interpreted correctly and not casually.

The simplest case to consider is one in which there are two competing substrates that individually give Michaelis–Menten kinetics when studied

---

*An exception arises if one substrate binds so tightly that it cannot be kinetically characterized by ordinary methods. The limiting rate for such a substrate can, however, be directly measured, and studying the kinetics in the presence of another substrate for which both Michaelis–Menten parameters are known allows the value of the missing parameter for the tight-binding substrate to be deduced (see Klyosov, 1996).

$$E + A \underset{k_{-1}}{\overset{k_1}{\rightleftharpoons}} EA \overset{k_2}{\longrightarrow} E + P$$

$$E + A' \underset{k'_{-1}}{\overset{k'_1}{\rightleftharpoons}} EA' \overset{k'_2}{\longrightarrow} E + P'$$

Figure 2.3. Competition between substrates for the same enzyme.

separately (Figure 2.3). (A slightly more complicated version will be used in Section 4.5 as an illustration.) It gives the following pair of rate equations:

$$v = \frac{dp}{dt} = \frac{Va}{K_m(1 + a'/K'_m) + a} = \frac{\dfrac{V}{K_m}a}{1 + \dfrac{a}{K_m} + \dfrac{a'}{K'_m}} \qquad (2.21)$$

$$v' = \frac{dp'}{dt} = \frac{V'a'}{K'_m(1 + a/K_m) + a'} = \frac{\dfrac{V'}{K'_m}a'}{1 + \dfrac{a}{K_m} + \dfrac{a'}{K'_m}} \qquad (2.22)$$

in which $V = k_2 e_0$ and $V' = k'_2 e_0$ are the limiting rates, and $K_m = (k_{-1} + k_2)/k_1$ and $K'_m = (k'_{-1} + k'_2)/k'_1$ are the Michaelis constants of the two reactions in isolation. As we shall see in Section 5.2.1, each equation is exactly of the form of equation 5.1, the equation for competitive inhibition, so if one measures the specific "inhibition constant" for a competitive substrate by treating it as if it were an inhibitor the value that results is its Michaelis constant. This is illustrated in Table 2.1, which shows K_m values for several poor substrates of fumarase measured both directly and in competing reactions. In each case the values of K_m measured in the two ways agree to within experimental error.

A more important point about equations 2.21–22, also illustrated by the data in Table 2.1, is that they provide the basis for a rigorous definition of enzyme specificity. Consider the parameters for fluorofumarate and for fumarate. The value of k_{cat} for fluorofumarate is about three times that for fumarate; thus at high concentrations fluorofumarate appears to be a better substrate than fumarate, if the two reactions are considered in isolation. The reverse is true at low concentrations, however, because k_{cat}/K_m is about 60% greater for fumarate than for fluorofumarate. Which of these results is more fundamental? Which is the more specific substrate? The question may seem to be just one of definitions, but a clear and satisfying answer emerges when one realizes that it is artificial to consider the two substrates in isolation from one another. In most physiological discussions of specificity one ought to consider the proportion of reaction using each substrate when they are mixed together,

Table 2.1. Kinetic parameters for substrates of fumarase. The data (Teipel, Hass and Hill, 1968) refer to measurements at 25 °C in buffer of pH 7.3. Values of the catalytic constant k_{cat} (or V/e_0, see Section 2.3.2) and K_m were measured in conventional kinetic experiments. Values in the column labelled "K_i" are K_m values for the poor substrates measured by treating them as competitive inhibitors of the reaction with fumarate as substrate.

Substrate	k_{cat} (s^{-1})	K_m (mM)	"K_i" (mM)	k_{cat}/K_m (s^{-1} mM^{-1})
Fluorofumarate	2700	0.027	–	100 000
Fumarate	800	0.005	–	160 000
Chlorofumarate	20	0.11	0.10	180
Bromofumarate	2.8	0.11	0.15	25
Iodofumarate	0.043	0.12	0.10	0.36
Mesaconate	0.023	0.51	0.49	0.047
L-Tartrate	0.93	1.3	1.0	0.72

which can be determined by dividing equation 2.21 by equation 2.22:

$$\frac{v}{v'} = \frac{dp}{dp'} = \frac{(V/K_m)a}{(V'/K'_m)a'} = \frac{(k_{cat}/K_m)a}{(k'_{cat}/K'_m)a'} = \frac{k_A a}{k'_A a'} \tag{2.23}$$

This shows why the name *specificity constant* for the parameter $k_A = k_{cat}/K_m$ introduced in Section 2.3.4 is not arbitrary, because it is indeed the parameter that determines the ratio of rates for competing substrates when they are mixed together, and it thus expresses the ability of an enzyme to discriminate in favour of any substrate A in the presence of others. It follows, therefore, that in an equimolar mixture of fumarate and fluorofumarate, *at any concentration*, the rate of the fumarase-catalysed hydration of fumarate is 60% faster than the hydration of fluorofumarate, and that fumarate is therefore the more specific substrate.

2.5 Validity of the steady-state assumption

In deriving equation 2.13 it was assumed that a steady state would be reached in which $dx/dt = 0$. In fact, however, equation 2.10 is readily integrable if a is treated as a constant, and it is instructive to derive a rate equation without making the steady-state assumption, because this sheds some light on the validity of the assumption. Separating the two variables, x and t, we have

$$\int \frac{dx}{k_1 e_0 a - (k_1 a + k_{-1} + k_2)x} = \int dt \tag{2.24}$$

Despite its complicated appearance, the left-hand side is just the integral of dx divided by the sum of a constant and a term proportional to x, so it has the same simple form as several integrals we have encountered already (for example, in Section 1.2), and may be integrated in the same way:

$$\frac{\ln[k_1 e_0 a - (k_1 a + k_{-1} + k_2)x]}{-(k_1 a + k_{-1} + k_2)} = t + \alpha \qquad (2.25)$$

At the instant when the reaction starts there has not been enough time to produce any EA complex, so $x = 0$ when $t = 0$, and

$$\alpha = \frac{\ln(k_1 e_0 a)}{-(k_1 a + k_{-1} + k_2)} \qquad (2.26)$$

and so

$$\ln\left[\frac{k_1 e_0 a - (k_1 a + k_{-1} + k_2)x}{k_1 e_0 a}\right] = -(k_1 a + k_{-1} + k_2)t \qquad (2.27)$$

Taking exponentials of both sides, we have

$$1 - \frac{(k_1 a + k_{-1} + k_2)x}{k_1 e_0 a} = e^{-(k_1 a + k_{-1} + k_2)t} \qquad (2.28)$$

and solving for x we have

$$x = \frac{k_1 e_0 a[1 - e^{-(k_1 a + k_{-1} + k_2)t}]}{k_1 a + k_{-1} + k_2} \qquad (2.29)$$

The rate is given by $v = k_2 x$, and so, substituting $V = k_2 e_0$ and $K_m = (k_{-1} + k_2)/k_1$, we have

$$v = \frac{Va[1 - e^{-(k_1 a + k_{-1} + k_2)t}]}{K_m + a} \qquad (2.30)$$

When t becomes very large the exponential term approaches $e^{-\infty}$, which is zero, and so equation 2.30 becomes identical to equation 2.15, the Michaelis–Menten equation. How large t must be for this to happen depends on the magnitude of $(k_1 a + k_{-1} + k_2)$: if it is of the order of $1000\,s^{-1}$ (a reasonable value in practice), then the exponential term is less than 0.01 for values of t greater than 5 ms: in other words equation 2.30 should become indistinguishable from the Michaelis–Menten equation after a few milliseconds.

In deriving equation 2.30 the substrate concentration a was treated as a constant, which is not strictly accurate because it must change as the reaction proceeds. However, provided that a_0 is much greater than e_0, as it usually is in steady-state experiments, the variation of a during the time it takes to establish the steady state is trivial and can be neglected without significant inaccuracy. Laidler (1955) derived an equation similar to equation 2.30 as a special case of a more general treatment in which he allowed for a to decrease from its initial value a_0. He found that a steady state was achieved in which

$$v = \frac{V(a_0 - p)}{K_m + a_0 - p} \tag{2.31}$$

which is the same as equation 2.15 apart from the replacement of a with $a_0 - p$.

It may seem contradictory to refer to a steady state in which v must decrease as p increases, but this decrease in v is extremely slow compared with the rapid increase in v that occurs in the *transient phase*, the period in which equation 2.30 must be used, before the steady state is established. The argument of Briggs and Haldane is hardly altered by replacing the assumption that $dx/dt = 0$ with an assumption that dx/dt is very small: equation 2.11 becomes a good approximation instead of being exact. As Wong (1975) pointed out, what matters is not the absolute magnitude of dx/dt but its magnitude relative to $k_1 e_0 a$.

This example illustrates the relationship between mathematics and science. From the mathematical point of view equation 2.24 cannot be integrated unless a is treated as a constant. Nonetheless, equation 2.31, obtained without treating a as a constant, is so accurate in practice that it would be quite difficult to devise an experiment to detect any deviations from its predictions. The point is that it requires experience with enzymes and their kinetic behaviour, not mathematical expertise, to know what simplifying assumptions can safely be made.

2.6 Graphs of the Michaelis–Menten equation

2.6.1 Plotting v against a

When several initial rates are measured at different substrate concentrations, graphical display of the results allows the values of the kinetic parameters and the precision of the experiment to be visually assessed. The most obvious way is to plot v against a, as in Figure 2.1. This is unsatisfactory in practice, however, for several reasons: it is difficult to draw a rectangular hyperbola accurately; it is difficult to locate the asymptotes correctly (because one is tempted to place them too close to the curve); it is difficult to perceive the relationship between a family of hyperbolas; and it is difficult to detect deviations from the expected curve if they occur. These points were recognized by Michaelis and Menten (1913), who instead plotted v against $\log a$. Their plot has some advantages apart from its historical interest, and it is particularly useful for comparing the properties of different isoenzymes that catalyse the same reaction with widely

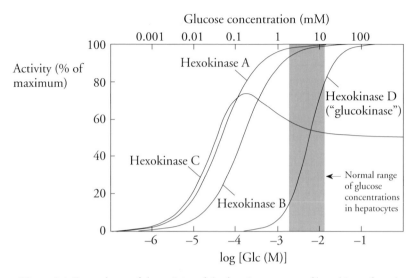

Figure 2.4. Dependence of the activity of the four isoenzymes of hexokinase found in rat liver on the glucose concentration. Using a logarithmic scale for the abscissa allows isoenzymes with very different kinetic properties to be compared. From Figure 6.1 of Cárdenas (1995).

different affinities for substrate. Consider, for example, the comparison shown in Figure 2.4 between the four isoenzymes of hexokinase found in the rat: as hexokinase C and hexokinase D differ by more than 300-fold in their affinity for glucose (Cárdenas, 1995) it would be impossible to choose scales for any of the other plots described in this chapter that would allow a direct comparison between the four isoenzymes. Even if hexokinase D were omitted the other three differ sufficiently from one another to make such a comparison difficult.

2.6.2   The double-reciprocal plot

Most workers since Lineweaver and Burk (1934) have preferred to rewrite the Michaelis–Menten equation in ways that allow the results to be plotted as points on a straight line. The way most commonly used is obtained from equation 2.15 by taking reciprocals of both sides:

$$\frac{1}{v} = \frac{1}{V} + \frac{K_m}{V} \cdot \frac{1}{a} \tag{2.32}$$

or, from equation 2.14,

$$\frac{e_0}{v} = \frac{1}{k_{cat}} + \frac{1}{k_A} \cdot \frac{1}{a} \tag{2.33}$$

The second of these emphasizes the correspondence between k_{cat} and k_A, which is somewhat obscured by the usual choice of V and K_m as the parameters

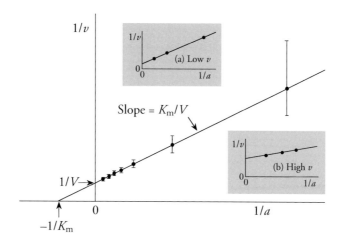

Figure 2.5. Plot of $1/v$ against $1/a$. Note the enormous variation in the lengths of the identically calculated error bars for v, each of which represents $0.05V$. Many authors refer to this plot as a *double-reciprocal plot* or a *Lineweaver–Burk plot*. The two insets illustrate how judicious choices of scales for the axes can disguise a poor design of experiment (see Section 2.6.4), whether (a) because the range of v values is too low, or (b) because it is too high. In each inset the three points shown are the three extreme points from the main plot.

of the Michaelis–Menten equation. Either way, a plot of $1/v$ against $1/a$ is a straight line with slope $1/k_A e_0$ $(= K_m/V)$ and intercepts $1/k_{cat} e_0$ $(=1/V)$ on the $1/v$ axis and $-k_A/k_{cat}$ $(= -1/K_m)$ on the $1/a$ axis. This plot is commonly known as the *Lineweaver–Burk* or *double-reciprocal plot*, and it is illustrated in Figure 2.5. Although it is by far the most widely used plot in enzyme kinetics, it cannot be recommended, because it gives a grossly misleading impression of the experimental error: for small values of v small errors in v lead to enormous errors in $1/v$; but for large values of v the same small errors in v lead to barely noticeable errors in $1/v$. This may be judged from the error bars shown in Figure 2.5, which are noticeably unsymmetrical even though they were calculated from the same symmetrical range of errors in v.

Dowd and Riggs (1965) drew attention to the capacity of the double-reciprocal plot to "launder" poor data, to minimize the *appearance* but not the reality of scatter; they suggested, probably rightly, that this accounted for its extraordinary popularity with biochemists, not much diminished nearly 40 years later.

In principle the problems with the double-reciprocal plot can be overcome by using suitable weights, but this solution is not altogether satisfactory because it often leads to a "best-fit" line that appears to the eye to fit badly. Incidentally, Lineweaver and Burk should not be blamed for the misuse of their plot by later workers: they were well aware of the need for weights and the methods to be used for determining them (see Lineweaver and Burk, 1934, and, especially,

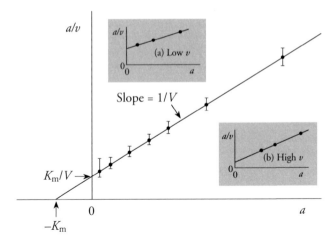

Figure 2.6. Plot of a/v against a, with error bars of $0.05V$, as in Figure 2.5. This is sometimes called a *Hanes plot* or a *Woolf plot*. The insets have the same significance as in Figure 2.5.

Lineweaver, Burk and Deming, 1934). One cannot assume, when authors report that they have used the method of Lineweaver and Burk, that they have actually read these papers or that they have used the method that Lineweaver and Burk used.

2.6.3   The plot of a/v against a

If we multiply both sides of equation 2.32 or 2.33 by a, we obtain the equation for a better plot:

$$\frac{a}{v} = \frac{K_m}{V} + \frac{1}{V} \cdot a \tag{2.34}$$

$$\frac{e_0 a}{v} = \frac{1}{k_A} + \frac{1}{k_{cat}} \cdot a \tag{2.35}$$

This shows that a plot of a/v against a should also be a straight line, with slope $1/k_{cat}e_0$ $(=1/V)$ and intercepts $1/k_A e_0$ $(=K_m/V)$ on the a/v axis and $-k_{cat}/k_A$ $(=-K_m)$ on the a axis. (Note that the slope and ordinate intercepts are the opposite way round from those of the double-reciprocal plot.) This plot, sometimes called a *Hanes plot* or a *Woolf plot*, is illustrated in Figure 2.6. Over a fair range of a values the errors in a/v provide a faithful reflection of those in v, as may be judged from the error bars in Figure 2.6: it is for this reason that the plot of a/v against a should be preferred over the other straight-line plots whenever the main objective is to illustrate how well the data agree with the interpretation.

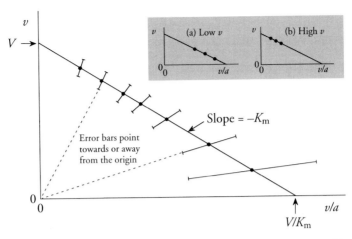

Figure 2.7. Plot of v against v/a, with error bars of $0.05V$, as in Figure 2.5. This plot is sometimes called an *Eadie–Hofstee plot*. The insets have the same significance as those in Figure 2.5, but now it is impossible to choose scales that disguise the poor experimental design in the insets.

2.6.4 The plot of v against v/a

Multiplying both sides of equation 2.32 by vV and rearranging, we obtain the equation for the third straight-line plot of the Michaelis–Menten equation:

$$v = V - K_m \frac{v}{a} \tag{2.36}$$

This shows that a plot of v against v/a should be a straight line with slope $-K_m$ and intercepts V on the v axis and V/K_m on v/a axis. This plot is often called an *Eadie–Hofstee plot*, and is illustrated in Figure 2.7. It gives fairly good results in practice, though the presence of v in both coordinates means that errors in v affect both of them and cause deviations towards or away from the origin rather than parallel with the ordinate axis.

The plot of v against v/a has the opposite character from the double-reciprocal plot: instead of making poor data look better it tends to make good data look worse, as it makes any deviations of the points from the line more difficult to hide, especially if there are systematic deviations (Dowd and Riggs, 1965). It follows that it is an excellent plot to use for detecting deviations from Michaelis–Menten behaviour. Moreover, the v axis from 0 to V corresponds to the entire observable range (all substrate concentrations from zero to infinity), so a poor experimental design covering only a small part of this range is likewise difficult to hide. In the world of advertising and public relations these characteristics would be seen as faults (and cynical biochemists may feel that this is indeed how they are seen by their colleagues who prefer to use the double-reciprocal plot). However, for researchers seriously interested in uncovering the behaviour of an enzyme they are clearly virtues.

All three of the straight-line plots were first ascribed in print to Woolf (1932), but were not published by him (see Haldane, 1957). Equation 2.34 was first published by Hanes (1932), but he did not present his results graphically. The plots became widely known and used as a result of the work of Lineweaver and Burk (1934), Eadie (1942) and Hofstee (1952), which is why these names have become associated with them.

2.6.5 The direct linear plot

The *direct linear plot* (Eisenthal and Cornish-Bowden, 1974) is a quite different way of plotting the Michaelis–Menten equation, which may be rearranged, most simply by taking equation 2.36 as a starting point, in yet another way to show the dependence of V on K_m:

$$V = v + \frac{v}{a} K_m \qquad (2.37)$$

If V and K_m are treated as variables, and a and v as constants, this equation defines a straight line of slope v/a and intercepts v on the V axis and $-a$ on the K_m axis. It may seem perverse to treat V and K_m as variables and a and v as constants, but in fact it is more logical than it appears: once a and v have been measured in an experiment, they are constants, because any honest analysis of the results will leave them unchanged, but until we have decided on best-fit values of V and K_m we can try any values we like, and in that sense they are variables.

For any pair of values a and v there is an infinite set of values of V and K_m that satisfy them exactly. For any arbitrary value of K_m, equation 2.37 defines the corresponding value of V. Consequently, the straight line drawn according to this equation relates all pairs of K_m and V values that satisfy one observation exactly. If a second line is drawn for a second observation (with different values of a and v), it will relate all pairs of K_m and V values that satisfy the second observation exactly. However, the two lines will not define the same K_m and V values except at the point of intersection. This point therefore defines the unique pair of K_m and V values that satisfies *both* observations exactly.

If there were no experimental error one could plot a series of such lines, each corresponding to a single determination of v at a particular value of a, and they would all intersect at a common point, which would specify the values of K_m and V that gave rise to the observations. This is illustrated in the main part of Figure 2.8. In reality, however, observations are never exact and so all the lines cannot be expected to intersect at exactly the same point, and a real

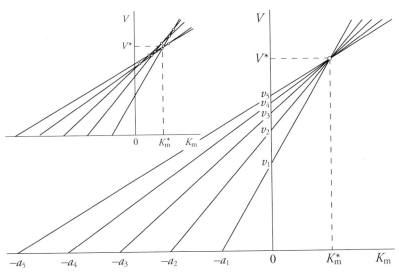

Figure 2.8. Direct linear plot of V against K_m. Each line represents one observation, and is drawn with an intercept of $-a$ on the abscissa and an intercept of v on the ordinate. In the idealized version (without experimental error) shown in the main part of the Figure, all the lines intersect at a unique point whose coordinates yield the values of K_m and V that fit the data. More realistically, as shown in the inset, experimental error causes this unique point to degenerate into a family of points, each of which yields an estimate of K_m and an estimate of V, and the best estimates can be taken as the medians (middle values) of the two series. For more detail see Figure 14.1 in Section 14.3.

plot is likely to resemble the one illustrated in the inset to Figure 2.8. Each intersection point provides one estimate of K_m and one of V, and in each series one can take the median (middle) estimate as the best one: the vertical line showing the best K_m value is drawn to the left of half of the individual intersection points and to the right of the other half; similarly, the horizontal line showing the best V value is drawn above half of the individual intersection points and below the other half.

Just as there are three ways of plotting the Michaelis–Menten equation as a straight line, there are three variants of the direct linear plot. The one shown in Figure 2.8 is the one described originally, but the variant shown in Figure 2.9 (Cornish-Bowden and Eisenthal, 1978), in which the intercepts are a_i/v_i and $1/v_i$ rather than $-a_i$ and v_i may be preferable in practice because the lines intersect at larger angles and the points of intersection are better defined. Moreover, as will be discussed in Section 14.3, it avoids the need for any special rules to deal with intersection points that occur outside the first quadrant, that is to say points that define negative values of one or other parameter.

The direct linear plot has a number of advantages over the other graphical methods described in this section, of which the most obvious is that in its original form (Figure 2.8) it requires no calculation at all. This allows it to

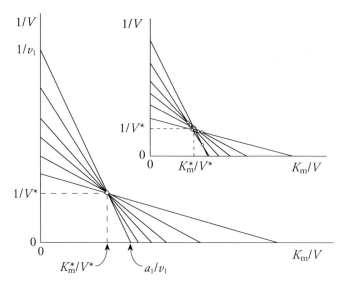

Figure 2.9. Modified direct linear plot, with $1/V$ plotted against K_m/V. Each observation is plotted as a straight line making an intercept of a/v on the abscissa and of $1/v$ on the ordinate. Experimental error causes the unique intersection point to degenerate into a family of points, as in Figure 2.8, and is handled in the same way.

be used at the laboratory bench while an experiment is proceeding, giving an immediate visual idea of the likely parameter values and of the design needed for defining them precisely. In other words one can obtain an immediate idea of the value of K_m from the first two or three measurements and use it to choose the appropriate substrate concentrations for the later ones. More important, perhaps, the direct linear plot (in any of its variants) has some desirable statistical properties that imply that it leads to reliable parameter values: this point will be taken up in Section 14.3. On the other hand, it is not a good plot for showing a large amount of data on the same graph, because it rapidly becomes crowded, and, like the plot of v against v/a, it tends to make any faults in the data rather obvious. These characteristics make it more suitable for use in the laboratory for the actual analysis of data than for subsequent presentation of the results.

2.7 The reversible Michaelis–Menten mechanism

2.7.1 The reversible rate equation

Many reactions of importance in biochemistry are reversible in the practical sense that significant amounts of both substrates and products exist in the reaction mixture when it has reached equilibrium. It is evident, therefore, that the Michaelis–Menten mechanism, as given, is incomplete, and that allowance should be made for the reverse reaction:

$$\text{E} + \text{A} \underset{k_{-1}}{\overset{k_1}{\rightleftharpoons}} \text{EA} \underset{k_{-2}}{\overset{k_2}{\rightleftharpoons}} \text{E} + \text{P} \qquad (2.38)$$
$$_{e_0-x \quad a x p}$$

The steady-state assumption is now expressed by

$$\frac{\mathrm{d}x}{\mathrm{d}t} = k_1(e_0 - x)a + k_{-2}(e_0 - x)p - (k_{-1} + k_2)x = 0 \qquad (2.39)$$

Gathering terms in x and rearranging, we obtain

$$x = \frac{k_1 e_0 a + k_{-2} e_0 p}{k_{-1} + k_2 + k_1 a + k_{-2} p} \qquad (2.40)$$

Because this is a reversible system of reactions, the *net* rate of release of P is obtained by subtracting the rate at which it is consumed in the reaction $\text{E} + \text{P} \rightarrow \text{EA}$ from the rate at which it is released in the reaction $\text{EA} \rightarrow \text{E} + \text{P}$:

$$v = k_2 x - k_{-2}(e_0 - x)p$$
$$= \frac{k_2(k_1 e_0 a + k_{-2} e_0 p)}{k_{-1} + k_2 + k_1 a + k_{-2} p} - k_{-2} e_0 p + \frac{k_{-2}(k_1 e_0 a + k_{-2} e_0 p)p}{k_{-1} + k_2 + k_1 a + k_{-2} p} \qquad (2.41)$$

Cross-multiplication to express everything over the same denominator gives an apparently complicated numerator with eight terms. However, six of these cancel, leaving

$$v = \frac{k_1 k_2 e_0 a - k_{-1} k_{-2} e_0 p}{k_{-1} + k_2 + k_1 a + k_{-2} p} \qquad (2.42)$$

The special case $p = 0$ gives the same equation as before, equation 2.13, except that a should be replaced by a_0, because only at zero time is it legitimate to put $p = 0$. It is important to realize that this simplification is possible because p is zero, *not* because of any assumption about the magnitude of k_{-2}: $k_{-2}p = 0$ if $p = 0$, regardless of the value of k_{-2}. The essential distinction is between a rate and rate constant, and as it is pervasive in kinetics it needs to be thoroughly understood: a rate can be zero regardless of the size of any individual factor in the rate expression if any of the other factors is zero.

When $a = 0$ equation 2.42 simplifies to a complementary special case for the initial rate of the reverse reaction:

$$v = \frac{-k_{-1} k_{-2} e_0 p}{k_{-1} + k_2 + k_{-2} p} \qquad (2.43)$$

The negative sign in this equation is a consequence of defining the rate as the rate of release of P, dp/dt; if it had been defined as da/dt the rate would have turned out to be positive. Apart from the sign, this equation is of the same form as the Michaelis–Menten equation, and by comparing it with equations 2.13–20 we can define parameters for the reverse reaction:

$$k_{cat}^P = k_{-1}; \quad k_P = \frac{k_{-1}k_{-2}}{k_{-1}+k_2}; \quad K_{mP} = \frac{k_{-1}+k_2}{k_{-2}} \tag{2.44}$$

which are analogous to the corresponding definitions for the forward reaction:

$$k_{cat}^A = k_2; \quad k_A = \frac{k_1k_2}{k_{-1}+k_2}; \quad K_{mA} = \frac{k_{-1}+k_2}{k_1} \tag{2.45}$$

Using the definitions, equation 2.42 can be rewritten as follows:

$$v = \frac{k_A e_0 a - k_P e_0 p}{1 + \dfrac{a}{K_{mA}} + \dfrac{p}{K_{mP}}} \tag{2.46}$$

This equation can be regarded as the general reversible form of the Michaelis–Menten equation. It has the advantage over equation 2.42 that it does not imply any particular mechanism and can be regarded as purely empirical: there are many mechanisms more complicated than the one defined by equation 2.38 that lead to rate equations equivalent to equation 2.46.

We shall need to consider more complicated mechanisms immediately because equation 2.38 as it stands is illogical, and the apparent symmetry in reading it from left to right or vice versa is an illusion. The problem is that although k_2 and k_{-1} look as if they are analogous to one another they are not, because k_{-1} is usually understood as the rate constant for a simple substrate-release step, whereas k_2 is understood to include not only product release but also the chemical transformation of enzyme-bound substrate into enzyme-bound product. This might not matter much if it were not that a major objective of mechanistic investigations of enzymes is often to understand, in as much detail as possible, the chemical processes that allow an enzyme to effect a chemical transformation, and one cannot understand this if one fails to separate processes that are conceptually distinct.

We therefore now consider the more realistic three-step mechanism in which the conversion of A into P in the catalytic site of the enzyme is represented as a process separate from release of P from the enzyme:

$$\underset{e_0-x-y}{E} + \underset{a}{A} \underset{k_{-1}}{\overset{k_1}{\rightleftharpoons}} \underset{x}{EA} \underset{k_{-2}}{\overset{k_2}{\rightleftharpoons}} \underset{y}{EP} \underset{k_{-3}}{\overset{k_3}{\rightleftharpoons}} \underset{p}{E + P} \tag{2.47}$$

In principle we can derive a rate equation for this mechanism by the same method as before. However, there are now two intermediates EA and EP, and both dx/dt and dy/dt must be set to zero. Two simultaneous equations in x and y must be solved and the derivation is more complicated than those made earlier. As I shall be describing a more versatile method in Chapter 4, I shall simply state here that the three-step mechanism again leads to equation 2.46, but the definitions of the parameters are now

$$k_{cat}^A = \frac{k_2 k_3}{k_{-2} + k_2 + k_3}; \quad k_A = \frac{k_1 k_2 k_3}{k_{-1} k_{-2} + k_{-1} k_3 + k_2 k_3}; \quad K_{mA} = \frac{k_{-1} k_{-2} + k_{-1} k_3 + k_2 k_3}{k_1 (k_{-2} + k_2 + k_3)}$$

$$(2.48)$$

for the forward reaction and

$$k_{cat}^P = \frac{k_{-1} k_{-2}}{k_{-1} + k_{-2} + k_2}; \quad k_P = \frac{k_{-1} k_{-2} k_{-3}}{k_{-1} k_{-2} + k_{-1} k_3 + k_2 k_3}; \quad K_{mP} = \frac{k_{-1} k_{-2} + k_{-1} k_3 + k_2 k_3}{(k_{-1} + k_{-2} + k_2) k_{-3}}$$

$$(2.49)$$

for the reverse reaction. In spite of their complicated appearance, the expressions for K_{mA} and K_{mP} simplify to the true substrate dissociation constants K_{sA} and K_{sP} of EA and EP respectively if the second step in the appropriate direction is rate-limiting. Thus,

$$K_{mA} = \frac{k_{-1}}{k_1} = K_{sA} \quad \text{if} \quad k_2 \ll (k_{-2} + k_3) \qquad (2.50)$$

$$K_{mP} = \frac{k_3}{k_{-3}} = K_{sP} \quad \text{if} \quad k_{-2} \ll (k_{-1} + k_2) \qquad (2.51)$$

Both simplifications apply simultaneously if both conditions are satisfied, that is to say if the interconversion of EA and EP is rate-limiting in both directions.

The question for which equations 2.50–51 give the answer is not often asked the other way round, but doing so helps to understand the points raised at the end of Section 2.3.1: are there any conditions in which the two Michaelis constants are *both* very different from the corresponding equilibrium constants? Atkinson (1977) is one of the few to have asked this question, and the answer is surprising. Clearly, from equations 2.50–51, we require k_2 to be large compared with $k_{-2} + k_3$ and k_{-2} to be *simultaneously* large compared with $k_{-1} + k_2$. As this is impossible, it follows that in at least one direction of reaction (and often in both) the Michaelis constant will be reasonably close to the corresponding dissociation constant.

Atkinson concludes that "we need not hesitate to infer probable affinities of enzymes for substrates from kinetic Michaelis constants". This, however, takes the argument too far, because there are many reactions in biochemistry that proceed in only one direction under all physiological conditions, and in which the equilibrium constant favours this physiological direction of reaction. For these reactions we are more likely to be interested in the Michaelis constant for the forward direction, and it is this Michaelis constant that is more likely to differ from the corresponding dissociation constant, because forward rate constants must on average be larger than comparable reverse rate constants for any reaction with an equilibrium constant that favours the forward direction.

It is also possible for both Michaelis constants to be equilibrium constants without either binding step being at equilibrium: if $k_{-1} = k_3$ (if A and P are released from their respective complexes with the same rate constant), then both expressions have common factors of $(k_{-2} + k_2 + k_3)$ in numerator and denominator, which cancel. Both of the simplifications given in equations 2.50–51 then apply without any assumptions about the magnitudes of k_{-2} or k_2 (Cornish-Bowden, 1976a).

Despite these various examples in which K_m is equivalent to a substrate dissociation constant, in general it is not safe to assume this unless there is definite evidence for it. K_m is best taken as an empirical quantity that describes the dependence of v on a, not as a measure of the thermodynamic stability of the enzyme–substrate complex.

2.7.2   The Haldane relationship

When a reaction is at equilibrium, the net rate must be zero, and so if a_∞ and p_∞ are the equilibrium values of a and p it follows from equation 2.46 that

$$k_A e_0 a_\infty - k_P e_0 p_\infty = 0 \tag{2.52}$$

and, as $k_A = k_{cat}^A / K_{mA}$ and $k_P = k_{cat}^P / K_{mP}$, it follows that

$$\frac{k_A}{k_P} = \frac{k_{cat}^A K_{mP}}{k_{cat}^P K_{mA}} = \frac{p_\infty}{a_\infty} = K_{eq} \tag{2.53}$$

where K_{eq} is the equilibrium constant of the reaction, and k_{cat}^A and k_{cat}^P are the catalytic constants for the forward and reverse reactions respectively. This result is known as the *Haldane relationship* (Haldane, 1930), and is true for any mechanism described by equation 2.46, not merely for the simple two-step Michaelis–Menten mechanism. More complicated rate equations, such as those that describe reactions of several substrates, lead to more complicated Haldane

John Burdon Sanderson Haldane (1892–1964)

J. B. S. Haldane was one of the great biologists of the twentieth century, and his work on enzymes, important though it was, formed just a small part of a career that included major contributions to physiology, genetics, mathematical statistics, popular science, journalism and politics. His book *Enzymes*, published in 1930, remains challenging and interesting more than 70 years later. In the preface to the paperback edition of 1964 he remarked that Frederick Hopkins had convinced him that enzymes were the central topic of biochemistry, and that his father, the physiologist J. S. Haldane, had shown in about 1910 that despite the large size of the haemoglobin molecule its properties could be understood in terms of the laws that hold for small molecules: two insights that have lost none of their importance in the era of genomic research. He retained his interest in enzymes to the end of his life, this preface being written only a few months before he died from cancer. He was almost unique in having enjoyed his service in the front lines during the First World War, and the verse that he wrote about his colostomy, which begins "I wish I had the voice of Homer, to sing of rectal carcinoma…", indicates an ability to remain cheerful in almost any circumstances.

relationships, but for all equations there is at least one relationship of this kind between the kinetic parameters and the equilibrium constant.

2.7.3 "One-way enzymes"

Some enzymes are much more effective catalysts for one direction of reaction than the other. As a striking example, the limiting rate of the forward reaction catalysed by methionine adenosyltransferase is about 2×10^5 greater than that for the reverse reaction, even though the equilibrium constant is close to unity (Mudd and Mann, 1963). Even after a thorough discussion of this type of behaviour by Jencks (1975), many biochemists remain rather uneasy about it, suspecting that it may violate the laws of thermodynamics. However, although these laws, in the form of the Haldane relationship, do indeed place limits on the kinetic behaviour that an enzyme can display, they still allow a wide range of behaviour.

　　The thermodynamic condition defined by equation 2.52 shows that the ratio of the specificity constants k_A/k_P is equal to the equilibrium constant K_{eq}. The specificity constants fully define the rates in the limit of low substrate concentrations, and so under these conditions adding a catalyst must increase the rates equally in the forward and reverse directions, since otherwise the equilibrium constant would be changed.

At higher concentrations, however, the enzyme can influence the rates differently in the two directions. The simplest case to consider concerns the rates at the limit of high substrate concentrations. They now depend not on k_A and k_P but on k_{cat}^A and k_{cat}^P, which may be written as $k_A K_{mA}$ and $k_P K_{mP}$. It follows that if K_{mA} is very different from K_{mP}, the degree of catalysis in one direction may be very different from that in the other. Suppose, for example, that the equilibrium constant is unity, and that K_{mA} is 10^5 times greater than K_{mP}. The catalytic constant $k_{cat}^A = k_A K_{mA}$ is then 10^5 times larger than $k_{cat}^P = k_P K_{mP}$, and in the limit of high substrate concentrations the enzyme affects the rate in the forward direction much more than that in the reverse direction.

To examine the range of behaviour possible, we can ignore the ratio of specificity constants, because this is set by thermodynamics, and we can also ignore the actual magnitudes of these constants because these just reflect the general level of catalytic activity of the enzyme. So, returning to equation 2.46, and setting $k_A e_0 = 1$, $k_P e_0 = 1$ and $a + p = 1$ (all in arbitrary units), it can be written as follows:

$$v = \frac{1 - p/a}{\dfrac{1}{a} + \dfrac{1}{K_{mA}} + \dfrac{p/a}{K_{mP}}} = \frac{1 - p/a}{1 + \dfrac{p}{a} + \dfrac{1}{K_{mA}} + \dfrac{p/a}{K_{mP}}} \qquad (2.54)$$

(The right-hand expression follows from dividing all terms in the definition $a + p = 1$ by a.) This equation now violates the principles of dimensional analysis (Section 1.3), so it cannot be the expression of a general kinetic property, but it is acceptable as long as its use is limited to numerical calculations of the effects of particular parameters on the kinetic behaviour.

Equation 2.54 shows that in the conditions considered the rate of approach to equilibrium is determined by the values of p/a and the Michaelis constants. As seen in Figure 2.10, if K_{mA} and K_{mP} are very different in magnitude a plot of v against $\ln(p/a)$ drawn at high reactant concentrations is highly unsymmetrical: if $K_{mP} \ll K_{mA}$ (if product binds much more tightly than substrate, as happens, for example, with numerous dehydrogenases that convert the oxidized form of NAD into the reduced form) the curve is much steeper when equilibrium is approached in the forward direction than when it is approached in the reverse direction; in the opposite conditions it is the reverse. An enzyme may therefore be much better at catalysing one direction of the reaction than the other, without violating any thermodynamic constraints. Study of Figure 2.10 confirms that the rate at equilibrium is zero, and that in any other state the reaction is always towards equilibrium: these are all that thermodynamics

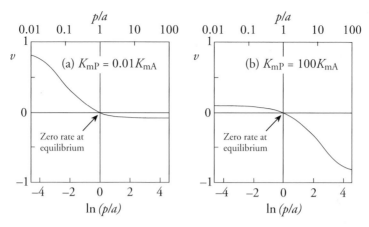

Figure 2.10. "One-way enzymes". Both curves are calculated from equation 2.54, which assumes an equilibrium constant of 1, or equilibrium at $p/a = 1$, with the total reactant concentration set at $a + p = 1$, but the ratio of Michaelis constants in the forward and reverse directions is very different in the two cases. (a) When $K_{mP} \ll K_{mA}$ the curve is steep as equilibrium is approached in the forward direction, but almost flat when it is approached in the reverse direction; but (b) the reverse is true when $K_{mP} \gg K_{mA}$. Note that these curves imply no violation of thermodynamic principles, because under all circumstances the *direction* of reaction is towards equilibrium.

requires; there is no requirement that the curves should be symmetrical about the equilibrium point or that equilibrium should be approached just as steeply from either direction.

The possibility of a "one-way enzyme" makes physiological sense, because many enzymes never need to catalyse a reaction in the reverse direction *in vivo*, and so there is no reason why efficient catalysis of the reverse reaction should have evolved. The absence of a disadvantage is not the same as an advantage, however, and one may still ask why such behaviour should have been selected. The answer probably lies in the limited amount of binding energy available for designing the enzyme, as discussed by Jencks (1975). If the active site is made strictly complementary to the transition state of the uncatalysed reaction that converts A into P, the enzyme will be optimized as a catalyst for both directions. However, if only one of the directions has any physiological importance the efficiency in this direction can be improved at the expense of the other by evolving an active site that binds one or other reactant better than it binds the transition state.

2.8   Product inhibition

Product inhibition is simply a special case of inhibition, which will be discussed in detail in Chapters 5 and 6, but because it follows naturally from the previous section it is convenient to mention it briefly here. When equation 2.46 applies, the rate must decrease as product accumulates, even if the decrease in substrate concentration is negligible, because the negative term in the numerator becomes increasingly important as equilibrium is approached, and because the third term in the denominator increases with p. In any reaction, the negative term in the numerator can only have a noticeable effect if the reaction is significantly reversible. However, product inhibition is observable in many essentially irreversible reactions, such as the classic example of the invertase-catalysed hydrolysis of sucrose. Such inhibition indicates that the product must be capable of binding to the enzyme, and is compatible with the simplest two-step mechanism only if the *first* step is irreversible and the second is not. Irreversible binding of substrate followed by reverse product release does not seem likely, at least as a general phenomenon. On the other hand, the three-step mechanism predicts that product inhibition can occur in an irreversible reaction if it is the second step, the chemical transformation, that is irreversible. When this is true the accumulation of product causes the enzyme to be sequestered as the EP complex, and hence unavailable for reacting with the substrate. For an irreversible reaction equation 2.46 then becomes

$$v = \frac{k_A e_0 a}{1 + \dfrac{a}{K_{mA}} + \dfrac{p}{K_{mP}}} = \frac{k_{cat} e_0 a}{K_{mA}(1 + p/K_{sP}) + a} \tag{2.55}$$

If the reaction is irreversible K_{mP} can legitimately be written as an equilibrium constant, K_{sP}, because if k_{-2} approximates to zero it must be small compared with $(k_{-1} + k_2)$. As will be seen in Section 5.2.1, equation 2.55 has exactly the form of the commonest type of inhibition, known as *competitive inhibition*.

Of course, the effect of added product should be the same as that of accumulated product, so one could measure initial rates with different concentrations of added product. For each product concentration, the initial rate for different substrate concentrations would obey the Michaelis–Menten equation, but with *apparent values* of k_{cat} and K_{mA}, given by $k_{cat}^{app} = k_{cat}$ and $K_{mA}^{app} = K_{mA}(1 + p/K_{sP})$, as may be seen by comparing equation 2.55 with equation 2.14. Thus k_{cat}^{app} is constant with the same value k_{cat} as for the uninhibited reaction, but K_{mA}^{app} is larger than K_{mA} and increases linearly with p.

The first thorough investigations of product inhibition were done by Michaelis and Rona (1914) on maltase and Michaelis and Pechstein (1914) on invertase, though they were anticipated to some degree by Henri (1903), who had already derived an equation equivalent to equation 2.55. With some products, such as fructose as an inhibitor of invertase, they observed competitive inhibition, but with others, such as glucose with the same enzyme, the behaviour was more complicated. There are anyway many substances apart from products that inhibit enzymes, so it is clear that a more complete theory is needed: this will be developed in later chapters (especially Chapter 5).

2.9　Integration of enzyme rate equations

2.9.1　Michaelis–Menten equation without product inhibition

As mentioned at the beginning of this chapter, the earliest students of enzyme kinetics encountered many difficulties because they followed reactions over extended periods of time, and then tried to explain their observations in terms of integrated rate equations similar to those commonly used in chemical kinetics. Michaelis and Menten (1913) then showed that the behaviour of enzymes could be studied much more simply by measuring initial rates, when the complicating effects of product accumulation and substrate depletion did not apply. An unfortunate by-product of this early history, however, has been that biochemists have been reluctant to use integrated rate equations even when they have been appropriate. It is not always possible to do steady-state experiments in such a way that the progress curve (the plot of p against t) is essentially straight during an extended period, and estimation of the initial slope of a curve, and hence the initial rate, is subjective and liable to be biassed. Much of this subjectivity can be removed by using an integrated form of the rate equation, as I shall now describe.

The Michaelis–Menten equation can be written as an equation in three variables, p, t and a, as follows (compare equation 2.15):

$$\frac{dp}{dt} = \frac{Va}{K_m + a} \tag{2.56}$$

As such it cannot be integrated directly, but one of the three variables can be removed by means of the stoicheiometric relationship $a + p = a_0$, which is accurate enough in the usual conditions with a_0 much larger than the enzyme concentration. Then we have

$$\frac{dp}{dt} = \frac{V(a_0 - p)}{K_m + a_0 - p} \tag{2.57}$$

which may be integrated by separating the two variables on the two sides of the equation:

$$\int \frac{(K_m + a_0 - p)dp}{a_0 - p} = \int V dt \tag{2.58}$$

The left-hand side of this equation is not immediately recognizable as a simple integral, but it can be separated into two terms:

$$\int \frac{K_m dp}{a_0 - p} + \int dp = \int V dt \tag{2.59}$$

The first is of the standard form used several times already in this book (for example, in Section 1.2):

$$\int \frac{K_m dp}{a_0 - p} = -K_m \ln(a_0 - p) \tag{2.60}$$

and the second is trivial, so

$$-K_m \ln(a_0 - p) + p = Vt + \alpha \tag{2.61}$$

in which α is a constant of integration. It may be evaluated as $\alpha = -K_m \ln a_0$ by means of the boundary condition $p = 0$ when $t = 0$. After substituting this value of α and rearranging, we have

$$Vt = p + K_m \ln \left(\frac{a_0}{a_0 - p} \right) \tag{2.62}$$

2.9.2   Effect of product inhibition on the progress curve

Although equation 2.62 is a valid integrated form of equation 2.56, it is highly misleading to regard it as the integrated form of the Michaelis–Menten equation, and it can lead to gross errors in the parameter estimates if it is used without taking account of violation of the initial-rate conditions that are fundamental to the Michaelis–Menten approach. The essential problem is that although product inhibition can usually be ignored in initial-rate studies, it can rarely* be ignored in studies of time courses, because whatever the product

---

*The main exceptions occur when the product released from the enzyme is transformed rapidly and completely into a different substance by a non-enzymic reaction. For example, an enzyme like aldehyde dehydrogenase releases its product as the free carboxylic acid, which would certainly inhibit the enzyme if it accumulated. In practice, however, it does not accumulate because it is deprotonated to the corresponding anion. If the anion acted as a significant inhibitor of the enzyme it would be a matter of definition whether to call this product inhibition, but for the purposes of this section it would have the same effects as product inhibition.

concentration at the beginning it will accumulate to positive values as the reaction proceeds. In the simplest case we need to replace equation 2.57 with an equation that allows for competitive inhibition by P with inhibition constant K_p:

$$\frac{\mathrm{d}p}{\mathrm{d}t} = \frac{V(a_0 - p)}{K_m\left(1 + \dfrac{p}{K_p}\right) + a_0 - p} \tag{2.63}$$

This has the same algebraic form as equation 2.57, which has two important consequences: first, we *cannot tell* from experimental time-course data that accord with equation 2.62 whether equation 2.57, as opposed to equation 2.63, is the correct starting point; second, the same logic that allowed integration of equation 2.57 can be applied with only trivial changes to the integration of equation 2.63, and the result is as follows:

$$\frac{Vt}{1 - K_m/K_p} = p + \frac{K_m(1 - a_0/K_p)}{1 - K_m/K_p} \ln\left(\frac{a_0}{a_0 - p}\right) \tag{2.64}$$

Henri (1903) obtained an equation equivalent to this in his studies of invertase. It has exactly the same form as equation 2.62 and can be written as follows:

$$V^{app}t = p + K_m^{app} \ln[a_0/(a_0 - p)] \tag{2.65}$$

with

$$V^{app} = \frac{V}{1 - K_m/K_p} \tag{2.66}$$

$$K_m^{app} = \frac{K_m(1 + a_0/K_p)}{1 - K_m/K_p} \tag{2.67}$$

Thus equation 2.62 has been written with V and K_m replaced by apparent values V^{app} and K_m^{app} respectively. Note that the denominator of $1 - K_m/K_p$ in both expressions means that the "apparent" values can differ from the real Michaelis–Menten parameters by enormous factors, unless the binding of the product to the enzyme is truly negligible, and they will be negative if it binds more tightly than the substrate (if K_p is less than K_m). To appreciate what is meant by "truly negligible" in this context, suppose that we want to analyse time courses with $a_0 = 2K_m$ and we need K_m^{app} to be within 5% of the true K_m. Making the appropriate substitutions in equation 2.67 we find that K_m needs to be at most $0.03K_p$, which means that the substrate needs to bind at least 33 times more tightly to the enzyme than the product. This is not impossible, but it is sufficiently unlikely to make it dangerous to use equation 2.62 directly for estimating the Michaelis–Menten parameters if no precautions are taken.

2.9.3   Accurate estimation of initial rates

Despite the complications just discussed, measurements of V^{app} and K_m^{app} can be useful if we recognize them for their true worth, not as direct measures of V and K_m but as vehicles for calculating highly accurate values of the initial rate v_0 by inserting them into the following equation:

$$v_0 = \frac{V^{app} a_0}{K_m^{app} + a_0} \tag{2.68}$$

This equation follows from equation 2.65 by differentiation, regardless of the meanings of V^{app} and K_m^{app}.

Rearranging equation 2.65, we have

$$\frac{t}{\ln[a_0/(a_0 - p)]} = \frac{1}{V^{app}} \left\{ \frac{p}{\ln[a_0/(a_0 - p)]} \right\} + \frac{K_m^{app}}{V^{app}} \tag{2.69}$$

which shows that a plot of $t/\ln[a_0/(a_0 - p)]$ against $p/\ln[a_0/(a_0 - p)]$ gives a straight line of slope $1/V^{app}$ and intercept K_m^{app}/V^{app} on the ordinate. V^{app} and K_m^{app} can readily be determined from such a plot, and v_0 can be calculated from them with equation 2.68. However, v_0 may also be found directly without evaluating V^{app} and K_m^{app} by a simple extrapolation of the line: rearrangement of equation 2.68 into the form of equation 2.34 shows that the point $(a_0, a_0/v_0)$ ought to lie on a straight line of slope $1/V^{app}$ and intercept K_m^{app}/V^{app} on the ordinate, the same straight line as that plotted from equation 2.69. This means that if that line is extrapolated back to a point at which $p/\ln[a_0/(a_0 - p)] = a_0$, the value of the ordinate must be a_0/v_0. The whole procedure is illustrated in Figure 2.11. The extrapolated point can then be treated as a point on an ordinary plot of a_0/v_0 against a_0 (compare Figure 2.6), and if several such points are found from several progress curves with different values of a_0, K_m and V may be found as described previously (Section 2.6.3).

This procedure, which originated with Jennings and Niemann (1955), may seem an unnecessarily laborious way of generating an ordinary plot of a_0/v_0 against a_0, but it provides more accurate values of a_0/v_0 than are available by more ordinary methods, and hence more accurate parameter values. The extrapolation required is short, and it can be carried out more precisely and less subjectively than estimating the tangent of a curve extrapolated back to

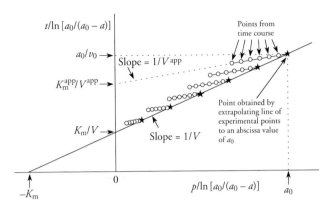

Figure 2.11. Determination of kinetic parameters from time courses. At each of several different initial substrate concentrations a_0 the values of $t/\ln[a_0/(a_0 - p)]$ are plotted as time t increases against $p/\ln[a_0/(a_0 - p)]$, and each set of such points ($\circ$) is extrapolated back to an abscissa value of a_0 to give a point ($\star$) that lies on a straight line of slope $1/V$ with intercepts $-K_m$ and K_m/V on the abscissa and ordinate respectively.

zero time. A study of subtilisin variants in which this approach was found to give satisfactory results is described by Brode and co-workers (1996).

A simpler method, involving only trivial calculations (no logarithms) may be obtained by using an approximation to the logarithmic term, as $-a_0 + p/2$ is almost exactly equal to $p/\ln[a_0/(a_0 - p)]$ over a sufficiently wide range, and applying the logic of the direct linear plot. This is described by Cornish-Bowden (1975).

Another method has been described by Boeker (1982), which also requires only trivial calculations, using a different approximation to the logarithmic term. To introduce this it is convenient to start with the integrated rate equation for a simple first-order reaction, equation 1.4, and note that it can be written as follows:

$$\ln(1 - p/a_0) = -kt \tag{2.70}$$

with p representing the concentration of product at time t and first-order rate constant k, for a starting reactant concentration of a_0. For small values of x, $2x/(2 + x)$ is an excellent approximation to $\ln(1 + x)$, with errors of less than 0.1% and 1% for x less than 0.1 and 0.41 respectively. So, for up to 40% of reaction, equation 2.70 can be written as follows with better than 1% accuracy:

$$\frac{2p/a_0}{2 - p/a_0} = kt \tag{2.71}$$

With a little rearrangement,

$$\frac{p}{t} = ka_0 - \frac{kp}{2} \qquad (2.72)$$

it is seen that a plot of p/t against p should give a straight line with an intercept on the ordinate of ka_0, the initial rate of reaction. Boeker (1982) showed that the intercept still gives the initial rate in more general cases of first-order kinetics, for example if p is not zero at $t = 0$, or there is less than 100% conversion when the reaction is complete. In such cases the *increase* in p must be plotted, in other words the ordinate variable needs to be $(p - p_0)/t$ rather than p/t. More important for enzyme kinetics, she also realized that although the theory does not apply quite as simply to enzyme reactions, there is no practical difference, because the departure from expectation is normally too small to detect.

Applying the same approximation for $\ln(1 + x)$ in equation 2.62, therefore, the result after rearrangement is as follows:

$$\frac{p}{t} = \frac{Va_0}{K_m + a_0} - \frac{Vp}{2(K_m + a_0)} + \frac{p^2}{2(K_m + a_0)t} \qquad (2.73)$$

Because of the third term on the right-hand side this no longer exactly defines a straight line, but there are two important points to consider. First, this third term is zero at the start of the reaction, so it has no effect on the ordinate intercept, which is still the initial rate. Second, it is in general small enough to produce only trivial deviations from linearity in the early part of the reaction, and consequently has only trivial effects on the use of a plot of p/t against p for estimating the initial rate.

A time course published by Schønheyder (1952) provides a convenient example for testing this (or any other) method of estimating initial rates, as it consists of 25 observations covering the reaction from 2.5% to 98.5% completion. The Boeker plot for this example is illustrated in Figure 2.12, with the unmodified progress curve in the inset. Notice that the Boeker plot is essentially straight for the first 15 minutes, up to about 40% of reaction. Notice also that, although systematic errors in the approximation tend to zero at zero time, the reverse is true of random errors, because, in the limit at zero time the value of p/t is 0/0, which means that it is undefined. The practical consequence is that the line should be drawn by eye, paying little attention to the scattered points at low t. Applying a naive linear regression to obtain the "best" line is likely to produce a line that is far from being the best because it will be excessively influenced by the least precise points.

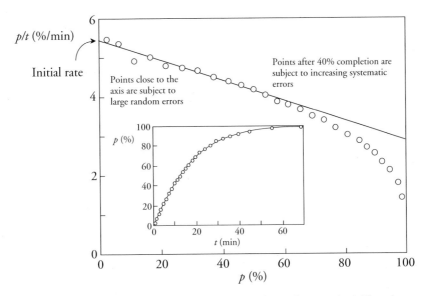

Figure 2.12. Accurate determination of initial rates by Boeker's method. The value of p/t extrapolated to $p = 0$ provides the initial rate. It is important to realize that random errors are large near the axis, so the extrapolation must be done by eye, not by linear regression. The inset shows the plain progress curve: note that it is curved over the whole time course, with no suggestion of an initial "linear phase".

2.9.4   Time courses for other mechanisms

In the initial development of methods for integrating rate equations and analysing time courses each mechanism tended to be treated in isolation; although numerous examples were analysed, including reactions subject to competitive product inhibition (Henri, 1903; Huang and Niemann, 1951; Schønheyder, 1952), reversible reactions (Alberty and Koerber, 1957) and some reactions with more than one substrate (Laidler and Bunting, 1973), these did not fit in any obvious way into a general framework. Later, however, Boeker (1984, 1985) developed a systematic approach that allows many of the mechanisms important in the study of enzymes to be handled in a consistent way. By this time, however, less efficient initial-rate methods of analysing the same mechanisms had become so widespread that her work has had less impact than it merited. An example of an application may be found, however, in a study of variants of aspartate aminotransferase (Schiller, Holmes and Boeker, 1996).

   In recent years the main work on the analysis of time courses of enzyme-catalysed reactions has been that of Duggleby and co-workers (Duggleby, 1995; Goudar, Sonnad and Duggleby, 1999), and his recent article (Duggleby, 2001) should be consulted for more information. The second of these articles is of particular interest as it describes how to express the progress of a reaction as a function of time, which is far more convenient for curve-fitting purposes than the usual forms (such as equation 2.62) that express time as a function of the progress.

2.10　Artificial enzymes, RNA enzymes and catalytic antibodies

2.10.1　"Alternative enzymes"

Nearly all known enzymes are naturally occurring proteins, often though not always containing non-protein cofactors. Important though this is, however, it has little direct impact on the kinetic description and analysis of enzyme-catalysed reactions, and little in this book would need to be rewritten if it were otherwise. For example, if natural enzymes were based on the structure of RNA, the main thing that would need to be changed in discussing enzyme kinetics would be the treatment of ionizing groups (Section 9.2). Increasing interest in recent years in various kinds of "alternative enzymes" has therefore scarcely affected the relevance of the classical material, as all of the classical methods apply to these as well as they do to the classical kinds of enzymes. Nonetheless, a brief description of these alternatives is appropriate.

Catalytic RNA has the same claim to our attention as classical protein enzymes: it is a natural component of living systems, and understanding its behaviour is a natural step in the understanding of life. Moreover, although at present there are far fewer RNA enzymes than protein enzymes, they have a particular importance for studies of evolution and the origin of life, because it is widely believed that they evolved earlier than protein enzymes (see for example Szathmáry and Maynard-Smith, 1993). Interest in the other alternative enzymes is primarily technological, as it is hoped that they can provide catalysts more suitable than natural enzymes for industrial purposes, offering either greater stability or specificity for reactions that do not occur in living organisms.

It is natural for people interested in these molecules to attribute to them an importance that is based more on an optimistic assessment of their potential than on anything that has actually been achieved. One of the ways that this is done is to use special names that obscure rather than illuminate their nature, such as *synzyme* (= *syn*thetic en*zyme*) for artificial enzyme, *ribozyme* (= *ribo*nucleic acid en*zyme*) for catalytic RNA, and *abzyme* (= *ab*tibody en*zyme*) for catalytic antibody. More serious is a tendency to present results in a misleading way, suggesting that they are much better catalysts than they are: although this tendency has decreased, exaggerated claims are still being made for catalytic antibodies (Wentworth, 2002), for example. One should be particularly cautious of results obtained with substrates chosen to illustrate specific claims rather than for the industrial importance of the reactions they can undergo.

2.10.2   Artificial enzymes

The first "artificial enzyme" was made from the bacterial proteinase subtilisin by using chemical methods to alter the active-site serine residue into a cysteine residue (Polgar and Bender, 1966; Neet and Koshland, 1966). As the resulting molecule was a protein, albeit an unnatural one, we probably would not nowadays call it an artificial enzyme, especially as unnatural proteins are now produced daily by genetic manipulation rather than by chemical modification. Nonetheless, it is interesting to compare the reactions of the two groups who independently prepared it, as they shed light on modern attitudes to artificial enzymes. Polgar and Bender were impressed that thiol-subtilisin could catalyse a reaction at all, and called their molecule a "new enzyme containing a synthetically formed active site". Neet and Koshland, noting that thiol groups are in general much more reactive than alcohol groups and that the change implied an extremely small conformational change, called it "a chemical mutation". They considered that the important question to ask was why it was so much worse a catalyst than one might have hoped, and the answer is the same today when one is faced with similarly unimpressive results: it is because we still lack a complete understanding of why natural enzymes are as good catalysts as they are.

Artificial enzymes can also be macromolecules produced by chemical methods that do not necessarily use any of the structural features found in real enzymes. As the long-term objective is to produce catalysts for industrial use that combine high stability (for example at high temperature or extreme pH) with the high catalytic effectiveness and specificity of enzymes, the backbone structures are often derived from polymers that are more stable and resistant to extreme conditions than proteins, such as cyclic peptides, poly(ethylene imine) and cyclodextrins (cyclic polymers of glucose). Polymerization of ethylene imine produces a highly branched molecule containing primary, secondary and tertiary amine groups, which confer high solubility in water and which can readily be modified by alkylation or acylation to produce a variety of microenvironments and functionalities.

One of the first successes claimed for this kind of polymer was the production of a catalyst for the hydrolysis of 2-hydroxy-5-nitrophenyl sulphate, about 100 times more efficient than the same reaction catalysed by the enzyme type IIA aryl sulphatase (Kiefer and co-workers, 1972). Nonetheless, this is not a particularly useful reaction to be able to catalyse; the value of

$k_{cat}/K_m = 1.1 \, \mathrm{M^{-1} \, s^{-1}}$ is many orders of magnitude smaller than the values observed with many enzymes with natural substrates, which can be as high as $3 \times 10^8 \, \mathrm{M^{-1} \, s^{-1}}$ (see Fersht, 1999); and the comparison with type IIA aryl sulphatase is irrelevant, because 2-hydroxy-5-nitrophenyl sulphate is not its natural substrate, so it did not evolve to catalyse this reaction and there is no reason to regard it as optimized for it.

Despite the disappointing nature of this first result, Pike (1987) could write fifteen years later that "no other synthetic polymer has been shown to provide a larger rate enhancement accompanied by substrate turnover". After fifteen more years, results with completely synthetic artificial enzymes continue to be unimpressive, but modifying existing protein structures, a return to the approach initiated by Polgar and Bender (1966), or "rational redesign", as it has been called (Cedrone, Ménez and Quéméneur, 2000), is beginning to yield genuinely useful new catalysts. For example, Ma and Penning (1999) modified the steroid binding site of 3α-hydroxysteroid dehydrogenase so that it became a 20α-hydroxysteroid dehydrogenase; in other words they changed the sex hormone specificity from androgen to progestin, with a change in specificity constant for the desired reaction by a factor of 2×10^{11}. In another example, Quéméneur and co-workers (1998) modified cyclophilin, an enzyme that normally catalyses *cis–trans* isomerization of aminoacyl-proline bonds in proteins, so that it would catalyse hydrolysis of such bonds; they did this by introducing mutations that would make the active site similar to that of well-studied proteinases like chymotrypsin, and achieved an increase of about 10^4-fold in the desired reaction.

Each of these enzymes had natural catalytic activity, so the effect was to alter the specificity of an existing enzyme. Attempts to convert a protein with no native catalytic action into a useful enzyme have not yet been correspondingly successful. For example, Nixon and co-workers (1999) attempted to create a scytalone dehydrogenase, an enzyme involved in melanin synthesis in fungi, by introducing catalytic groups into a non-catalytic protein involved in protein translocation. Although they were successful in producing a substantial activity in a previously inactive protein, this still fell far short, by a factor of around 10^5, of the activity of a natural scytalone dehydrogenase.

2.10.3 Catalytic RNA

The study of catalytic RNA started with the discovery by Zaug and Cech (1986) that an intron in the RNA of *Tetrahymena thermophila* catalysed a specific reaction in the processing of RNA and that it had all of the properties of an enzyme. As they noted, its efficiency as a catalyst, with a k_{cat}/K_m value of about $10^3 \, \mathrm{M^{-1} \, s^{-1}}$, is low by the standards of many protein enzymes (such as those tabulated by Fersht, 1999), but within the range found with other enzymes that recognize specific nucleic acid sequences. High specificity is much more important than high activity for the biological role of such enzymes.

This field has developed substantially since the original discovery, and many examples of similar catalytic RNA molecules have now been sequenced and studied (Doudna and Cech, 2002). Although they will not supplant proteins as the primary catalysts of metabolic reactions, their importance for a better understanding of the origin of life is great. One of the major problems with the older view of proteins as the sole enzymes was that it seemed barely possible to synthesize enzymes without nucleic acids, or nucleic acids without proteins, but even more difficult to imagine that the two kinds of polymer could have originated independently. The existence of catalytic RNA thus allows a more credible picture of the origin of life in which nucleic acids existed before proteins and fulfilled some of the functions now associated with proteins. They undoubtedly fulfilled these functions quite badly, as RNA molecules have a much more limited range of functional groups available to them than proteins, but this was of little importance in a world where more efficient competing catalysts did not exist. Some of the early functions of RNA enzymes still survive in the present world to fulfil major biological roles; for example, the ribosome appears to be an RNA enzyme (Lilley, 2001), as is the structure that splices out the major class of introns (Doudna and Cech, 2002).

Catalytic RNA molecules are finding important applications for manipulating the levels of genes, and hence of gene products, in living organisms. This is because the basis of their specificity for particular base sequences is well understood, as it derives from the usual base pairing that underlies the operation of the genetic code. Consequently the specificity of a particular catalytic RNA with ribonuclease activity can be modified by introducing appropriate base sequences complementary to sequences in the mRNA of the gene one wishes to manipulate. This approach has, for example, allowed Efrat and co-workers (1994) to obtain transgenic mice with activities of hexokinase D in the pancreatic islets much below the normal, while leaving the activities of other hexokinase isoenzymes unaffected.

2.10.4 Catalytic antibodies

A catalytic antibody is a sort of natural artificial enzyme: on the one hand it is a natural protein synthesized by the usual biological processes; on the other hand it is intended to catalyse a reaction for which no real enzyme is available. The essential idea is to raise antibodies to a molecule considered to mimic the transition state of the reaction that is to be catalysed, that is to say a molecule resembling a strained structure intermediate between the substrate and product, believed to occur on the reaction pathway (compare Section 1.7.3). The hope is that some of the antibodies produced will happen to possess groups capable of promoting the reaction. Reviews of this field (for example Blackburn and co-workers, 1989; Wentworth, 2002) tend to be written in the same breathless style that one associates with other articles on artificial enzymes, sometimes with abundant use of exclamation marks for readers who might otherwise miss the points being made, but the achievements to date have hardly justified the excitement. For example, aminoacylation of the 3′-hydroxyl group of thymidine with an alanyl ester can be catalysed by an antibody raised against a phosphonate diester, "a remarkably efficient catalyst", with a k_{cat}/K_m value of only about $10^3 \, M^{-1} s^{-1}$ (Jacobsen and co-workers, 1992).

Problems

2.1 For an enzyme obeying the Michaelis–Menten equation, calculate (a) the substrate concentration relative to K_m at which $v = 0.1V$, (b) the substrate concentration relative to K_m at which $v = 0.9V$, and (c) the ratio between the two.

2.2 In Henri's time it was not considered unreasonable that an enzyme might act merely by its presence, without necessarily entering into combination with its substrate. Derive a rate equation for the following mechanism

$$\begin{array}{c} EA \\ \Big\updownarrow K \\ E + A \xrightarrow{\;\;k\;\;} E + P \end{array}$$

in which EA is formed but is not on the pathway from A to P, and show that it leads to a rate equation of the same form as the Michaelis–Menten equation. What are the definitions of V and K_m in terms of K and k?

2.3 The curve defined by the Michaelis–Menten equation is described in Section 2.3.4 (and in other books) as a rectangular hyperbola. However, it is not immediately obvious how the biochemists' equation $v = Va/(K_m + a)$ is related to the equation $xy = A$ (for variables x and y and constant A) used as the standard representation of a rectangular hyperbola in mathematics texts. Show that the equation $xy = A$ can be rearranged into the form of the Michaelis–Menten equation after replacing x by $K_m + a$ and y by $V - v$. What value for A does this imply? The equations for the two asymptotes are $x = 0$ and $y = 0$: what do these become when expressed in terms of the usual kinetic symbols?

2.4 An activity assay for an enzyme should be designed so that the measured initial rate is insensitive to small errors in the substrate concentration. How large must a/K_m be if a 10% error in a is transmitted to v as an error of less than 1%? (Assume that the Michaelis–Menten equation is obeyed.)

2.5 In an investigation of the enzyme fumarase from pig heart, the kinetic parameters for the forward reaction were found to be $K_m = 1.7\,\text{mM}$, $V = 0.25\,\text{mM s}^{-1}$, and for the reverse reaction they were found to be $K_m = 3.8\,\text{mM}$, $V = 0.11\,\text{mM s}^{-1}$. Estimate the equilibrium constant for the reaction between fumarate and malate. For a sample of fumarase from a different source, the kinetic parameters were reported to be $K_m = 1.6\,\text{mM}$, $V = 0.024\,\text{mM s}^{-1}$ for the forward reaction, and $K_m = 1.2\,\text{mM}$, $V = 0.012\,\text{mM s}^{-1}$ for the reverse reaction. Comment on the plausibility of this report.

2.6 The table below shows values of product concentration p (in mM) at various times t (in min), for five different values of the initial substrate concentration a_0 as indicated. Estimate the initial rate v_0 at each initial substrate concentration from plots of p against t (do *not* use the more elaborate methods described in Section 2.9.3). Hence estimate K_m and V, assuming that the initial rate is given by the Michaelis–Menten equation, by each of the methods illustrated in Figures 2.1 and 2.3–7. Finally, estimate K_m and V by the method of Jennings and Niemann (Section 2.9.3). Account for any differences you observe between the results given by the different methods.

t	$a_0 = 1$	$a_0 = 2$	$a_0 = 5$	$a_0 = 10$	$a_0 = 20$
1	0.095	0.18	0.37	0.56	0.76
2	0.185	0.34	0.71	1.08	1.50
3	0.260	0.49	1.01	1.57	2.20
4	0.330	0.62	1.29	2.04	2.88
5	0.395	0.74	1.56	2.47	3.50
6	0.450	0.85	1.80	2.87	4.12
7	0.505	0.95	2.02	3.23	4.66
8	0.555	1.04	2.22	3.59	5.24
9	0.595	1.12	2.40	3.92	5.74
10	0.630	1.20	2.58	4.22	6.24

2.7 If a reaction is subject to product inhibition according to equation 2.55, the progress curve obeys an equation of the form

$$k_{cat}e_0 t = (1 - K_{mA}/K_{sP})(a_0 - a) + K_{mA}(1 + a_0/K_{sP}) \ln(a_0/a)$$

where a_0 is the value of a when $t = 0$ and the other symbols are as defined in equation 2.55. (a) Show that equation 2.55 is the differentiated form of this equation. (b) Compare the equation with equation 2.65 and write down expressions for V^{app} and K_m^{app} (defined as in equation 2.65). (c) Under what conditions will V^{app} and K_m^{app} be negative?

2.8 Atassi and Manshouri (1993) report synthetic peptides (or "pepzymes") symbolized ChPepz and TrPepz that catalyse the hydrolysis of N-benzoyl-L-tyrosine ethyl ester and N-tosyl-L-arginine methyl ester with kinetic constants comparable with those of the enzymes α-chymotrypsin and trypsin respectively, as shown in the following table[*]:

Substrate	Catalyst	K_m (mM)	k_{cat} (s^{-1})
N-benzoyl-L-tyrosine ethyl ester	ChPepz	1.11	147
	α-Chymotrypsin	1.07	185
N-tosyl-L-arginine methyl ester	TrPepz	2.42	85
	Trypsin	2.56	221

How far do these results support the claim that "clearly, pepzymes should have enormous biological, clinical, therapeutic, industrial and other applications"?

2.9 Three isoenzymes catalysing the same reaction with K_m values of 0.04, 0.2 and 5 mM are found to occur in a cell extract with V values of 0.7, 1.2 and 0.8 (arbitrary units) respectively. How could this information be most conveniently presented in the form of a graph with substrate concentrations covering at least the range 0.01 to 20 mM?

Additional exercises concerned primarily with specificity are included at the end of Chapter 5.

---

[*]The factual accuracy of the information in the table has been questioned (Corey and Phillips, 1994; Wells and co-workers, 1994; Corey and Corey, 1996). However, although this underlines the importance of maintaining a healthy scepticism about new results, it does not directly affect the question posed in this problem, which is intended to be answered on the assumption that the results are correct.

Chapter 3

Practical Aspects of Kinetic Studies

3.1 Enzyme assays

3.1.1 Discontinuous and continuous assays

All enzyme kinetic investigations rest ultimately on assays of catalytic activity, and these are essential also in enzyme studies that are not primarily kinetic. If it is unavoidable one can use a *discontinuous assay*, in which samples are removed at intervals from the reaction mixture and analysed to determine the extent of reaction. It is more convenient to use a *continuous assay*, in which the progress of the reaction is monitored continuously with automatic recording apparatus. If the reaction causes a change in absorbance at a conveniently accessible wavelength it can easily be followed in a recording spectrophotometer (John, 2002). For example, many reactions of biochemical interest involve the interconversion of the oxidized and reduced forms of NAD, and for these one can usually devise a spectrophotometric assay that exploits the large absorbance of reduced NAD at 340 nm. Even if no such convenient spectroscopic change occurs, and no suitable fluorescence change is available as an alternative, it may well be possible to "couple" the reaction of interest to one that is more easily assayed, as will be discussed in Section 3.1.4.

Reactions for which no spectrophotometric assay is suitable may still often be followed continuously by taking advantage of the release or consumption of protons that many enzyme-catalysed reactions involve. Such reactions may be followed in unbuffered solutions with a "pH-stat", an instrument that adds base or acid automatically and maintains a constant pH (Brocklehurst, 2002). Because of the stoicheiometry of the reaction, a record of the amount of base or acid added provides a record of the progress of the reaction.

3.1.2   Estimating the initial rate

The subject matter of this section may appear old-fashioned, as it deals with methods that have been almost entirely superseded by automated ways of doing the same thing. Rather than estimating initial rates by analysing a curve on paper with a ruler, it is now standard to obtain the initial rate of a reaction from the software supplied with the spectrophotometer or other recording instrument. In principle there is little objection to this, but in practice there may be serious objections. Before investing trust in the numbers that emerge from a machine an experimenter needs to be confident that the numbers are properly calculated and that they are appropriate. To be "appropriate", the number interpreted as an initial rate must have been computed as the slope of the tangent to the progress curve at time zero and not, for example, as the slope of a chord that approximates the curve during a finite period of time. Use of an instrument with high standards of optics and photometry does not guarantee that the built-in software is of corresponding quality: for an example of an instrument supplied with software far inferior to the quality of the instrument itself, see Cárdenas and Cornish-Bowden (1993). Thus even if one rarely needs to estimate an initial rate manually, one still needs to know how to do it when necessary, and one always needs an appreciation of the principles involved.

Ideally one must try to find conditions in which the progress curve is virtually straight during the period of measurement. Strictly speaking this is impossible, because regardless of the mechanism of the reaction one expects the rate to change — usually to decrease — as the substrates are consumed, the products accumulate and, sometimes, the enzyme loses activity. A simple example of such slowing down is considered in Section 2.9, and some more complicated but more realistic ones in Cornish-Bowden (1975). However, if the assay can be arranged so that less than 1% of the complete reaction is followed the progress curve may then be indistinguishable from a straight line. This happy situation is less common than one might think from reading the literature, because many experimenters are reluctant to admit the inherent difficulty of drawing an accurate tangent to a curve, and prefer to persuade themselves that their progress curves are biphasic, with an initial "linear" period followed by a tailing off. This nearly always causes the true initial rate to be underestimated, for reasons that should be clear from Figure 3.1. Even if the line is constrained to pass through the origin, it will tend to underestimate the true slope (albeit by less than is shown in Figure 3.1) if the first few time points are treated as points on a straight line.

To avoid the bias evident in Figure 3.1 one must be aware of the problem and remember that one is trying to find the *initial* rate, not the average rate during the first few minutes of reaction, because it is the initial rate that appears in the

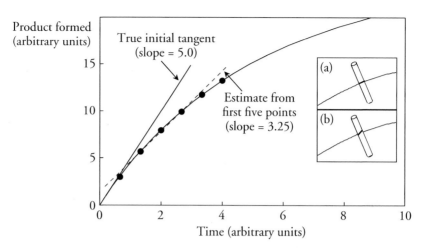

Figure 3.1. Bias in estimating an initial rate. The straight line through the origin is a true initial tangent, with a slope more than 50% greater than that of the broken line, which was drawn by treating the first five experimental points as if they occurred during a supposed "linear" phase of reaction. In reality there is no linear phase and the line is curved over its entire range. The insets show the effect of laying a glass rod over a curve: (a) if it is laid exactly at right angles to the tangent to the curve at the crossing point, the line appears continuous across the rod, but (b) if it makes a slight angle to this perpendicular there are breaks in the line as it passes under the rod.

kinetic equations. So one must try to draw a tangent to the curve extrapolated back to zero time, not a chord. For following the progress of an enzyme purification more refinement than this is hardly needed, as there is no need for highly precise initial rates in this context. Subsequent kinetic study of the purified enzyme, however, may well need better-defined initial rates than one can hope to get by drawing initial tangents by hand. A method based on an integrated rate equation, as discussed in Section 2.9, is then likely to be useful.

If the curve itself is clean and well defined one can use a glass rod as a ruler to draw a straight line *at right angles* to the curve (from which the actual tangent can then easily be found by drawing a second straight line at right angles to the first). This exploits a refractive property of the glass rod such that any deviation from the correct angle is magnified when one looks at the curve through the glass. This property is illustrated in the insets to Figure 3.1.

3.1.3   Increasing the straightness of the progress curve

Provided that the rate of an enzyme-catalysed reaction is proportional to the total enzyme concentration (as one usually tries to ensure), the curvature of the progress curve cannot be altered by using more or less enzyme: this just alters the scale of the time axis and any apparent change in curvature is an illusion;

indeed, this property forms the basis of Selwyn's test for enzyme inactivation as described in Section 3.2. Nonetheless, it may still be advantageous to use less enzyme as it will decrease the amount of product formed during the period between mixing the reagents and starting to observe the reaction; in other words it can decrease the proportion of the curve that cannot be directly observed.

One can, however, improve the linearity of an assay by increasing the substrate concentration, so long as the products do not bind more tightly than the substrates to the enzyme (as often happens, for example, in reactions in which the oxidized form of NAD is converted to the reduced form). To illustrate this effect of substrate concentration I shall consider the simplest possible reaction, one that obeys the Michaelis–Menten equation and is not subject to product inhibition or retardation due to any cause apart from depletion of substrate. The progress curve is then described by equation 2.62, the naive integrated form of the Michaelis–Menten equation (or equation 2.65, with $V^{app} = V$ and $K_m^{app} = K_m$). If the initial substrate concentration a_0 is $5K_m$, the initial rate v_0 is $0.83V$, and is largely unaffected by small errors in a_0. Even if a_0 is doubled to $10K_m$, v_0 increases only by 9%, to $0.91V$. So it might seem that the assay would be only trivially improved, despite being made considerably more expensive, by using the higher initial substrate concentration. But if curvature is a prime concern this conclusion is mistaken, as we may see by making some simple calculations with equations 2.57 and 2.62. For $a_0 = 5K_m$, with initial rate $0.833V$, a decrease by 1% gives a rate of $0.825V$, corresponding (from equation 2.57) to $p = 0.23K_m$, which gives $t = 0.34K_m/V$ when substituted into equation 2.62. For $a_0 = 10K_m$, however, the same calculation gives 99% of the initial rate for $p = K_m$, at $t = 1.11K_m/V$. Doubling the initial substrate concentration thus increases more than threefold the time required for a 1% decrease in rate.

In practice this calculation will usually be an oversimplification, because nearly all enzyme-catalysed reactions are subject to product inhibition, but the principle still applies qualitatively: increasing the initial substrate concentration usually extends the "linear" period, unless products bind more tightly than substrates. Do not forget, however, that the property of NAD-dependent dehydrogenases noted above makes this last qualification an important one.

Another reason for using as high a substrate concentration in an enzyme assay as cost, solubility and substrate inhibition allow is that this minimizes the sensitivity of the assay to small errors in the substrate concentration, not only during the course of the reaction, as just discussed, but also from one experiment to another. Working with $a_0 = 0.1K_m$, for example, requires precisely prepared solutions and great care, because a 10% error in a_0 generates almost a 10% error in the measured rate; when $a_0 = 10K_m$, however, much less precision is needed because a 10% error in a_0 generates less than a 1% error in the measured rate.

Whether or not steps are taken to increase the straightness of a progress curve, it is a good idea to observe a reaction mixture for a long enough period for the curvature to be easily visible. Even if the later stages are not used in the analysis, their presence in the trace will tend to decrease the probability of the error of treating any part of the curve as if it is straight, making the faults illustrated in Figure 3.1 less likely.

3.1.4 Coupled assays

When it is not possible or convenient to follow a reaction directly in a spectrophotometer it may still be possible to follow it indirectly by "coupling" it to another reaction. Consider, for example, the hexokinase-catalysed transfer of a phosphate group from ATP to glucose:

$$\text{glucose} + \text{ATP} \rightarrow \text{glucose 6-phosphate} + \text{ADP} \qquad (3.1)$$

This important reaction is not accompanied by any convenient spectrophotometric change, but it may nonetheless be followed spectrophotometrically by coupling it to the following reaction, catalysed by glucose 6-phosphate dehydrogenase:

$$\text{glucose 6-phosphate} + \text{NAD}_{\text{oxidized}} \rightarrow \text{6-phosphogluconate} + \text{NAD}_{\text{reduced}}$$
$$(3.2)$$

Provided that the activity of the coupling enzyme is high enough for the glucose 6-phosphate to be oxidized as fast as it is produced, the rate of NAD reduction recorded in the spectrophotometer will correspond exactly to the rate of the reaction of interest.

The requirements for a satisfactory coupled assay may be expressed in simple but general terms by means of the scheme

$$A \xrightarrow{\;v_1\;} B \xrightarrow{\;v_2\;} C \qquad (3.3)$$
$$a \qquad b \qquad c$$

in which the conversion of A into B at rate v_1 is the reaction of interest and the conversion of B into C is the coupling reaction, with a rate v_2 that can readily be measured. For measurements of v_2 to provide accurate information about the initial value of v_1 a steady state in the concentration of B must be reached before v_1 decreases perceptibly from its initial value. Some treatments of this system assume that v_2 must have a first-order dependence on b, but this is both unrealistic and unnecessary, and can lead to the design of assays that use more materials than necessary, which may not only be wasteful but may have undesirable side effects as well. As the coupling reaction is usually enzyme-catalysed, it is more appropriate to suppose that v_2 depends on b according to the Michaelis–Menten equation:

$$v_2 = \frac{V_2 b}{K_{m2} + b} \tag{3.4}$$

in which the subscripts 2 are to emphasize that V_2 and K_{m2} are the Michaelis–Menten parameters of the second (coupling) enzyme. If v_1 is a constant (as it is approximately during the period of interest, the early stages of reaction), the equation expressing the rate of change of b with time,

$$\frac{db}{dt} = v_1 - v_2 = v_1 - \frac{V_2 b}{K_{m2} + b} \tag{3.5}$$

can readily be integrated (for details, see Storer and Cornish-Bowden, 1974). It leads to the conclusion that the time t required for v_2 to reach any specified fraction of v_1 is given by an equation of the form

$$t = \phi K_{m2} / v_1 \tag{3.6}$$

in which ϕ is a dimensionless number that depends only on the ratios v_2/v_1 and v_1/V_2. For example, suppose that v_1 is 0.1 mM min^{-1} and V_2, the limiting rate of the coupling enzyme, is 0.5 mM min^{-1}; then $v_1/V_2 = 0.2$ and one is trying to measure a rate that is 20% of the limiting rate of the coupling reaction. For these conditions the value of ϕ tabulated by Storer and Cornish-Bowden (1974) for $v_2/v_1 = 0.99$ is 1.31: this means that the time required to reach 99% of the target rate is $1.31 \times 0.2/0.1$ min, or 2.62 min. One needs to use discretion to decide what value of v_2/v_1 is appropriate, as using 0.99 when 0.9 will do will produce an unnecessarily expensive assay, and using 0.9 when 0.99 is needed will produce an inadequate one. There is no universal answer to this question. To follow the progress of a purification of an enzyme high accuracy is not needed, and general physiological characterization of an enzyme is hardly more demanding, even in modern proteomic studies. On the other hand, analysing the kinetic variations of a series of mutant forms of an enzyme to study the relations between structure and function require the highest accuracy available: if the numbers that emerge from such a study are not accurate they are hardly worth having at all.

The validity of this treatment can be checked by following the coupling reaction over a period of time and showing that the value of v_2 does increase in the way expected. An example of such a check is shown in Figure 3.2. In that

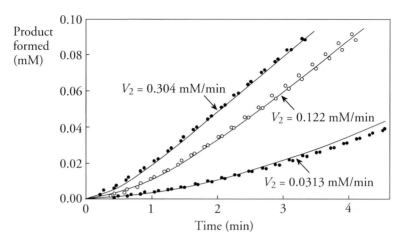

Figure 3.2. Acceleration phase of a coupled assay. The data are from Storer and Cornish-Bowden (1974) and refer to the assay of hexokinase D with glucose 6-phosphate dehydrogenase as coupling enzyme. The experimental points show (for duplicate time courses at each value of V_2) the concentrations of reduced NAD, the product of the coupling reaction, at various times after the start of the reaction and at three different values of V_2 as indicated. The three curves drawn were not fitted to the data but were calculated independently from the known values of V_2, according to the theory outlined in the text.

experiment V_2 was deliberately made rather smaller than would be appropriate for a satisfactory assay in order to make the period of acceleration clearly visible.

Even if the reaction of interest can be assayed directly it is sometimes advantageous to couple it to a second reaction. For example, it may be difficult to measure the initial rate accurately if one of the products of the first reaction is a powerful inhibitor, or if a reversible reaction is being studied in the less favoured direction and equilibrium is reached after only a small proportion of substrate has reacted. Problems of this kind can often be overcome by coupling the reaction to an irreversible reaction that removes the inhibitory product or displaces the equilibrium. Much the same analysis as before can be used, but with a stricter definition of the steady state of the reaction. The steady-state value of b in the scheme of equation 3.3 is obtained by setting $v_2 = v_1$ and solving equation 3.4 for b, which gives $b = K_{m2}v_1/(V_2 - v_1)$. It is then simple to decide how large V_2 must be if the steady-state value of b is not to be large enough to create problems.

Coupling enzymes, used especially as ATP-regenerating systems, can also be valuable for maintaining the concentration of a reactant as constant as possible. Adding pyruvate kinase to an assay mixture, for example, with excess ADP and phospho*enol*pyruvate, can ensure that ATP is regenerated as fast as it is consumed, as a result of the following reaction:

$$\text{ADP} + \text{phospho}enol\text{pyruvate} \rightarrow \text{pyruvate} + \text{ATP} \qquad (3.7)$$

In a recent variant of this approach, Chittock and co-workers (1998) showed how to use it as an amplification system for detecting and measuring small concentrations of ATP.

Sometimes it is necessary to couple a reaction with two or more coupling enzymes. For example, the coupled assay for hexokinase based on the reaction in equation 3.2 would be useless for studying inhibition of hexokinase by glucose 6-phosphate, because the coupling system would remove not only the glucose 6-phosphate released in the reaction but also any added by the experimenter. To avoid this, the production of ADP needs to be coupled to the oxidation of NAD, which can be done with two enzymes, pyruvate kinase (equation 3.7) and lactate dehydrogenase:

$$\text{pyruvate} + \text{NAD}_{\text{reduced}} \rightarrow \text{lactate} + \text{NAD}_{\text{oxidized}} \qquad (3.8)$$

Rigorous kinetic analysis of systems with two or more coupling enzymes is difficult, but qualitatively they resemble the simple example we have considered: one must ensure that the activities of the coupling enzymes are high enough for the measured rate to reach an appropriate percentage of the required rate within the period that the required rate remains effectively constant. This can most easily be checked by experiment: if the concentrations of the coupling enzymes are high enough the observed rate should be proportional to the concentration of the enzyme of interest over the whole range to be used. The times required for the two or more coupling enzymes to reach their steady states are additive (Easterby, 1981), and so a reasonably accurate total time can be calculated quite easily.

With all of these examples there may be reasons other than expense for not using more coupling enzyme than is absolutely necessary for an efficient assay. Unless the coupling enzyme is as carefully purified as the target enzyme one has the danger that impurities may catalyse unwanted reactions that interfere with the assay, and even if it is quite pure it may have unwanted activities of its own that are best minimized. For example, glucose 6-phosphate dehydrogenase not only catalyses the oxidation of glucose 6-phosphate (equation 3.2) but also has a weak activity towards glucose. Similar potential dangers apply in all coupled assays as there is always the risk that an enzyme chosen for its capacity to react with the product of a reaction may also react with its substrate.

3.2   Detecting enzyme inactivation

Many enzymes are much more stable at high concentrations than at low, so it is not uncommon for an enzyme to lose activity rapidly when it is diluted from a stable stock solution to the much lower concentration used in the assay. This can obviously lead to errors in the estimate of the general level of activity, but, less obviously, it may also produce errors in the *type* of behaviour reported. It often happens that an enzyme–substrate complex is more stable than the corresponding free enzyme, and consequently enzymes often lose activity more slowly at high substrate concentrations. If this effect is not noticed, the abnormally low activity observed at low substrate concentrations can be falsely attributed to cooperativity (Chapter 11), that is to say to deviations of the initial rate from Michaelis–Menten kinetics. Even if there are no effects as serious as this, assay conditions that minimize inactivation are likely to give results more reproducible than they would otherwise be, and it is anyway of interest to know whether the decrease in rate that occurs during the reaction is caused wholly or partly by loss of enzyme activity (rather than by substrate depletion or accumulation of products, for example). Fortunately Selwyn (1965) has described a simple test of this.

He pointed out that as long as the rate dp/dt at all times during a reaction is proportional to the initial total enzyme concentration e_0 then it can be expressed as the product of the constant e_0 times some function of the instantaneous concentrations of the substrates, products, inhibitors and any other species (apart from the enzyme) that may be present. But because of the stoicheiometry of the reaction, these concentrations can in principle be calculated from p, the concentration of one product at any time. So the rate equation can be written in the simple form

$$\frac{dp}{dt} = e_0 f(p) \tag{3.9}$$

where f is a function that can in principle be derived from the rate equation. It is of no importance that f may be difficult to derive or that it may be a complicated function of p, because its exact form is not required. It is sufficient to know that it is independent of e_0 and t, and so the integrated form of equation 3.9 must be

$$e_0 t = F(p) \tag{3.10}$$

where F is another function. The practical importance of this equation is that it shows that the value of $e_0 t$ after a specified amount of product has been formed is independent of e_0. Consequently if progress curves are obtained with various values of e_0 but otherwise identical starting conditions, plots of p against $e_0 t$ for the various e_0 values should be superimposable. If they are not, the initial assumption that the rate is proportional throughout the reaction to the initial total enzyme concentration must be incorrect. Figure 3.3 shows two examples

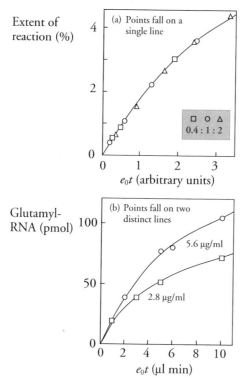

Figure 3.3. Selwyn's test of inactivation. In a reaction in which there is no apprecia-ble enzyme inactivation during the period of observation, plots of the extent of reaction against e_0t should be superimposable, as in plot (a), which shows data of Michaelis and Davidsohn (1911) for invertase at three different enzyme concentra-tions in the ratio 0.4:1:2 as shown. If the enzyme becomes inactivated during the reaction, or if the rate is not strictly proportional to e_0, the plots are not superimposable, as in plot (b), which shows data of Deutscher (1967) for glutamyl ribonucleic acid synthetase with 5.6 and 2.8 $\mu g\,ml^{-1}$ enzyme.

of this plot, one in which the results are as expected for a satisfactory assay, the other in which they are not.

The simplest reason why Selwyn's test may fail, as in Figure 3.3b, is that e_0 varies because the enzyme loses activity during the reaction. Selwyn (1965) lists several other possibilities, all of which indicate either that the assay is unsatisfactory or that it is complicated in some way that ought to be invest-igated before it is used routinely. Problem 3.3 at the end of this chapter provides an example of this. A recent example of the use of the test to compare the effectiveness of different anions in stabilizing a bacterial arsenate reductase is described by Messens and co-workers (2002).

Selwyn's test has not been widely applied. During the 1970s and 1980s this could perhaps be justified by the increasing knowledge and use of conditions for maintaining natural enzymes in a stable condition. Since then, however, the enormous growth in the use of genetic techniques for producing mutant enzymes, which are usually less stable than their natural counterparts, has recreated a need for a reliable way of recognizing if they become inactivated during assay. Without this, comparisons between reasonably stable natural enzymes and much less stable variants are often likely to be invalidated by artefacts.

The principle embodied in Selwyn's test was widely known in the early years of enzymology: the data used for constructing Figure 3.3a were taken from Michaelis and Davidsohn (1911), and similar data are given by Hudson (1908); it is clear, moreover, from the discussion given by Haldane (1930) that similar tests were applied to many enzymes. As early as 1890, O'Sullivan and Tompson commented that "the time necessary to reach any given percentage of inversion [hydrolysis of sucrose] is in inverse proportion to the amount of the inverting preparation present; that is to say, the time is in inverse proportion to the inverting agent". In spite of this, the test was largely forgotten until Selwyn (1965) adapted the treatment of Michaelis and Davidsohn (1911) and discussed the various reasons why it might fail.

3.3   Experimental design

3.3.1   Choice of substrate concentrations

A full account of the design of enzyme kinetic experiments would require a great deal of space, and this section will provide only a brief and simplified guide. In general, the conditions that are optimal for assaying an enzyme, or determining the amount of catalytic activity in a sample, are unlikely to be ideal for determining its kinetic parameters. This is because an enzyme assay is ideally run in conditions where the measured rate depends *only* on the enzyme concentration, so that slight variations in other conditions have little effect; but investigating the kinetic properties of an enzyme implies learning how it responds to changes in conditions. It is then essential to work over a wide range of substrate concentrations in which the rate varies appreciably. In practice, for an enzyme that obeys the Michaelis–Menten equation this means that the a values should extend from well below K_m to well above K_m.

If one is confident that the enzyme obeys the Michaelis–Menten equation, it is sufficient to consider what range of a values will define K_m and V precisely. It is easy to decide how to define V precisely, by recalling that v approaches V as a becomes very large (Section 2.3); obviously therefore, one should include some a values as large as cost, solubility and other constraints (such as high specific absorbance in the spectrophotometer) permit. In principle, the larger the largest a value the better, but in reality there are two reasons why this may not be so. First, one's confidence that the Michaelis–Menten equation is obeyed may be misplaced: many enzymes show substrate inhibition at high a values, and as a result the v values measured at very high a may not be those expected from the K_m and V values that define the kinetics at low and moderate a. Second, even if the Michaelis–Menten equation is accurately obeyed, the advantage of including a values greater than about $10K_m$ is slight and may well be outweighed by the added cost in materials. Moreover, if the substrate is an ion it may become difficult to avoid varying the ionic strength if excessively high concentrations are used.

One limit to the maximum substrate concentration that can be used is imposed by the specific absorbance of the substrate or product used for the assay. For example, many assays depend on the absorbance of reduced NAD at 340 nm, about 0.9 in a 1 cm cuvette for a 0.15 mM solution. This might suggest that it would be difficult to use assays with a much higher concentration, but in fact use of 0.2 cm cuvettes allows a fivefold extension of the range. Curiously, 1 cm cells have become so familar that many modern experimenters appear not to know (or to have forgotten) that other pathlengths are possible. Longer pathlengths, such as 5 or 10 cm, are useful for assaying solutions that absorb weakly, but these may be less convenient to use as they are more likely to require special modifications to the spectrophotometer itself.

Just as the rate at high a is largely determined by V, so the rate at low a is largely determined by V/K_m (see Section 2.3). So for V/K_m to be well defined it is necessary to have some observations at a values less than K_m. Defining K_m itself needs accurate values of *both* V and V/K_m; thus a should range from about $0.1K_m$ to about $10K_m$ or as high as conveniently possible.

It is not necessary to go to the lowest a values for which measurements are possible, however, because the need for v to be zero when a is zero provides a fixed point on the plot of v against a through which the curve must pass. As a result there is little advantage in using a values less than about $0.1K_m$. Theoretical studies made under rather idealized conditions (Endrenyi, 1981) suggest that if v has a constant standard deviation the optimum value for the low end of the range may be as high as $0.4K_m$ (the exact value depending on the high end of the range), but if v has a constant coefficient of variation the optimum value for the low end of the range is zero. However, one should be cautious about taking the first part of this result too literally, because it was derived for conditions that may not be satisfied in a real experiment. In practice one rarely knows with sufficient certainty that the curve passes through the origin or that the Michaelis–Menten equation is obeyed with sufficient accuracy to justify omitting observations at low substrate concentration, regardless of what the theoretical analysis may suggest.

It cannot be overemphasized that the above remarks were prefaced with the condition that one must be confident that the Michaelis–Menten equation is obeyed, or at least that one does not care whether it is obeyed or not outside the range of the experiment. If the motivation is primarily physiological there is no reason to want to know about deviations from simple behaviour at grossly unphysiological concentrations; but for studying an enzyme mechanism as wide a range of conditions as possible should certainly be explored, because deviations from the expected behaviour at the extremes of the experiment may well provide clues to the mechanism. Hill, Waight and Bardsley (1977) argued that there may be few enzymes (if indeed there are any at all) that truly obey the Michaelis–Menten equation. They thought that excessively limited experimental designs, coupled with an unwillingness to take note of deviations from expected behaviour, had led to an unwarranted belief in almost universal adherence to the Michaelis–Menten equation.

The Michaelis–Menten equation will undoubtedly remain useful as a first approximation in enzyme kinetics, even if it may sometimes need to be rejected after careful measurements, but it is always advisable to check for the commonest deviations. Is the rate truly zero in the absence of substrate (and enzyme, for that matter)? If not, is the discrepancy small enough to be accounted for by instrumental drift or other experimental error? If there is a significant "blank rate" in the absence of substrate or enzyme, can it be removed by careful purification? Does the rate approach zero at a values appreciably greater than zero? If so one should check for evidence of cooperativity (Chapter 11). Is there any evidence of substrate inhibition, that is to say of a decrease in v as a increases? Even if there is no decrease in v at high a values, failure to increase as much as predicted by the Michaelis–Menten equation (see Figure 2.1) may indicate substrate inhibition.

3.3.2   Choice of pH, temperature and other conditions

Even if there is no intention to study the pH and temperature dependence of an enzyme-catalysed reaction, care must still be taken in choosing the pH and temperature. For many purposes it will be appropriate to work under approximately physiological conditions — pH 7.2, 37 °C, ionic strength 0.15 mol l^{-1} for most mammalian enzymes, for example — but there may be good reasons for deviating from these in a mechanistic study. Many enzymes become denatured appreciably fast at 37 °C and may be much more stable at 25 °C (though there are exceptions, so this should not be taken as a universal rule). It is also advisable to choose a pH at which the reaction rate is insensitive to small changes in pH. This is sometimes expressed as a recommendation to work at the pH "optimum", but, as will become clear in Chapter 9, this may well be meaningless advice if K_m varies with pH: if so, then even though the Michaelis–Menten equation may be obeyed the maximum value of V/K_m will not occur at the same pH as the maximum value of V, and the "optimum" pH will be different at different substrate concentrations.

In studies of reactions with more than one substrate, the experimental design must obviously be more elaborate than that required for one-substrate reactions, but the principles are similar. Each substrate concentration should be varied over a wide enough range for its effect on the rate to be manifest. If the Michaelis–Menten equation is obeyed when any single substrate concentration is varied under conditions that are otherwise constant, the measured values of the Michaelis–Menten parameters are *apparent* values, and are likely to change when the other conditions are changed. To obtain the maximum information, therefore, one needs to choose a range of substrate concentrations in relation to the appropriate apparent K_m, not the limiting K_m as the other substrate or substrates approach saturation, which may not be relevant. I shall return to this topic in Chapter 7 after introducing the basic equation for a two-substrate reaction. Similar considerations apply in studies of inhibition, and are discussed in Chapter 5.

3.3.3   Use of replicate observations

At the end of a kinetic study one always finds that the best equation one can determine fails to fit every observation exactly. The question then arises whether the discrepancies are small enough to be dismissed as experimental error, or whether they indicate the need for a more complicated rate equation. Answering this requires an idea of the magnitude of the random error in the experiment, which is most easily obtained in a clear way by comparing replicate observations. If these agree with one another much better than they agree with the fitted line there are grounds for rejecting the fitted line and perhaps introducing more terms into the equation. If, on the other hand, there is about as much scatter within each group of replicates as there is between the fitted line and the points, there are no grounds for rejecting the equation until more precise observations become available.

This approach is possible because in a repeated experiment one knows what the degree of agreement ought to be if there were no random error. Hence such an experiment measures only random error, often called *pure error* in this context to distinguish it from the *lack of fit* that results from fitting an inadequate equation. The disagreement between an observation and a fitted line, on the other hand, may be caused either by error in the observation, or inadequacy of the theory, or, most likely, a combination of the two; it does not therefore measure pure error.

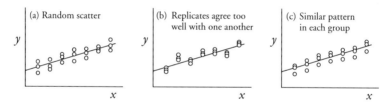

Figure 3.4. Use of repeated observations. When observations are properly repeated, the scatter of points about the fitted line should be irregular, as in (a). Regular scatter, as in (b) and (c), suggests that the experiment has not been properly done, as discussed in the text.

The use of replicate observations is not without its pitfalls. To give a meaningful result the disagreement between two replicates must be truly representative of the random error in the experiment as a whole. This will be true only if the repeated measurements are made just like any others, and not in any special way. This is perhaps best understood by examining the three examples shown in Figure 3.4. In Figure 3.4a the points are scattered within each group of replicates to about the same extent as all the points are scattered about the line; this is what one expects if there is no lack of fit and the repeated measurements have been made correctly, just like any others. In Figure 3.4b the scatter within each group of replicates is much less than the scatter about the line, even though the latter scatter does appear to be random rather than systematic. This is an unsatisfactory result, which can arise from various kinds of design fault: perhaps the most common is to measure all of the observations within a group in succession, so that the average time between them is small compared with the average for the experiment. If this is done, then any error caused by slow changes during the whole experiment — for example, instrumental drift, deterioration of stock solutions, increase in ambient temperature, fatigue of the experimenter — will not be properly reflected by the repeats.

Figure 3.4c shows the opposite problem, where the arrangement of each group of replicates is suspiciously regular, with a spread that is noticeably larger than the spread of points about the fitted line. This suggests that the repeats are overestimating the actual random error, perhaps because the figure actually represents three separate experiments done on three different days or with three different samples of enzyme.

The question of how many repeats there ought to be in a kinetic experiment is not one that can be answered dogmatically. For any individual study the answer must depend on how much work is needed for each measurement, how long the enzyme and other stock solutions can be kept in an essentially constant state, how large the experimental error is, and how complicated the equation to be fitted is. The first essential is to include as many *different* concentrations of substrate (and any other relevant components of the system, such as inhibitors) as are needed to characterize the shape of the curve adequately. A one-substrate enzyme that obeyed straightforward Michaelis–Menten kinetics might be adequately characterized with as few as five substrate concentrations in the range $0.5K_m$ to $5K_m$; but a two-substrate enzyme, again with straightforward kinetics, might well require a minimum of 25 different combinations of concentrations; and enzymes that showed deviations from simple kinetics would certainly require these numbers to be increased.

These estimates assume that full analysis of the data will involve graphs as well as statistical calculations in the computer (see Chapter 14). So far as the latter are concerned there is no need for the concentrations to be organized as a grid, even approximately, as it does not affect the validity of the calculations if a line calculated as one of a family of lines has only one point on it; thus a two-substrate experiment could in principle give satisfactory results with as few as eight different combinations of concentrations. The important point is to concentrate observations in the regions that give the most information about the validity of the model proposed. However, the human eye can readily spot unexpected behaviour on a graph that would pass unnoticed by even the best of computer programs, and for this reason it is unwise to rely wholly on the computer. This places a constraint on the experimental design, because a satisfactory graph needs to have a reasonable number of points on every line.

Only when the number of *different* combinations of concentrations to be used has been decided can one make any intelligent decision about the number of replicates. Suppose that one has decided that 25 different combinations are necessary and that it is possible and convenient to measure 60 rates in the time available for the experiment, or the time during which deterioration of the enzyme is negligible. It would then be appropriate to do ten sets of triplicates — spread over the whole experiment, not concentrated in one part of it — and the rest as duplicates. If, on the other hand, one could only manage 30 measurements one would have to decrease the number of repeats. To advocate a universal rule, that each measurement should be done in triplicate, for example, seems to me to be silly, not only because it oversimplifies the problem, but also because it may lead to experiments in which too few different sets of conditions are studied to provide the information sought.

3.4   Treatment of ionic equilibria

The substrates of many reactions of biochemical interest are not well-characterized compounds with directly measurable concentrations, but ions in equilibrium with other ions, some of which have their own interactions with the enzymes catalysing the reactions. Most notable of these ions is $MgATP^{2-}$, the true substrate of most of the enzymes that are loosely described as ATP-dependent. It is impossible to prepare a solution of pure $MgATP^{2-}$, because any solution that contains $MgATP^{2-}$ must also contain numerous other ions; for example, an equimolar mixture of ATP and $MgCl_2$ at pH 7 contains appreciable proportions of $MgATP^{2-}$, ATP^{4-}, $HATP^{3-}$, Mg^{2+} and Cl^-, as well as traces of $MgHATP^-$, Mg_2ATP and $MgCl^+$. Moreover, their proportions vary with the total ATP and $MgCl_2$ concentrations, the pH, the ionic strength and the concentrations of any other species (such as buffer components) that may be present.

For example, studying the effect of $MgATP^{2-}$ on an enzyme obviously requires some assurance that effects attributed to $MgATP^{2-}$ are indeed due to that ion and not to variations in the Mg^{2+} and ATP^{4-} concentrations that accompany variations in the $MgATP^{2-}$ concentration. Failure to take account of this possibility may lead to quite spurious suggestions of cooperativity or other deviations from Michaelis–Menten kinetics with respect to $MgATP^{2-}$. It is necessary, therefore, to have some method of calculating the composition of a mixture of ions, and it is desirable to have some way of varying the concentration of one ion without concomitant large variations in the concentrations of other ions.

The stability constants of many of the ions of biochemical interest have been measured. It is thus a simple matter to calculate the concentration of any complex if the concentrations of the free components are known. Unfortunately, however, the problem usually appears in inverse form: given the total concentrations of the components of a mixture how can one calculate the free concentrations? For example, given the total ATP and $MgCl_2$ concentrations, the pH and all relevant equilibrium constants, how can the concentration of $MgATP^{2-}$ be calculated? A simple and effective way is to proceed as follows:

1. Assume initially that no complexes exist and that all ionic components are fully dissociated. For example, assume that a mixture of 1 mM ATP, 2 mM $MgCl_2$ and 100 mM KCl contains 1 mM ATP^{4-}, 2 mM Mg^{2+}, 100 mM K^+ and 104 ($=100 + 2 \times 2$) mM Cl^-.
2. Use these free concentrations and the association constants to calculate the concentrations of all complexes that contain Mg^{2+}. When added up these will give a total Mg^{2+} concentration that exceeds (probably by a large amount in the first step of the calculation) the true total Mg^{2+} concentration.
3. Correct the free Mg^{2+} concentration by multiplying it by the true total Mg^{2+} concentration divided by the calculated total Mg^{2+} concentration.
4. Repeat with each component replacing Mg^{2+} in turn, with ATP^{4-}, K^+ and Cl^- in this example. In principle, H^+ may be treated in the same way, but

in usual experimental practice the free H^+ concentration is controlled and measured directly and so the free H^+ concentration should not be changed during the calculation but maintained throughout at its correct value.

5. Repeat the whole cycle, steps 2–4, until the results are self-consistent, that is to say until the concentrations do not change from cycle to cycle.

This procedure is modified slightly from one described by Perrin (1965) and Perrin and Sayce (1967). Although the number of cycles required for self-consistency is likely to be too large for convenient calculation by hand, the method is simple to express as a computer program (Storer and Cornish-Bowden, 1976a), and is then easy and efficient to apply to most problems that arise in enzyme kinetics; Kuzmic (1998) has developed a more powerful method for ones that present difficulty.

Experience in using a program of this sort has led to a simple experimental design for varying the concentration of $MgATP^{2-}$ while keeping variations in the concentrations under control. Three designs are in common use, of which one gives good results and the others give unacceptably poor results (Figure 3.5). The "good" design is to keep the total $MgCl_2$ concentration in constant excess over the total ATP concentration. The best results are obtained with an excess of about $5\,mM$ $MgCl_2$, but if the enzyme is inhibited by free Mg^{2+} or if there are other reasons for wanting to minimize the concentration of free Mg^{2+}, the excess can be lowered to $1\,mM$ with only small losses of efficiency. If the excess is greater than $10\,mM$ there may be complications due to the presence of significant concentrations of Mg_2ATP. With this design the ATP concentration may be varied over a wide range ($1\,\mu M$ to $0.1\,M$ at least)

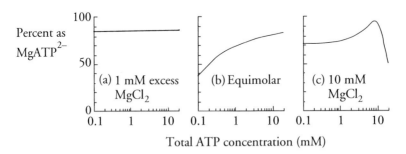

Figure 3.5. Three designs for controlling the concentration of $MgATP^{2-}$. (a) If the total $MgCl_2$ concentration is maintained in $1\,mM$ excess over the total ATP concentration, the percentage of ATP existing as $MgATP^{2-}$ is high and nearly constant over the whole range plotted, 0.1–$20\,mM$ ATP. (b) If the total $MgCl_2$ and ATP concentrations are kept equal this percentage varies over the whole range, and is not even approximately constant except at high concentrations. (c) If the total $MgCl_2$ concentration is kept at $10\,mM$ for all ATP concentrations the percentage of ATP existing as $MgATP^{2-}$ is reasonably constant at low ATP concentrations, but this design fails badly at high concentrations. Based on Figure 3.11 of Cárdenas (1995).

with a high and almost constant proportion of the ATP existing as $MgATP^{2-}$ and a nearly constant concentration of free Mg^{2+}. Thus effects due to variation in the concentration of $MgATP^{2-}$ may be clearly separated from effects due to variation in the free concentration of Mg^{2+}.

The two "bad" designs are unfortunately both common in practice. The first is to vary the total concentrations of ATP and $MgCl_2$ in constant ratio. Whether this ratio is $1:1$ or any other, this design leads to wild variations in the proportion of ATP existing in any particular form, and cannot be recommended. The second, also not to be recommended, is to keep the total $MgCl_2$ concentration constant at a value that exceeds the highest ATP concentration by about 2–5 mM. Although this design does ensure that ATP exists largely as $MgATP^{2-}$, it can produce undesirably large variations in the concentrations of free Mg^{2+} and of Mg_2ATP. Moreover, as illustrated in Figure 3.5c, it fails badly if the total ATP concentration is made too high.

Although the conclusions outlined in the preceding paragraphs depend to some degree on the numerical values of the equilibrium constants for complexes of Mg^{2+}, ATP^{4-} and H^+, the principles are general. As a rough guide, a component A of a binary complex AB exists largely in complexed form if B is maintained in excess over A by an amount about 100 times the dissociation constant of AB.

In this discussion I have simplified the problem by ignoring the fact that ionic equilibrium constants strictly define ratios of *activities* rather than concentrations (see Alberty, 2003). In practice, therefore, to avoid the complication of dealing with activity coefficients one must work at constant ionic strength. A value of about $0.15 \, \text{mol} \, l^{-1}$ is appropriate, both because it is close to the ionic strength of many living cells and because many of the equilibria of biochemical interest are insensitive to variations in the ionic strength near this value.

I have also simplified the problem in a different sense, by considering only some of the ions that occur in typical biochemical mixtures. Alberty (2003) gives a full account of the theory of handling biochemical equilibria, with details of the numerical values of the equilibrium constants involved, as originally tabulated by Alberty and Goldberg (1992).

Problems

3.1 Hexokinase A from mammalian brain is strongly inhibited by glucose 6-phosphate at concentrations above 0.1 mM. What must the limiting rate V_2 of glucose 6-phosphate dehydrogenase ($K_m = 0.11$ mM for glucose 6-phosphate) be if it is required as the coupling enzyme in an assay in which rates v_1 not exceeding $0.1 \, \text{mM min}^{-1}$ are to be measured and the concentration of glucose 6-phosphate is never to exceed 0.1 mM?

3.2 Maintaining a total $MgCl_2$ concentration 5 mM in excess of the total ATP concentration ensures that effects due to $MgATP^{2-}$ and Mg^{2+} can be clearly separated, because it allows the $MgATP^{2-}$ concentration to be varied with little concomitant variation in the free Mg^{2+} concentration. However, it does not permit unequivocal distinction between effects of $MgATP^{2-}$ and of ATP^{4-}, because it keeps their concentrations almost in constant ratio. Suggest a design that would allow the $MgATP^{2-}$ concentration to be varied with little variation in the ATP^{4-} concentration.

3.3 The following values of product concentrations p_a and p_b (in μM) and time t (in min) refer to two assays of the same enzyme, with identical reaction mixtures except that twice as much enzyme was added for values of p_b than for values of p_a. Suggest a cause for the behaviour observed.

t	p_a	p_b
0	0.0	0.0
2	10.5	4.3
4	18.0	8.3
6	23.7	11.7
8	27.9	14.5
10	31.3	16.8
12	34.0	19.0

Chapter 4

Deriving Steady-State Rate Equations

4.1 Introduction

In principle, the steady-state rate equation for any mechanism can be derived in the same way as that for the two-step Michaelis–Menten mechanism: first write down expressions for the rates of change of concentrations of all but one of the enzyme forms; next set them all to zero; write down an additional equation to express the sum of all these concentrations as a constant; finally solve the simultaneous equations that result. In practice this method is extremely laborious and liable to error for all but the simplest mechanisms, because it generates many terms that subsequently need to be cancelled from the final result. Fortunately, King and Altman (1956) described a schematic method that is simple to apply to any mechanism consisting of a series of reactions between different forms of one enzyme. It is not applicable to non-enzymic reactions, to mixtures of enzymes, or to mechanisms that contain non-enzymic steps. Nonetheless, it is applicable to most of the situations met with in enzyme catalysis and is useful in practice. It is described and discussed in this chapter.

It is not necessary to understand the theory of the King–Altman method in order to apply it, and indeed the theory is much more difficult than the practice. Readers unfamiliar with the properties of determinants may therefore prefer to proceed directly to the description in Section 4.3. However, if one understands a method that one uses one can better appreciate its scope and limitations, and for this reason the theory of the King–Altman method is given in the next section.

4.2 The principle of the King–Altman method

Consider a mechanism involving n different enzyme forms, $E_1, E_2 \ldots E_n$. Suppose that reversible first-order (or pseudo-first-order) reactions are possible between all pairs of species E_i and E_j, and let the rate constant for $E_i \rightarrow E_j$ be k_{ij} and that for $E_j \rightarrow E_i$ be k_{ji}, and so on. Then the rate of production of any particular enzyme form E_i is $k_{1i}e_1 + k_{2i}e_2 + \ldots + k_{ni}e_n$, where the sum includes every enzyme form except E_i itself; and the rate of consumption of E_i is $(k_{i1} + k_{i2} + \ldots + k_{in})e_i$, which we shall represent as $\Sigma k_{ij}e_i$. Then, the rate of change of e_i is

$$\frac{de_i}{dt} = k_{1i}e_1 + k_{2i}e_2 + \ldots + k_{i-1,i}e_{i-1} - \Sigma k_{ij}e_i + k_{i,i+1}e_{i+1} + \ldots + k_{ni}e_n = 0 \quad (4.1)$$

In the steady state this expression is equal to zero, and there are n expressions of the same type, one for each of the n species. However, only $(n-1)$ of these equations are independent, because any one of them can be obtained by adding the other $(n-1)$ together. To solve the equations for the n unknowns, it is necessary to have one further equation: this is provided by the condition that the sum of concentrations of all the enzyme forms must be the total enzyme concentration e_0:

$$e_1 + e_2 + \ldots + e_n = e_0 \quad (4.2)$$

It does not matter which of the original n equations is replaced with equation 4.2, but it is convenient when solving for e_m to replace the mth equation. The complete set of n simultaneous equations is then as follows:

$$
\begin{aligned}
-\Sigma k_{1j}e_1 + k_{21}e_2 + \ldots + k_{m1}e_m + \ldots + k_{n1}e_n &= 0 \\
k_{12}e_1 - \Sigma k_{2j}e_2 + \ldots + k_{m2}e_m + \ldots + k_{n2}e_n &= 0 \\
\vdots \qquad\qquad\qquad\qquad\qquad\qquad & \qquad\qquad (4.3) \\
e_1 + e_2 + \ldots + e_m + \ldots + e_n &= e_0 \\
\vdots \qquad\qquad\qquad\qquad\qquad\qquad & \\
k_{1n}e_1 + k_{2n}e_2 + \ldots + k_{mn}e_m + \ldots - \Sigma k_{nj}e_n &= 0
\end{aligned}
$$

These n simultaneous equations can in principle be solved by any ordinary algebraic method. The determinant method known as "Cramer's rule" is extremely inefficient as a numerical method for solving any but the most trivial sets of simultaneous equations, but it remains valuable for expressing the formal structure of the solution. For equation 4.3 it gives the following expression for e_m:

$$
e_m = \begin{vmatrix} -\Sigma k_{1j} & k_{21} & \ldots 0 \ldots & k_{n1} \\ k_{12} & -\Sigma k_{2j} & \ldots 0 \ldots & k_{n2} \\ \vdots & \vdots & \vdots & \vdots \\ 1 & 1 & \ldots e_0 \ldots & 1 \\ \vdots & \vdots & \vdots & \vdots \\ k_{1n} & k_{2n} & \ldots 0 \ldots & -\Sigma k_{nj} \end{vmatrix} \Big/ \begin{vmatrix} -\Sigma k_{1j} & k_{21} & \ldots k_{m1} \ldots & k_{n1} \\ k_{12} & -\Sigma k_{2j} & \ldots k_{m2} \ldots & k_{n2} \\ \vdots & \vdots & \vdots & \vdots \\ 1 & 1 & \ldots 1 \ldots & 1 \\ \vdots & \vdots & \vdots & \vdots \\ k_{1n} & k_{2n} & \ldots k_{mn} \ldots & -\Sigma k_{nj} \end{vmatrix} \tag{4.4}
$$

Inspection of the numerator $\mathcal{N}_m e_0$ of this expression shows that the mth column consists entirely of zeros apart from e_0 in the mth row. This element can be brought into the first column of the first row by m switches of rows and m switches of columns, leaving the rest of the determinant unchanged; $2m$ must be even whether m is odd or even, and so these switches leave the sign of the determinant unchanged. As the first row now consists of zeros apart from e_0 in the first column it follows that e_0 can be taken out as a factor leaving a determinant of order $(n-1)$. The normalized numerator ($\mathcal{N}_m e_0$ divided by e_0) can therefore be written as follows:

$$
\mathcal{N}_m = \begin{vmatrix} -\Sigma k_{1j} & k_{21} & \ldots & k_{n1} \\ k_{12} & -\Sigma k_{2j} & \ldots & k_{n2} \\ \vdots & \vdots & & \vdots \\ k_{1n} & k_{2n} & \ldots & -\Sigma k_{nj} \end{vmatrix} \tag{4.5}
$$

Careful examination of this determinant shows that it has the following properties:

1. It contains no constants k_{mj} with m as first index. Therefore, its expansion cannot anywhere contain a constant k_{mj} with m as first index.
2. Every constant with the same first index occurs in the same column. As every product of constants in the expansion must contain one element from each column, it follows that no product can contain two or more constants with the same first index, and every index other than m must occur once as first index in every product.
3. Every constant k_{ij}, where $i \neq m$ and $j \neq m$, occurs twice in the determinant, once as a non-diagonal element and once as one of the terms in a $-\Sigma k_{ij}$ summation. Every product containing a cycle of indices, such as $k_{12}k_{23}k_{31}$, which contains the cycle $1 \to 2 \to 3 \to 1$, must therefore cancel out when the determinant is expanded (notice that within any cycle each of a set of indices occurs once as a first index and once as a second index). To see why this should be so, it is simplest to look at a specific example, such as $k_{12}k_{23}k_{31}$.

This product will occur as $-k_{12}k_{23}k_{31}$, or $(-k_{12})(-k_{23})(-k_{31})$, as a term in the expansion of $(-\Sigma k_{1j})(-\Sigma k_{2j})(-\Sigma k_{3j})$, and also as $+k_{12}k_{23}k_{31}$ from the non-diagonal elements. Both of these products will be multiplied by the same combinations of elements from the rest of the determinant in the full expansion, because whether one constructs $k_{12}k_{23}k_{31}$ from diagonal or from non-diagonal elements one uses the same rows and columns, and therefore one leaves the same rows and columns available for choosing elements for the rest of the product.

We now need to consider the signs of these two sets of products. If the number of rate constants in the cycle is odd, as here, the product from the $-\Sigma k_{ij}$ summations is negative, because it contains an odd number of negative elements from the main diagonal. However, the product from the non-diagonal elements is positive, because it contains positive elements multiplied together, which require an even number of column switches to bring them to the main diagonal (switch columns 1 and 2, then columns 1 and 3, for the specific example considered). Thus these products are of opposite sign and cancel from the expansion.

On the other hand if the number of rate constants in the cycle is even the situation is reversed: the product from the diagonal elements is positive, because it contains an even number of negative elements, whereas the product from the non-diagonal elements is negative, because an odd number of switches are needed to bring these elements to the main diagonal. Either way, *all* products containing cycles cancel from the final expansion.

4. Any product containing a non-diagonal element must contain at least one other non-diagonal element, because selection of any non-diagonal element removes *two* diagonal elements from the choice of elements available for the rest of the product (for example, if the third element of the fourth row is used, both the third and fourth diagonal elements are excluded by the requirement that each product must contain one element only from each row and one element only from each column). Then selection of further non-diagonal elements can only be terminated by selecting an element with a first index that is already used as second index and a second index that is already used as first index; in other words, a cycle must be completed. However, we have seen that all products that contain cycles must cancel out. Consequently, all products that appear in the final expansion must be derived solely from diagonal elements. As all of the constants in the diagonal are negative, it follows that all of the products in the expansion must have the same sign: positive if $(n - 1)$ is even; negative if $(n - 1)$ is odd.

5. We have seen, under point (2), that every index except m must occur at least once as first index. Each product contains $(n-1)$ constants and so m must occur at least once as a second index because, if it did not, every index

that occurred as a second index would also occur as a first index, and the product would inevitably contain at least one cycle.

6. Every diagonal element $-\Sigma k_{ij}$ contains every possible second index except i. Consequently, every product that is not forbidden by the preceding rules must appear in the final expansion.

The conclusions from this discussion can be summarized by saying that the expansion of the normalized numerator from equation 4.4 contains a sum of products of $(n - 1)$ constants k_{ij}, of which each has the following properties: (1) m does not occur as first index; (2) every other index occurs once only as first index; (3) there are no cycles of indices; (4) every product has the same sign; (5) m occurs at least once as second index; (6) every allowed product occurs.

Fortunately the denominator in equation 4.4 does not require discussion in such complicated detail: it has the same value for every enzyme form and, because the total concentration of all the enzyme forms must be e_0, the denominator must be the sum of all the normalized numerators. Because of this, and because all numerators must have the same sign, the denominator must have the same sign also, so the fraction as a whole must be positive. This, of course, merely confirms the physical necessity that all concentrations be positive. It has the convenient consequence that we do not have to consider the signs of the numerators at all: whether they are positive or negative they can be written as positive.

To this point we have assumed that a reaction exists between every pair of enzyme forms. This is, of course, unrealistic, but it presents no problem, because any absent reaction can be treated as a reaction with zero rate constants. Products containing such constants must be zero and can be omitted.

Another objection to the above analysis is that two or more parallel reactions may connect the same pair of enzyme forms. The total forward rate is then the sum of the individual forward rates, and the total reverse rate is the sum of the individual reverse rates. So, in the above discussion, any k_{ij} can be considered to be the sum of a number of constants for parallel reactions.

Every product of rate constants discussed in this section can be regarded as a "tree" or pathway leading to one particular species from each of the other species. Consequently the discussion leads naturally and automatically to the method now to be described.

4.3 The method of King and Altman

The method is most easily described in terms of an example, and for this I shall take a mechanism for a reaction with two substrates and two products (to be considered again in Section 7.2):

$$E + A \underset{k_{-1}}{\overset{k_1}{\rightleftarrows}} EA$$

$$EA + B \underset{k_{-2}}{\overset{k_2}{\rightleftarrows}} EAB$$

$$EAB \rightleftarrows EPQ \qquad\qquad (4.6)$$

$$EPQ \underset{k_{-3}}{\overset{k_3}{\rightleftarrows}} EQ + P$$

$$EQ \underset{k_{-4}}{\overset{k_4}{\rightleftarrows}} E + Q$$

No rate constants are shown for the third reaction, because steady-state measurements provide no information about isomerizations between intermediates that react only in first-order reactions. For analysing the steady-state rate equation, therefore, we must treat EAB and EPQ as a single species, even though it may be mechanistically more meaningful to regard them as distinct.

The first step is to incorporate the whole of the mechanism into a closed scheme that shows all of the enzyme forms and the reactions between them, as in Figure 4.1a. All of the reactions must be treated as first-order reactions: this means that second-order reactions, such as the reaction $E + A \rightarrow EA$, must be given pseudo-first-order rate constants (Section 1.2.1); for example, the second-order rate constant k_1 is replaced by the pseudo-first-order rate constant k_1a by including the concentration of A.

Next, a *master pattern* (Figure 4.1b) is drawn representing the skeleton of the scheme, in this case a square. It is then necessary to find every pattern that (i) consists only of lines from the master pattern, (ii) connects every pair of enzyme forms and (iii) contains no closed loops. Each pattern contains one line fewer than the number of enzyme forms, and in this example there are four such patterns, as shown in Figure 4.1c.

In this example the application of these rules is fairly obvious, but in a more complicated mechanism it might not be, and to avoid any misunderstanding it may be helpful to give three examples of improper patterns, each of which satisfies two but violates one of the rules. These are shown in Figure 4.1d.

For each enzyme form, arrowheads are then drawn on the lines of the legitimate patterns, in such a way that each pattern leads to the enzyme form considered regardless of where in the pattern one starts. For the free enzyme E, the four patterns in Figure 4.1c are drawn as shown in Figure 4.1e. Each pattern is then interpreted as a product of the rate constants specified by the

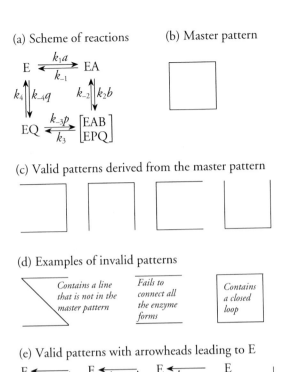

Figure 4.1. King–Altman method of deriving rate equations. (a) Scheme of reactions; (b) master pattern derived from the scheme; (c) complete set of King–Altman patterns derived from the master pattern; (d) examples of patterns that do not satisfy the rules given in the text; (e) patterns with arrowheads added so that each pattern terminates at E, with the corresponding rate constants marked; each of these directed patterns corresponds to the product of rate constants written underneath it.

arrows with reference to the complete mechanism in Figure 4.1a, and the whole set of four patterns is interpreted as the sum of four such products. For example, in Figure 4.1e, the first pattern gives $k_{-1}k_{-2}k_{-3}p$, and the complete set of patterns leading to E represents $(k_{-1}k_{-2}k_{-3}p + k_{-1}k_{-2}k_4 + k_{-1}k_3k_4 + k_2k_3k_4b)$. This sum is then the numerator of an expression that represents the fraction of the total enzyme concentration e_0 that exists as E in the steady state:

$$\frac{[E]}{e_0} = \frac{k_{-1}k_{-2}k_{-3}p + k_{-1}k_{-2}k_4 + k_{-1}k_3k_4 + k_2k_3k_4b}{D} \tag{4.7}$$

in which the denominator D will be defined shortly. Proceeding in exactly the same way for each of the other three enzyme forms provides three more fractions as follows:

$$\frac{[EA]}{e_0} = \frac{k_1 k_{-2} k_{-3} ap + k_1 k_{-2} k_4 a + k_1 k_3 k_4 a + k_{-2} k_{-3} k_{-4} pq}{D} \tag{4.8}$$

$$\frac{[EAB]+[EPQ]}{e_0} = \frac{k_1 k_2 k_{-3} abp + k_1 k_2 k_4 ab + k_{-1} k_{-3} k_{-4} pq + k_2 k_{-3} k_{-4} bpq}{D} \tag{4.9}$$

$$\frac{[EQ]}{e_0} = \frac{k_1 k_2 k_3 ab + k_{-1} k_{-2} k_{-4} q + k_{-1} k_3 k_{-4} q + k_2 k_3 k_{-4} bq}{D} \tag{4.10}$$

As these are the only four enzyme forms the sum of the four fractions must be 1, which means that the denominator D in each fraction must be the sum of the numerators, the sum of all 16 products obtained from the patterns.

The rate of the reaction is then the sum of the rates of the steps that generate one particular product, minus the sum of the rates of the steps that consume the same product. In this example, there is one step only that generates P, $(EAB + EPQ) \rightarrow EQ + P$, and one step only that consumes P, $EQ + P \rightarrow (EAB + EPQ)$, so

$$\begin{aligned}
v &= \frac{dp}{dt} \\
&= k_3([EAB] + [EPQ]) - k_{-3}[EQ]p \\
&= e_0(k_1 k_2 k_{-3} k_3 abp + k_1 k_2 k_3 k_4 ab + k_{-1} k_{-3} k_3 k_{-4} pq \\
&\quad + k_2 k_{-3} k_3 k_{-4} bpq - k_1 k_2 k_{-3} k_3 abp - k_{-1} k_{-2} k_{-3} k_{-4} pq \\
&\quad - k_{-1} k_{-3} k_3 k_{-4} pq - k_2 k_{-3} k_3 k_{-4} bpq)/D \\
&= (k_1 k_2 k_3 k_4 e_0 ab - k_{-1} k_{-2} k_{-3} k_{-4} e_0 pq)/D \tag{4.11}
\end{aligned}$$

This completes the King–Altman method for deriving steady-state rate equations.

For most purposes it is more important to know the *form* of the steady-state rate equation than to know its detailed expression in terms of rate constants. It is often therefore convenient to express a derived rate equation in *coefficient form*, which permits a straightforward prediction of the experimental properties of a given mechanism. For the example we have been examining, the coefficient form of the rate equation is

$$v = \frac{e_0(N_1 ab - N_{-1} pq)}{D_0 + D_1 a + D_2 b + D_3 p + D_4 q + D_5 ab + D_6 ap + D_7 bq + D_8 pq + D_9 abp + D_{10} bpq} \tag{4.12}$$

where the coefficients have the following values: $N_1 = k_1 k_2 k_3 k_4$; $N_{-1} = k_{-1} k_{-2} k_{-3} k_{-4}$; $D_0 = k_{-1}(k_{-2}+k_3)k_4$; $D_1 = k_1(k_{-2}+k_3)k_4$; $D_2 = k_2 k_3 k_4$; $D_3 = k_{-1} k_{-2} k_{-3}$; $D_4 = k_{-1}(k_{-2}+k_3)k_{-4}$; $D_5 = k_1 k_2(k_3+k_4)$; $D_6 = k_1 k_{-2} k_{-3}$; $D_7 = k_2 k_3 k_4$; $D_8 = (k_{-1}+k_{-2})k_{-3}k_{-4}$; $D_9 = k_1 k_2 k_{-3}$; $D_{10} = k_2 k_{-3} k_{-4}$.

4.4   The method of Wong and Hanes

Although the King–Altman method avoids generating any denominator terms that will subsequently be cancelled by subtraction, and thus avoids most of the wasted labour (and mistakes) implied by the simpler approach, it does not avoid cancellation of terms from the numerator of the rate equation. In equation 4.11, for example, eight numerator terms were initially written down, of which six were then cancelled by subtraction.

It is noticeable that the two numerator terms that survive have a rather tidy appearance compared with the six that disappeared: the positive numerator term consists of the product of the total enzyme concentration with all substrate concentrations for the forward reaction and the four rate constants for a complete forward cycle; the negative numerator term consists of the product of the total enzyme concentration with all substrate concentrations for the reverse reaction and the four rate constants for a complete backward cycle. All this is clear if we draw the corresponding arrow patterns in a sort of reverse King–Altman method, as in Figure 4.2.

This property was generalized by Wong and Hanes (1962) into a *structural rule* for the numerator of the rate equation. Initially the master pattern is drawn, exactly as in the King–Altman method (Figure 4.1b). Next every pattern is drawn that (i) consists only of lines from the master pattern, (ii) connects every pair of enzyme forms, (iii) contains one arrow leaving every enzyme form, and (iv) contains exactly one cycle capable of accomplishing a complete reaction in either the forward or the reverse direction. In a simple mechanism such as

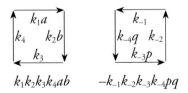

Figure 4.2. Wong–Hanes method of determining the numerator. Patterns are drawn and labelled as described in the text. Cycles that accomplish a complete forward reaction are positive; those that accomplish a complete reverse reaction are negative, and the numerator $\mathcal{N}e_0$ of the rate expression is the total enzyme concentration multiplied by the algebraic sum $\mathcal{N}$ of all the products of rate constants generated in this way, so $\mathcal{N} = k_1 k_2 k_3 k_4 ab - k_{-1} k_{-2} k_{-3} k_{-4} pq$ for the example illustrated.

the one considered already, the cycle will consist of the entire master pattern. However, in more complicated examples there may be additional enzyme forms outside the catalytic cycle: the lines connecting these to the cycle must have arrowheads that lead into the cycle. Finally one writes the numerator terms as an algebraic sum, such that every pattern accomplishing the forward reaction gives a positive product and every pattern accomplishing the reverse reaction gives a negative product. If the cycle converts more than one stoicheiometric equivalent it is weighted accordingly. Thus if there were a cycle in Figure 4.1b representing the reaction $2A + 2B \rightarrow 2P + 2Q$ it would be inserted in the rate equation with a stoicheiometric factor of 2.

4.5   Modifications to the King–Altman method

The King–Altman method as described is convenient and easy to apply to any of the simpler enzyme mechanisms. However, complicated mechanisms often require large numbers of patterns to be found. The derivation is then laborious and liable to mistakes, and one can easily overlook patterns or write down incorrect terms. Although in principle it is possible to calculate the total number of patterns (King and Altman, 1956; Chou and co-workers, 1979), this is tedious unless the mechanism is simple, because corrections are needed for all the cycles in the mechanism. Moreover, knowing how many patterns are to be found hardly helps finding them, and does not decrease the labour involved in writing down terms. In general it is better, for complicated mechanisms, to search for ways of simplifying the procedure. Volkenstein and Goldstein (1966) gave some rules for doing this, of which the simplest are the following:

1. If two or more steps interconvert the same pair of enzyme forms, these steps can be condensed into one by adding the rate constants for the parallel reactions. For example, the Michaelis–Menten mechanism is represented in the King–Altman method as shown in Figure 4.3a, which gives two patterns, ⌢ and ⌣. Because the two reactions connect the same pair of enzyme forms, they can be added to give Figure 4.3b. The resulting master pattern is itself the only pattern, so the expressions for [E] and [EA] can be written down immediately:

$$\frac{[E]}{e_0} = \frac{k_{-1}+k_2}{k_{-1}+k_2+k_1a+k_{-2}p} \qquad (4.13)$$

$$\frac{[EA]}{e_0} = \frac{k_1a+k_{-2}p}{k_{-1}+k_2+k_1a+k_{-2}p} \qquad (4.14)$$

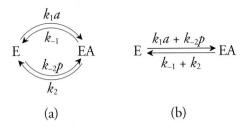

(a) (b)

Figure 4.3. Collapsing a pair of parallel steps (a) connecting the same pair of enzyme forms by adding the corresponding rate constants to give a single step (b).

2. If the mechanism contains different enzyme species that have identical properties, the procedure is greatly simplified by treating them as single species. For example, if an enzyme contained two *identical* active sites, the mechanism might be represented as shown in Figure 4.4a. Treated in this way the mechanism requires 32 patterns, but if one takes advantage of the symmetry about the broken line to write it as shown in Figure 4.4b it simplifies to one of only four patterns, which the addition of parallel steps (rule 1) allows to be simplified still further, to give Figure 4.4c. As this generates only one pattern, the expressions for the concentrations can be written down immediately:

$$\frac{[E]}{e_0} = \frac{2(k_{-1}+k_2)(k_{-3}+k_4)}{2(k_{-1}+k_2)(k_{-3}+k_4) + 4k_1(k_{-3}+k_4)a + 2k_1k_3a^2} \tag{4.15}$$

and so on.

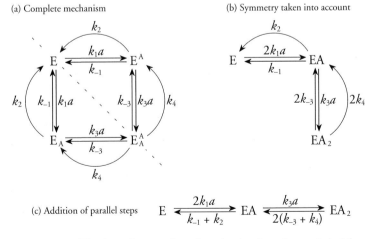

Figure 4.4. Simplification of a complicated mechanism by using statistical factors to replace multiple identical steps. The mechanism shown in (a) is symmetrical about the broken line and has exactly the same steady-state properties as the simpler mechanism shown in (b).

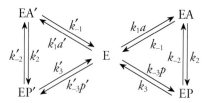

Figure 4.5. Mechanism in which two otherwise independent parts are connected through a unique enzyme form (E, in this example).

Statistical factors always appear whenever advantage is taken in this way of symmetry in the master pattern. In this example there are two ways for the reaction $E \rightarrow EA$ to occur, and so the total rate is the sum of the two rates, giving a rate constant $2k_1a$ that is twice the rate constant for either of the two paths on its own. The reverse reaction $EA \rightarrow A$, on the other hand, can only occur in one way, with a statistical factor of 1, so its rate constant remains k_{-1}.

3. If the master pattern consists of two or more distinct parts touching at single enzyme forms, it is convenient to treat the different parts separately. A simple example of this is provided by *competitive substrates* (Figure 4.5), in which a single enzyme simultaneously catalyses two separate reactions with different substrates. The expression for each enzyme form is then the product of the appropriate sums for the left and right halves of the master pattern:

$$\frac{[E]}{e_0} = (k'_{-1}k'_{-2} + k'_{-1}k'_3 + k'_2k'_3)(k_{-1}k_{-2} + k_{-1}k_3 + k_2k_3)/\mathcal{D} \qquad (4.16)$$

$$\frac{[EA]}{e_0} = (k'_{-1}k'_{-2} + k'_{-1}k'_3 + k'_2k'_3)(k_1k_{-2}a + k_1k_3a + k_{-2}k_{-3}p)/\mathcal{D} \qquad (4.17)$$

$$\frac{[EP]}{e_0} = (k'_{-1}k'_{-2} + k'_{-1}k'_3 + k'_2k'_3)(k_1k_2a + k_{-1}k_{-3}p + k_2k_{-3}p)/\mathcal{D} \qquad (4.18)$$

$$\frac{[EA']}{e_0} = (k'_1k'_{-2}a' + k'_1k'_3a' + k'_{-2}k'_{-3}p')(k_{-1}k_{-2} + k_{-1}k_3 + k_2k_3)/\mathcal{D} \qquad (4.19)$$

$$\frac{[EP']}{e_0} = (k'_1k'_2a' + k'_{-1}k'_{-3}a' + k'_2k'_{-3}p')(k_{-1}k_{-2} + k_{-1}k_3 + k_2k_3)/\mathcal{D} \qquad (4.20)$$

The numerator of the expression for $[E]/e_0$ is the product of two sums: the first of these comes from the patterns that lead to E in the left half of the master pattern, and it appears unchanged in the numerators for EA and EP; the second sum in the numerator for E comes from the patterns that lead to E in the right half of the master pattern, and it appears unchanged in the numerators for EA′ and EP′. The second sum in the numerator for EA comes from the patterns that lead to EA in the right half of the master pattern, and similarly for EP. The first sum in the numerator for EA′ comes from the patterns that lead to EA′ in the left half of the master pattern, and similarly for EP′.

4.6 Reactions containing steps at equilibrium

Some mechanisms are important enough to be worth analysing in detail, but so complicated that even with the aid of the methods described above they produce unmanageable rate equations. Simplifying assumptions cannot then be avoided, and there are often great advantages in assuming that suitable steps, such as protonation steps, are at equilibrium in the steady state. Such assumptions may, of course, turn out on further investigation to be false, but they are useful as a first approximation.

Cha (1968) described a method for analysing mechanisms that contain steps at equilibrium that is much simpler than the full King–Altman analysis because each group of enzyme forms at equilibrium is treated as a single species. As an example, consider the mechanism shown in Figure 4.6, for an enzyme in which the unprotonated enzyme E^0 and the protonated form HE^+ are both catalytically active but with different kinetic constants (in the usual treatment of the pH behaviour of enzymes, discussed in Chapter 9, only one protonation

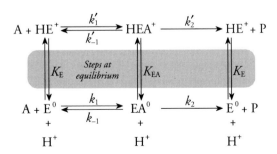

Figure 4.6. Treatment of steps at equilibrium. If the protonation steps (vertical in the example) are treated as equilibria, the mechanism can be simplified by treating each group of enzyme forms at equilibrium as a single species.

state is assumed to be catalytically active, but here we shall assume that both EA^0 and HEA^+ can react to give products as otherwise the example is too trivial to serve as an illustration of Cha's method). If equilibrium between E^0 and HE^+ is maintained in the steady state, E^0 forms a fraction $K_E/(K_E + h)$ of the composite species E, where the hydrogen-ion concentration is written as h, and HE^+ forms a fraction $h/(K_E + h)$. The rate of binding of A to E^0 is $k_1[E^0]a$ and can thus be written as $k_1[E]aK_E/(K_E + h)$; that of A to HE^+ can similarly be written as $k_1'[E]ah/(K_E + h)$, and the total rate of binding of A to the composite species E is $[E]a(k_1K_E + k_1'h)/(K_E + h)$. In effect, the individual rate constants k_1 and k_1' have been replaced by a composite rate constant $(k_1K_E + k_1'h)/(K_E + h)$ for the composite species: more generally, this composite rate constant is the sum of the individual rate constants weighted according to the fraction of the relevant enzyme form in the equilibrium mixture. The release of A from the composite species EA and the conversion of the same species to products can be dealt with in the same way.

All of this may seem to be making the analysis more complicated rather than simpler. However, what has happened is that the original mechanism that would require the King–Altman method to analyse has been replaced by a mechanism for which the solution is already known: in effect it is the Michaelis–Menten mechanism, albeit with unusually complicated expressions for the component rate constants. The rate equation can thus be written down immediately, replacing k_1 in equation 2.13 by $(k_1K_E + k_1'h)/(K_E + h)$, and similarly for k_{-1} and k_2.

Cha's method is applicable to any mechanism that contains steps at equilibrium. In general any number of enzyme forms in equilibrium with one another can be treated as one species, and each rate constant k_i for a component becomes a term k_if_i in the summation for the rate constant of the composite species, where f_i is the fraction of the component in the equilibrium mixture. Although the general validity of the method has been called into question (Segel and Martin, 1988), Topham and Brocklehurst (1992) have shown that the objections raised were unfounded[*]. Their analysis has more than a passing interest, because, as they point out, other authors have made similar errors and it is important to understand how to avoid them. The essential point is that whenever a pathway contains cycles one must take account of

---

[*]Varón and co-workers (1992) later suggested that Topham and Brocklehurst (1992) had given an incorrect account of the errors in the work of Segel and Martin (1988), but their arguments included several mistakes (Selwyn, 1993; Brocklehurst and Topham, 1993) and the numerical calculations offered in support of them failed to take account of the need for the product of ratios of rate constants around a cycle to agree with the equilibrium constant for the cycle. All of this underlines the point that proper analysis of rate equations cannot be done without proper care, and that incorrect ideas are widespread in the literature.

all of the parallel routes that connect any pair of species in the mechanism. More generally, whatever rate constants are assumed for any of the steps in a mechanism, it is essential that the product of all the ratios of forward and reverse rate constants for the steps connecting a pair of species must be the equilibrium constant for the net process. It ought to be superfluous to add that this equilibrium constant must be the same for every parallel route that accomplishes the same reaction. It follows that a cycle that accomplishes no net change, such as the cycle $E^0 \rightarrow HE^+ \rightarrow HEA^+ \rightarrow EA^0 \rightarrow E^0$ in Figure 4.6, must have a net equilibrium constant of 1.

This important principle is known as the *principle of microscopic reversibility* or the *principle of detailed balance*. The two names are almost equivalent, but microscopic reversibility emphasizes that a reaction proceeding in the reverse direction follows the same mechanism as it does in the forward direction, whereas detailed balance emphasizes that when an entire process is at equilibrium all of its component steps are individually at equilibrium as well. Occasionally there have been deliberate challenges to the validity of the principle by authors well aware of its existence, for example in the work of Weber and Anderson (1965) mentioned later (Section 11.6.3), but far more often it is violated out of ignorance or carelessness, as with the criticisms of Cha's method just noted. Failure to understand it also led to some erroneous ideas about how kinetic cooperativity could arise in monomeric enzymes; only later were the valid models described in Section 11.8 developed.

A word of warning should be given about the use of Cha's method in relation to mechanisms that contain parallel binding pathways, such as a mechanism in which two substrates A and B can bind in either order to arrive at the same ternary complex EAB. Because the full rate equation for such a mechanism is often too complicated to be manageable, it is common practice to use equations derived on the assumption that binding steps are at equilibrium. Mechanistically, there is no more basis for making any such assumption than there is for a one-substrate reaction (Section 2.3.1), yet the resulting equations are often found to be in good agreement with experiment. The problem is that the additional terms present in the rigorous steady-state rate equation may be numerically negligible, or they may vary so nearly in proportion with other terms, that they are almost impossible to detect, as discussed by Gulbinsky and Cleland (1968) in relation to a study of galactokinase. This is an instance of an important general point for the correct application of kinetics: more than one set of assumptions can lead to the same rate equation, or to a family of rate equations that are experimentally indistinguishable; experimental adherence to a particular rate equation is not a proof that the assumptions used to derive the equation were correct.

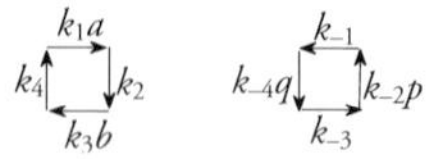

$$v = \cfrac{\cfrac{V_+ab}{K_{iA}K_{mB}} - \cfrac{V_-pq}{K_{iP}K_{mQ}}}{\cfrac{a}{K_{iA}} + \cfrac{K_{mA}b}{K_{iA}K_{mB}} + \cfrac{p}{K_{iP}} + \cfrac{K_{mP}q}{K_{iP}K_{mQ}} + \cfrac{ab}{K_{iA}K_{mB}} + \cfrac{ap}{K_{iA}K_{iP}} + \cfrac{K_{mA}bq}{K_{iA}K_{mB}K_{iQ}} + \cfrac{pq}{K_{iP}K_{mQ}}}$$

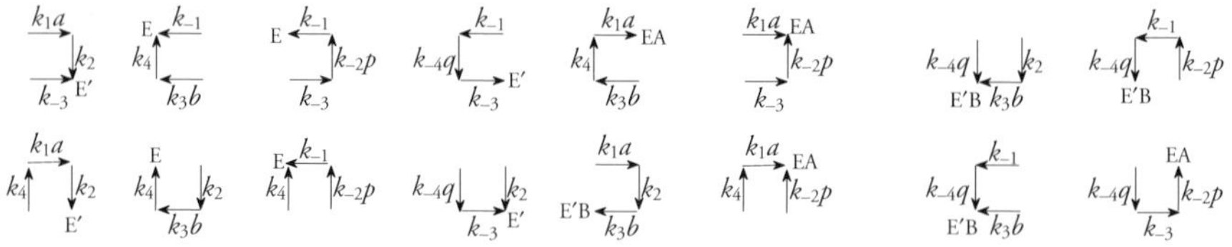

Figure 4.7. Structure of a rate equation revealed by inspection. The equation for the substituted-enzyme mechanism to be discussed in Section 7.3.3 (equation 7.7) is analysed in terms of all the King–Altman and Wong–Hanes patterns that account for it.

4.7   Analysing mechanisms by inspection

4.7.1   Topological reasoning

One might think — indeed, it seems too obvious to be worth saying — that the main role of the King–Altman method in modern enzymology was for deriving rate equations. However, this would be a mistake, because outside the context of teaching one rarely has occasion to derive rate equations, those of greatest importance being mostly known already: many examples, both of rate equations and of their derivation, may be found in the books of Plowman (1972), Segel (1975) and Schulz (1994). The real importance of the King–Altman method is that once one is thoroughly conversant with it one can deduce important conclusions about the steady-state properties of enzyme mechanisms without having to do any explicit algebra, a process Wong (1975) has aptly called *topological reasoning*. Figure 4.7 illustrates this idea: it shows equation 7.7, which we shall encounter in Chapter 7, together with all the King–Altman patterns that account for the different denominator terms, as well as the two Wong–Hanes patterns for the two numerator terms. With a proper understanding of the methods of this chapter the analysis of the structure of the equation that is made explicit in the Figure can be done mentally, so that one can recognize immediately where all of the terms in the equation have come from. Not only that, with no great effort one can also see why the different constants in the equation have the definitions in terms of rate constants that are listed in the right-hand column of Table 7.1.

In the King–Altman method every pattern generates a positive term, and every term appears in the denominator of the rate equation. As there are no negative terms, no terms can cancel by subtraction, and so every term for which a pattern exists must appear in the rate equation. The only exception to this rule is that sometimes the numerator and denominator share a common factor that can be cancelled by division, but this can only happen if there are symmetries in the mechanism such that certain rate constants appear more than once.

4.7.2 Mechanisms with alternative routes

In mechanisms with no relationships between rate constants apart from those required by thermodynamics, it is usually safe to assume that no cancellation between numerator and denominator will be possible, and that any term for which a pattern exists must appear in the rate equation. Consider, for example, the mechanism shown above in Figure 4.6, but without assuming the protonation steps to be at equilibrium: is there any conclusion we can draw about how the rate equation will differ from the one obtained with the equilibrium assumption, without actually deriving it? If no equilibrium steps are assumed, the King–Altman method must produce terms in a^2 and h^2, because one can easily find patterns that contain two A-binding steps or two H^+-binding steps, for example, the pattern terminating at EA^0 consisting of $E^0 \rightarrow EA^0$ and $HE^+ \rightarrow HEA^+ \rightarrow EA^0$. Thus simply by inspecting the mechanism one can deduce that the rate equation is second-order in a and in h. Note that this can be done without writing anything on paper, neither drawing patterns nor doing any algebra (though the process would perhaps be made a little easier by redrawing the mechanism in a closed form, as in Figure 4.1a, so that E^0 and HE^+ do not appear twice each).

4.7.3 Dead-end steps

Inspection is not only valuable for studying particular mechanisms; it also leads to important and far-reaching conclusions about mechanisms in general. Consider, for example, the effect of adding a step to the mechanism of Figure 4.1a, in which B binds to EQ to form a *dead-end complex* EBQ that is incapable of any reaction apart from release of B to re-form EQ, as shown in Figure 4.8. For every enzyme form of the original mechanism, that is to say every enzyme form apart from EBQ, every King–Altman pattern must contain the reaction $EBQ \rightarrow EQ$, and so every enzyme form apart from EBQ must have the same expression as before apart from being multiplied by k_{-5}. As the only route to EBQ is through EQ, it is obvious that every pattern for EBQ must be the same as a pattern for EQ with the reaction $EBQ \rightarrow EQ$ replaced

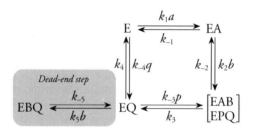

Figure 4.8. Mechanism with a dead-end step. A dead-end step (interconversion of EQ and EBQ in this example) is always at equilibrium in the steady state.

by $EQ \rightarrow EBQ$, so the expression for EBQ must be the same as that for EQ with k_{-5} replaced by k_5b, that is to say the ratio $[EBQ]/[EQ]$ must be k_5b/k_{-5}. As k_5/k_{-5} is the equilibrium association constant this means that the step must be at equilibrium in the steady state. The rate equation as a whole must be the same as for the mechanism without dead-end inhibition except that all terms for EQ in the denominator are multiplied by $(1+k_5b/k_{-5})$.

A little reflection will show that this conclusion is not particular to the mechanism considered, but applies to any mechanism with dead-end steps, even if these form a series, that is to say even if several steps are needed to connect a dead-end complex to the rest of the mechanism: in any mechanism, dead-end reactions are at equilibrium in the steady state. One can understand in mechanistic terms (without algebra) why this conclusion applies by realizing that the reason why ordinary steady-state reactions are not at equilibrium is that flux of reactants through a reaction is a process of continual unbalancing: constant replenishment of reactants on one side and removal of products from the other. As there is no net flux of reactants through a dead-end reaction, there is no corresponding unbalancing to interfere with the establishment of equilibrium.

4.8 Derivation of rate equations by computer

Deriving a rate equation, whether by the method of King and Altman or in any other way, is a purely mechanical process, and success or failure depends on the avoidance of mistakes rather than on making correct intellectual decisions about how to proceed at any point. As such it is ideal for computer implementation, and a number of computer programs for deriving rate equations have been described (Rhoads and Pring, 1968; Hurst, 1967, 1969; Fisher and Schulz, 1969; Rudolph and Fromm, 1971; Kinderlerer and Ainsworth, 1976; Cornish-Bowden, 1977; Olavarría, 1986). Unfortunately all of these were written in Fortran or other languages that are no longer widely used in

Table 4.1. Automatic derivation of rate equations.

(a) Matrix of rate constants

	$\rightarrow$E	$\rightarrow$EA	$\rightarrow$(EAB + EPQ)	$\rightarrow$EQ
E$\rightarrow$	(none)	k_1a	(none)	$k_{-4}q$
EA$\rightarrow$	k_{-1}	(none)	k_2b	(none)
(EAB + EPQ)$\rightarrow$	(none)	k_{-2}	(none)	k_3
EQ$\rightarrow$	k_4	(none)	$k_{-3}p$	(none)

(b) Identification of cyclic products

Product	$\rightarrow$E?	Cyclic?
$k_{-1}k_{-2}k_4$	Yes	No
$k_{-1}k_{-2}k_{-3}p$	Yes	No
$k_{-1}k_3k_4$	Yes	No
$k_{-1}k_3k_{-3}p$	Yes	Yes
$k_2bk_{-2}k_4$	Yes	Yes
$k_2bk_{-2}k_{-3}p$	No	Yes
$k_2bk_3k_4$	Yes	No
$k_2bk_3k_{-3}p$	No	Yes

biochemical computing, but adaptation to modern systems should not present any insurmountable difficulties once the methodology is clearly understood.

For implementing the method of King and Altman in the computer, the derivation is best treated as a series of operations on a table of rate constants rather than as the algebraic equivalent of an exercise in pattern recognition, because operating on a table of numbers is much easier to program than interpreting geometric relationships. Table 4.1 represents the same example as the one considered in Section 4.3, that of equation 4.6. To obtain a valid product of rate constants terminating at E we can take the first available rate constant from each row except the first, giving $k_{-1}k_{-2}k_4$: omission of the E$\rightarrow$ row ensures that the product has no step leading away from E, and including every other row ensures that there is exactly one step leading out of every other intermediate. However, this is only the first of the possible products. To obtain the others, we replace the rate constant from the EQ$\rightarrow$ row by each other possibility from that row, giving $k_{-1}k_{-2}k_{-3}p$ as the only new product. We then move to the second position of the previous row, the (EAB + EPQ)$\rightarrow$ row, and again choose all possibilities from the EQ$\rightarrow$ row, giving $k_{-1}k_3k_4$ and $k_{-1}k_3k_{-3}p$. If there were other possibilities in the $(EAB + EPQ) \rightarrow$ row we would do the same for them, but there are not, so we proceed to the next element of the $EA \rightarrow$ row, taking all possibilities in turn from the $(EAB + EPQ) \rightarrow$ and $EQ \rightarrow$ rows.

This procedure generates eight products in total, as listed in the lower part of Table 4.1. However, not all of these are valid, because some do not contain a step to E, and some are cyclic. In principle it is not necessary to check whether a product contains a step to the target species, because any non-cyclic product must satisfy this requirement. However, it is much easier to check if a product contains a step to E than it is to check if it is cyclic, because a product that leads to E must contain at least one rate constant from the $\rightarrow E$ column. It is best therefore to delete products that do not contain a rate constant from the $\rightarrow E$ column before checking if any of the others are cyclic. In Table 4.1 it is trivially easy to recognize cyclic products, because the only kind of cycle possible with this mechanism is one in which both forward and reverse rate constants from the same step occur, as in $k_{-1}k_3k_{-3}p$, which contains both the forward and reverse rate constants of step 3. More complicated cycles can occur in more complicated mechanisms, however, and a valid program must be able to recognize these.

Conceptually the simplest (though not the most efficient) way of determining if a product is cyclic is to start from each rate constant in turn that it contains and follow through at most $n - 1$ steps until the target species is found. For example, consider $k_2bk_3k_4$, and start with k_2b: this comes from the $\rightarrow(EAB + EPQ)$ column, and so we have to find the rate constant from the $(EAB + EPQ) \rightarrow$ row, which is k_3; this is from the $\rightarrow EQ$ column so we look for the entry from the $EQ \rightarrow$ row, which is k_4; this is in the $\rightarrow E$ column, so the process is terminated successfully in three steps ($3 = n - 1$ in this example). As the other two elements in this product were checked during the course of checking k_2b they do not need to be checked again. However, if we had started with k_3 the check would not have passed through k_2b and so it would have had to be checked separately.

This method of checking for cyclic products may seem tedious, but one should be careful not to be seduced by appealing shortcuts that may not always give correct results. For example, for many mechanisms any product that contains the same rate constant more than once, or contains the forward and reverse versions of the same rate constant, will be cyclic. However, one cannot assume that this will always be true, because certain kinds of symmetry in the enzyme mechanism can result in having one or more rate constants appearing more than once, as for example in Figure 4.4a. Some of the published methods for analysing mechanisms by computer (not among those cited above) have failed to include the necessary precautions and consequently give incorrect results for such mechanisms (Cornish-Bowden, 1976b).

After eliminating the invalid products the sum of the others gives the expression for $[E]/e_0$, as in equation 4.7:

$$[E]/e_0 = (k_{-1}k_{-2}k_4 + k_{-1}k_{-2}k_{-3}p + k_{-1}k_3k_4 + k_2bk_3k_4)/D \tag{4.21}$$

Conversion of this into equation 4.8, that is to say the corresponding expression for $[EA]/e_0$, now requires analysis of each product in the sum to identify the pathway from EA to E that it contains, reversing the rate constants for this part of the product and leaving the rest unchanged. For $k_{-1}k_{-2}k_4$, for example, the pathway from EA to E requires just k_{-1}, so this is replaced by k_1a and $k_{-2}k_4$ is left unchanged and the whole product becomes $k_1ak_{-2}k_4$. Doing this for each product in turn, and then for (EAB + EPQ) and EQ in the same way, provides equations equivalent to equations 4.8–10.

This then provides the basis for a computer equivalent of the method of King and Altman. The numerator can be dealt with by applying the same sort of logic to the rules of Wong and Hanes (Section 4.4), and hence it is straightforward to obtain the complete rate equation.

Problems

4.1 In the opening sentence of this chapter the second step in deriving a rate equation was defined as writing down "expressions for the rates of change of concentrations of all but one of the enzyme forms". Why is the number of expressions to be written down one fewer than the number of enzyme forms? Why not all?

4.2 Derive a rate equation for the following mechanism,

$$E + A \rightleftarrows EA \rightleftarrows E'P \rightleftarrows E' + P; \ E' + B \rightleftarrows E'B \rightleftarrows EQ \rightleftarrows E + Q$$

treating EA and E′P as a single enzyme form, and E′B and EQ as a single enzyme form. Ignoring terms containing the product concentrations p and q, how does the coefficient form of the rate equation differ from equation 4.12, for the mechanism of equation 4.6?

4.3 Without carrying out a complete derivation, write down a rate equation for the mechanism of Problem 4.2 modified by supposing that B can bind to E to give a dead-end complex EB with a dissociation constant K_{siB}.

4.4 Consider an enzyme that catalyses two reactions simultaneously, inter-conversion of A + B with P + Q and of A + B′ with P′ + Q; that is to say A and Q are common to the two reactions but B, B′, P and P′ are not.

Assuming that both reactions proceed individually by the mechanism shown in equation 4.6, draw a master pattern for the system and find all valid King–Altman patterns.

In this system the rate equation for v defined as dp/dt is not the same as that for v defined as $-da/dt$. Why not?

4.5 For a mechanism that contains two or more enzyme forms that react only in bimolecular reactions (never in unimolecular reactions), show that every term in the rate equation must contain at least one reactant concentration. (Problem 4.2 provides an example of such a mechanism.)

An exercise in the use of Cha's method is given in Problem 9.4.

Chapter 5

Reversible Inhibition and Activation

5.1 Introduction

A substance that decreases the rate of an enzyme-catalysed reaction when it is present in the reaction mixture is called an *inhibitor*. Inhibition can arise in a wide variety of ways, however, and there are many different types of inhibitor. This chapter deals with *reversible inhibitors*, which are substances that form dynamic complexes with the enzyme that have different catalytic properties from those of the uncombined enzyme. (Chapter 6 deals with irreversible inhibitors.) If the enzyme has no catalytic activity at all when saturated with inhibitor the inhibition may be described as *complete inhibition*, but is more often referred to as *linear inhibition*, in reference to the linear dependence of the apparent values of K_m/V and $1/V$ on the inhibitor concentration, or just *simple inhibition*. Objections can be raised to all of these terms, the first being likely to be misunderstood if used out of context, and the third because it is too vague. If "linear" is taken to mean that plots of the apparent values of K_m/V and $1/V$ against the inhibitor concentration are straight this may also be misleading, especially as these plots are less used today than they once were. However, linearity is a mathematical concept that is defined without any reference to graphs: a relationship between variables is linear, or not, regardless of whether one chooses to plot it as a straight line. In this book the term *linear inhibition* will be used for this simplest class of inhibition.

Less often considered (though probably not as rare in nature as is usually assumed) is the kind of inhibition that occurs when the enzyme–inhibitor complex has some residual catalytic activity: this is called *hyperbolic inhibition*, from the shapes of the curves obtained by plotting apparent Michaelis–Menten parameters against inhibitor concentration, or *partial inhibition*, from the survival of some activity when the enzyme is saturated with inhibitor (see Section 5.7.3).

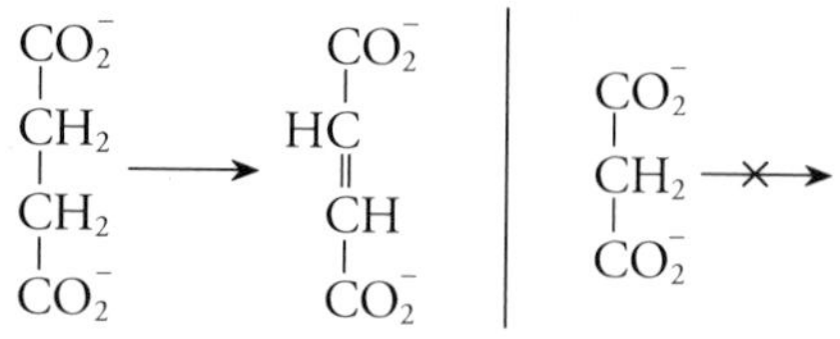

$$
\begin{array}{ccc}
\text{CO}_2^- & \text{CO}_2^- & \\
| & | & \text{CO}_2^- \\
\text{CH}_2 & \text{HC} & | \\
| \longrightarrow & \| & \text{CH}_2 \longrightarrow\!\!\!\times\!\!\!\longrightarrow \\
\text{CH}_2 & \text{CH} & | \\
| & | & \text{CO}_2^- \\
\text{CO}_2^- & \text{CO}_2^- & \\
\end{array}
$$

succinate   fumarate   malonate

Figure 5.1. Succinate dehydrogenase reaction. Although succinate and malonate are sufficiently similar in structure to be able to bind at the same site on the enzyme, malonate lacks the dimethylene group that would allow it to be dehydrogenated.

Both linear and hyperbolic inhibition may in principle be sub-classified according to the particular apparent Michaelis–Menten parameters they affect, but in practice the common terms always imply linear inhibition unless otherwise stated. The commonest type is usually called *competitive inhibition*, and is characterized by decreased apparent V/K_m with no change in apparent V. Because competitive inhibition was known before it was realized that V and V/K_m are the fundamental parameters of the Michaelis–Menten equation, the behaviour is often described as an increase in K_m. However, as will be clear later (see Table 5.1 on page 120) the classification of inhibitors is much simpler and more straightforward if expressed in terms of V and V/K_m.

The opposite extreme of inhibition is *uncompetitive inhibition*, in which apparent V is decreased, with apparent V/K_m unchanged: this can also be expressed by saying that apparent V and K_m are both decreased by the same factor. Spanning the range between these two is *mixed inhibition*, in which apparent V and V/K_m are both decreased. Another type, known as *pure non-competitive inhibition* (apparent V is decreased, with apparent K_m unchanged), was once thought to be equally important, but it is actually quite rare. All of these will now be described.

5.2   Linear inhibition

5.2.1   Competitive inhibition

In the reaction catalysed by succinate dehydrogenase, succinate is oxidized to fumarate, as illustrated in Figure 5.1. As this is a reaction of the dimethylene group it cannot occur with malonate, which has no dimethylene group. In other respects malonate has almost the same structure as succinate, and, not surprisingly, it can bind to the substrate-binding site of succinate dehydrogenase to give an abortive complex that is incapable of reacting. This is an example of *competitive inhibition*, so-called because the substrate and the inhibitor compete for the same site.

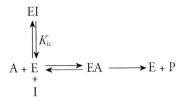

Figure 5.2. Mechanism that produces competitive inhibition.

The mechanism may be represented in general terms as shown in Figure 5.2, in which EI is a dead-end complex, as the only reaction that it can undergo is reformation of $E + I$. By the argument of Section 4.7.3, therefore, its concentration is given by a true equilibrium constant, $K_{ic} = [E][I]/[EI]$, called the *inhibition constant*, or, more explicitly, the *competitive inhibition constant*. It is often symbolized simply as K_i when the competitive character is taken as assumed. It is important to note, however, that Figure 5.2 is not the only possible way in which competitive inhibition can occur, and the inhibition constant is an equilibrium constant because of the particular mechanism assumed and not because the inhibition is competitive: it is perfectly possible for inhibition to be competitive but for the inhibition constant not to be an equilibrium constant. In many of the more complicated types of inhibition, including most types of product inhibition, the inhibition constant is not a true equilibrium constant because the enzyme–inhibitor complex is not a dead-end complex.

The defining equation for linear competitive inhibition, which applies not only to the mechanism of Figure 5.2 but to any mechanism, is

$$v = \frac{Va}{K_m(1 + i/K_{ic}) + a} \tag{5.1}$$

in which i is the free inhibitor concentration and V and K_m have their usual meanings as the limiting rate and the Michaelis constant. The equation is of the form of the Michaelis–Menten equation, and can be written as

$$v = \frac{V^{app}a}{K_m^{app} + a} \tag{5.2}$$

where V^{app} and K_m^{app} are the *apparent* values of V and K_m, the values they appear to have when measured in the presence of the inhibitor. They are given by

$$V^{app} = V \tag{5.3}$$

$$K_m^{app} = K_m(1 + i/K_{ic}) \tag{5.4}$$

$$V^{app}/K_m^{app} = \frac{V/K_m}{1 + i/K_{ic}} \qquad (5.5)$$

Hence the effect of a competitive inhibitor is to decrease the apparent value of V/K_m by the factor $(1 + i/K_{ic})$ while leaving that of V unchanged.

As equations 5.3–5 define the meaning of competitive inhibition regardless of the underlying mechanism, and as they show that the essential effect of a competitive inhibitor is to decrease the apparent specificity constant, it would appear that the name *specific inhibition* would be preferable. In contexts where activation (Section 5.7) is also being considered, this would be a major improvement, but for general use it is quite unrealistic to suppose that biochemists will abandon a name sanctioned by 90 years of use in order to avoid confusion when the much less important phenomenon of specific activation is under discussion. For this reason I continue to use the term *competitive inhibition* in this book, though I emphasize that it is defined *operationally**, not mechanistically, in accordance with the recommendations of the International Union of Pure and Applied Chemistry (1981) and the International Union of Biochemistry and Molecular Biology (International Union of Biochemistry, 1982) for kinetic terminology.

5.2.2   Mixed inhibition

Most elementary accounts of inhibition discuss two types only, competitive inhibition and *non-competitive inhibition*. Competitive inhibition is of genuine importance in nature, but non-competitive inhibition is a phenomenon found mainly in textbooks and it need not be considered in detail here. It arose originally because Michaelis and his collaborators, who were the first to study enzyme inhibition, assumed that certain kinds of inhibitor acted by decreasing the apparent value of V with no corresponding effect on that of K_m. Such an effect seemed at the time to be the obvious alternative to competitive inhibition, and was termed "non-competitive inhibition". It is difficult to imagine a reasonable explanation for such behaviour, however: one would have to suppose that the inhibitor interfered with the catalytic properties of the enzyme but that it had no effect on the binding of substrate; expressed somewhat differently, it would mean that two molecules (the free enzyme and the enzyme–substrate complex) with quite different properties in other respects had equal binding

---

*An operational definition is one that expresses what is observed, without regard to how it may be interpreted, whereas a mechanistic definition implies an interpretation. Mechanistic definitions of kinetic behaviour are frequently used, but they have the disadvantage that observations cannot be described until they have been interpreted, even though in research one often needs to describe observations before one has a satisfactory way of explaining them.

constants for the inhibitor. This is possible for very small inhibitors, such as protons, metal ions and small anions such as Cl⁻, but is unlikely otherwise.

Non-competitive inhibition by protons is, in fact, common. There are also several instances of non-competitive inhibition by heavy-metal ions, though some (maybe all) of these are really examples of partial irreversible inactivation, which will be discussed shortly. Non-competitive inhibition by other species is rare, and some supposed examples, such as the inhibition of β-fructofuranosidase ("invertase") by α-glucose (Nelson and Anderson, 1926) and the inhibition of arginase by various compounds (Hunter and Downs, 1945), prove, on re-examination of the original data, to be examples of *mixed inhibition*, or else they result from confusion between reversible non-competitive inhibition and *partial irreversible inactivation*. In general, it is best to regard non-competitive inhibition as a special, and not very interesting, case of mixed inhibition.

In partial irreversible inactivation it does not greatly matter whether the primary effect of the inactivation is on V, V/K_m or K_m, because whichever it is the net result is a loss of enzyme molecules from the system, with the unaffected molecules behaving just as before. However, as $V = k_{cat}e_0$ is the product of the catalytic constant k_{cat} and the concentration of active enzyme e_0, it will decrease regardless of whether the apparent value of k_{cat} is decreased (genuine non-competitive inhibition), or whether e_0 is decreased (irreversible inactivation).

Linear mixed inhibition is the type of inhibition in which both specific and catalytic effects are present, that is to say both V^{app}/K_m^{app} and V^{app} vary with the inhibitor concentration, according to the following equations:

$$V^{app} = \frac{V}{1 + i/K_{iu}} \tag{5.6}$$

$$K_m^{app} = \frac{K_m(1 + i/K_{ic})}{1 + i/K_{iu}} \tag{5.7}$$

$$V^{app}/K_m^{app} = \frac{V/K_m}{1 + i/K_{ic}} \tag{5.8}$$

The simplest formal mechanism for this behaviour is one in which the inhibitor can bind both to the free enzyme to give a complex EI with dissociation constant K_{ic}, and also to the enzyme–substrate complex to give a complex EAI with dissociation constant K_{iu}, as shown in Figure 5.3. As EI and EAI both exist in this scheme there is no obvious mechanistic reason why A should not bind directly to EI to give EAI, as shown, so the inhibitor-binding steps are not dead-end reactions and it does not therefore follow automatically that

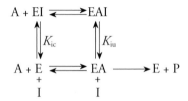

Figure 5.3. Mechanism that produces mixed inhibition.

they must be at equilibrium (Section 4.7.3). The complete steady-state rate equation is thus more complicated than equation 5.2 with the definitions given in equations 5.6–8: it includes terms in a^2 and i^2 that do not cancel unless all of the binding reactions are equilibria. However, the predicted deviations from linear kinetics are difficult to detect experimentally, and so adherence to linear kinetics is not adequate evidence that K_m, K_{ic} and K_{iu} are true equilibrium dissociation constants. In practice inhibitor binding is generally treated as an equilibrium nonetheless, but it should be understood that that is an *assumption* that does not follow directly from the mechanism: in this respect mixed inhibition differs from competitive inhibition (Section 5.2.1) and uncompetitive inhibition (considered in the next section), for which binding of the inhibitor at equilibrium is a necessary consequence of the way the mechanism is written.

Although it is formally convenient to define mixed inhibition in terms of Figure 5.3, it actually occurs mainly as an important case of product inhibition. If a product is released in a step that generates an enzyme form other than the one to which the substrate binds, product inhibition is predicted to be in accordance with equations 5.6–8. This conclusion does not depend on any equilibrium assumptions, being a necessary consequence of the steady-state treatment, as can readily be shown by the methods of Chapter 4. The simplest of many mechanisms of this type is one in which the product is released in the second of three steps, as shown in Figure 5.4. More complicated examples abound in reactions that involve more than one substrate or product, as will be seen in Chapter 7. In these cases, identification of K_{ic} and K_{iu} with dissociation constants is not very useful. Even in such a simple example as Figure 5.4, in which the product P acts as a mixed inhibitor rather than as a competitive inhibitor because it binds to a different form of enzyme from A, the inhibition

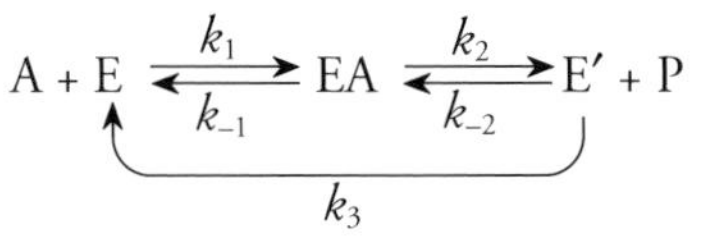

$$A + E \underset{k_{-1}}{\overset{k_1}{\rightleftharpoons}} EA \underset{k_{-2}}{\overset{k_2}{\rightleftharpoons}} E' + P$$
$$k_3$$

Figure 5.4. Mechanism that produces mixed inhibition by product. In this mechanism neither of the two inhibition constants is an equilibrium constant.

constants are $K_{ic} = (k_{-1} + k_2)k_3/k_{-1}k_{-2}$ and $K_{iu} = (k_2 + k_3)/k_{-2}$, and neither of them is an equilibrium constant except in special cases, such as $k_3 \ll k_2$.

The rareness of genuine non-competitive inhibition has led to a tendency to generalize the term to embrace mixed inhibition. There seems to be no advantage in doing this: apart from anything else it replaces a short unambiguous word with a longer ambiguous one, thereby adding to the confusion of an already confused nomenclature. To avoid the ambiguity one must refer to non-competitive inhibition in the classical sense as *pure non-competitive inhibition* or *true non-competitive inhibition*.

5.2.3   Uncompetitive inhibition

In uncompetitive inhibition, at the other extreme from competitive inhibition, the inhibitor decreases the apparent value of V with no effect on that of V/K_m:

$$V^{app} = \frac{V}{1 + i/K_{iu}} \tag{5.9}$$

$$K_m^{app} = \frac{K_m}{1 + i/K_{iu}} \tag{5.10}$$

$$V^{app}/K_m^{app} = V/K_m \tag{5.11}$$

The uncompetitive inhibition constant K_{iu} may be symbolized simply as K_i when the uncompetitive character can be assumed, but as this will be rather unusual it is best to use the more explicit symbol. Comparison of equations 5.9–11 with equations 5.6–8 shows that uncompetitive inhibition is a limiting case of mixed inhibition in which K_{ic} approaches infinity, i/K_{ic} being negligible at all values of i and hence disappearing from equations 5.6–8; it is thus the converse of competitive inhibition, the other limiting case of mixed inhibition, in which K_{iu} approaches infinity.

Uncompetitive inhibition is also, at least in principle, the mechanistic converse of competitive inhibition, because it is predicted for mechanisms in which the inhibitor binds only to the enzyme–substrate complex and not to the free enzyme. One example of clinical importance is the inhibition of *myo*-inositol monophosphatase by Li^+. This ion is used to treat manic depression, and its selectivity for cells with excessive signal-transduction activity is consistent with the uncompetitive character of the inhibition (Pollack and co-workers, 1994). Another example is discussed in Section 12.3.4.

The name *uncompetitive inhibition* is by no means as firmly established as its opposite, and there is a much stronger argument for abandoning it in favour of *catalytic inhibition*: the word "uncompetitive" is not used in ordinary

Table 5.1. Characteristics of linear inhibitors.

Type of inhibition	V^{app}	$V^{\mathrm{app}}/K_{\mathrm{m}}^{\mathrm{app}}$	$K_{\mathrm{m}}^{\mathrm{app}}$
Competitive	V	$\dfrac{V/K_{\mathrm{m}}}{1+i/K_{\mathrm{ic}}}$	$K_{\mathrm{m}}(1+i/K_{\mathrm{ic}})$
Mixed	$\dfrac{V}{1+i/K_{\mathrm{iu}}}$	$\dfrac{V/K_{\mathrm{m}}}{1+i/K_{\mathrm{ic}}}$	$\dfrac{K_{\mathrm{m}}(1+i/K_{\mathrm{ic}})}{1+i/K_{\mathrm{iu}}}$
Pure non-competitive[1]	$\dfrac{V}{1+i/K_{\mathrm{iu}}}$	$\dfrac{V/K_{\mathrm{m}}}{1+i/K_{\mathrm{ic}}}$	K_{m}
Uncompetitive	$\dfrac{V}{1+i/K_{\mathrm{iu}}}$	V/K_{m}	$\dfrac{K_{\mathrm{m}}}{1+i/K_{\mathrm{iu}}}$

[1]As pure non-competitive inhibition is mixed inhibition with the two inhibition constants identical, $K_{\mathrm{ic}} = K_{\mathrm{iu}}$, it is not strictly necessary to use both symbols in this line of the Table. This is done, however, to preserve the regularity of the columns for V^{app} and $V^{\mathrm{app}}/K_{\mathrm{m}}^{\mathrm{app}}$.

language and is easy to confuse with "non-competitive", itself a term that has been used in various different ways in the biochemical literature; uncompetitive inhibition itself is much less common than competitive inhibition, and the name cannot claim to be sanctioned by universal use — indeed, a major text-book by Laidler and Bunting (1973) used the term *anti-competitive inhibition*: this better suggests that it is at the other end of the spectrum from competitive inhibition, but it has failed to achieve any broad acceptance, and in this book I shall continue to use the older name.

5.2.4   Summary of linear inhibition types

The properties of the various types of linear inhibition are summarized in Table 5.1. They are easy to memorize as long as the following points are noted:

1. The two limiting cases are competitive (specific) and uncompetitive (cata-lytic) inhibition; pure non-competitive inhibition is simply a special case of mixed inhibition in which the two inhibition constants K_{ic} and K_{iu} are equal.
2. The effects of inhibitors on $V^{\mathrm{app}}/K_{\mathrm{m}}^{\mathrm{app}}$ and V^{app} are simple and regular: if $V^{\mathrm{app}}/K_{\mathrm{m}}^{\mathrm{app}}$ is decreased at all by the inhibitor it is decreased by a factor of $(1+i/K_{\mathrm{ic}})$; if V^{app} is decreased at all it is decreased by a factor of $(1+i/K_{\mathrm{iu}})$.
3. The effects of inhibitors on $K_{\mathrm{m}}^{\mathrm{app}}$ are confusing; they are most easily remembered by thinking of $K_{\mathrm{m}}^{\mathrm{app}}$ as the result of dividing V^{app} by $V^{\mathrm{app}}/K_{\mathrm{m}}^{\mathrm{app}}$ and not as a parameter in its own right.

5.3 Plotting inhibition results

5.3.1 Simple plots

Any of the plots described in Section 2.6 can be used to diagnose the type of inhibition, as they all provide estimates of the apparent values of the kinetic parameters. For example, if plots of a/v against a are made at several values of i, the intercept on the ordinate (K_m^{app}/V^{app}) varies with i if there is a specific component in the inhibition, and the slope $(1/V^{app})$ varies with i if there is a catalytic component in the inhibition. Alternatively, if direct linear plots of V^{app} against K_m^{app} are made at each value of i, the common intersection point shifts in a direction that is characteristic of the type of inhibition: for competitive inhibition, the shift is to the right; for uncompetitive inhibition it is directly towards the origin; and for mixed inhibition, it is intermediate between these extremes.

Other plots are needed for determining the actual values of the inhibition constants. The simplest approach is to estimate the apparent kinetic constants at several values of i, by the methods of Section 2.6, and to plot K_m^{app}/V^{app} and $1/V^{app}$ against i. Either plot should show a straight line, with intercept $-K_{ic}$ on the i axis if K_m^{app}/V^{app} is plotted and intercept $-K_{iu}$ on the i axis if $1/V^{app}$ is plotted. It may seem more natural to determine K_{ic} by plotting K_m^{app} rather than K_m^{app}/V^{app} against i, but this is not advisable, for two reasons. It is valid only if the inhibition is competitive, and it gives a curve instead of a straight line if the inhibition is mixed; even if the inhibition is strictly competitive, it is much less accurate, because K_m^{app} can never be estimated as accurately as K_m^{app}/V^{app} can (see Section 14.4).

Another method of estimating K_{ic}, introduced by Dixon (1953b), is also in common use. The full equation for mixed inhibition may be written as follows:

$$v = \frac{Va}{K_m(1 + i/K_{ic}) + a(1 + i/K_{iu})} \tag{5.12}$$

Taking reciprocals of both sides, and introducing subscripts 1 and 2 to distinguish measurements made at two different substrate concentrations a_1 and a_2, this becomes

$$\frac{1}{v_1} = \frac{K_m + a_1}{Va_1} + \frac{i\left(\dfrac{K_m}{K_{ic}} + \dfrac{a_1}{K_{iu}}\right)}{Va_1} \tag{5.13}$$

$$\frac{1}{v_2} = \frac{K_m + a_2}{Va_2} + \frac{i\left(\dfrac{K_m}{K_{ic}} + \dfrac{a_2}{K_{iu}}\right)}{Va_2} \tag{5.14}$$

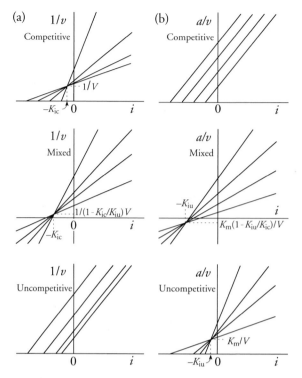

Figure 5.5. Determination of competitive and uncompetitive inhibition constants. (a) K_{ic} is given by plots of $1/v$ against i at various a values (Dixon, 1953b); (b) K_{iu} is given by plots of a/v against i at various a values (Cornish-Bowden, 1974). In mixed inhibition the point of intersection can be above the axis in the first plot and below it in the second, or vice versa, or it can be on the axis in both plots if $K_{ic} = K_{iu}$.

Both of these equations indicate that a plot of $1/v$ against i at a constant value of a is a straight line (Figure 5.5). If two such lines are drawn, from measurements at two different a values, the point of intersection can be found by setting $1/v_1 = 1/v_2$:

$$\frac{K_m + a_1}{Va_1} + \frac{i\left(\dfrac{K_m}{K_{ic}} + \dfrac{a_1}{K_{iu}}\right)}{Va_1} = \frac{K_m + a_2}{Va_2} + \frac{i\left(\dfrac{K_m}{K_{ic}} + \dfrac{a_2}{K_{iu}}\right)}{Va_2} \qquad (5.15)$$

which may be simplified as follows when substrate concentrations are cancelled (when appropriate) between numerators and denominators, terms identical on the two sides of the equation are omitted, and the common factor K_m/V is omitted:

$$\left(\frac{1}{a_1} - \frac{1}{a_2}\right)\left(1 + \frac{i}{K_{ic}}\right) = 0 \qquad (5.16)$$

and so $i = -K_{ic}$ and $1/v = (1 - K_{ic}/K_{iu})/V$ at the point of intersection. In principle, if several lines are drawn at different a values they should all intersect at a common point; in practice, experimental error will usually ensure some variation.

Notice that terms containing K_{iu} cancelled out in going from equation 5.15 to equation 5.16. Consequently, the Dixon plot provides the value of K_{ic} regardless of the value of K_{iu}, measuring the specific component of the inhibition regardless of whether the inhibition is competitive, mixed or pure non-competitive. The horizontal coordinate of the point of intersection does not distinguish between these three possibilities, but as both inhibition constants contribute to the vertical coordinate the location of the point in relation to the abscissa axis provides some qualitative information: intersection above the axis indicates that the competitive component is stronger than the uncompetitive component, and vice versa. In uncompetitive inhibition, when K_{ic} is infinite, the Dixon plot generates parallel lines.

Although the Dixon plot does not give the value of the uncompetitive inhibition constant K_{iu}, an exactly similar derivation shows that it can be found by plotting a/v against i at several a values (Cornish-Bowden, 1974). For this plot a different set of straight lines is obtained, which intersect at the point where $i = -K_{iu}$ and $a/v = K_m(1 - K_{iu}/K_{ic})/V$. Both types of plot are shown schematically for the various types of inhibition in Figure 5.5, in which one should note that both plots are needed for determining both inhibition constants for an enzyme showing mixed inhibition. An experimental example showing competitive inhibition is shown in Figure 5.6.

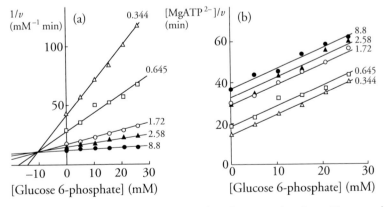

Figure 5.6. Inhibition of hexokinase D by glucose 6-phosphate (Storer and Cornish-Bowden, 1977) plotted as in Figure 5.5. The concentrations of $MgATP^{2-}$ were 8.8 mM (●), 2.58 mM (▲), 1.72 mM (○), 0.645 mM (□) and 0.344 mM (△). The combination of intersecting lines in (a) with parallel lines in (b) indicate competitive inhibition with $K_{ic} = 11.5$ mM.

5.3.2 Combination plots

All of the plots described in the preceding section imply constraints on the experimental design for inhibitor experiments (Section 5.8), because to have enough points on each line there must be several inhibitor concentrations at each substrate concentration, or several substrate concentrations at each inhibitor concentration, or both. This is not a constraint for computer analysis (Chapter 14), because a properly written program has no requirement for the pairs of concentrations to be points in a well-defined grid, but even then one may want to display the results of the analysis graphically. Hunter and Downs (1945) pointed out many years ago that it is possible to plot all of the observations at haphazard combinations of substrate and inhibitor concentrations in such a way that the resulting points fall on a single curve, provided that the uninhibited rate at each substrate concentration is known or can be calculated. In reciprocal form equation 5.12 for mixed inhibition may be written as follows:

$$\frac{1}{v_i} = \frac{K_m + a}{Va} + \frac{i(K_m/K_{ic} + a/K_{iu})}{Va} \tag{5.17}$$

where v_i is the rate at inhibitor concentration i. As the first fraction on the right-hand side is the reciprocal of the uninhibited rate v_0, this equation may be written as follows:

$$\frac{1}{v_i} = \frac{1}{v_0} + \frac{i(K_m/K_{ic} + a/K_{iu})}{Va} \tag{5.18}$$

With a simple rearrangement, this becomes

$$\frac{iv_i}{v_0 - v_i} = \frac{Va/v_0}{K_m/K_{ic} + a/K_{iu}} \tag{5.19}$$

and as v_0 is the same as $Va/(K_m + a)$ it can be removed from the right-hand side by writing the equation as follows:

$$\frac{iv_i}{v_0 - v_i} = \frac{K_m + a}{K_m/K_{ic} + a/K_{iu}} \tag{5.20}$$

Thus in a plot of $iv_i/(v_0 - v_i)$ against a all of the points should lie on a single curve if equation 5.12 is obeyed. Although the values of v_0 are needed for this plot they do not need to be individually measured if the Michaelis–Menten parameters are known, as they can easily be calculated.

    This plot of Hunter and Downs may be useful if one needs to analyse observations that were originally made without subsequent analysis in mind and with a design inappropriate for the more commonly used plots. Chan (1995) provides further discussion, including description of some variants that generate straight lines for the common inhibition types.

5.4   Multiple inhibitors

In pharmacological applications it is often necessary to consider the cumulative effects of two or more inhibitors that are simultaneously present. In the simplest case, when the binding is *exclusive*, with no more than one inhibitor molecule able to be bound to the same enzyme form, the rate $v_{1,2\ldots n}$ for a mixture of n mixed inhibitors follows a simple generalization of equation 5.12:

$$v_{1,2\ldots n} = \frac{Va}{K_m(1 + i_1/K_{ic1} + i_2/K_{ic2} + \ldots) + a(1 + i_1/K_{iu1} + i_2/K_{iu2} + \ldots)} \tag{5.21}$$

As Chou and Talalay (1977) pointed out, the structure of this equation is clearer when it is written in reciprocal form, and it then allows a straight-forward classification of multiple inhibition without requiring knowledge of the inhibition constants:

$$\frac{1}{v_{1,2\ldots n}} = \frac{K_m + a}{Va} + \frac{i_1(K_m/K_{ic1} + a/K_{iu1})}{Va} + \frac{i_2(K_m/K_{ic2} + a/K_{iu2})}{Va} + \ldots$$

$$= \frac{1}{v_0} + \left(\frac{1}{v_1} - \frac{1}{v_0}\right) + \left(\frac{1}{v_2} - \frac{1}{v_0}\right) + \ldots$$

$$= \frac{1}{v_1} + \frac{1}{v_2} + \ldots + \frac{1}{v_n} - \frac{n-1}{v_0} = \sum_{j=1}^{n} \frac{1}{v_j} - \frac{n-1}{v_0} \tag{5.22}$$

In practice, of course, this equation is not necessarily obeyed, because it is not necessarily true that the binding is exclusive. If the measured reciprocal rate proves to be larger than predicted by equation 5.22, that is to say if the rate is smaller, then Chou and Talalay define the cumulative effects of the inhibitors as *synergism*; if the reciprocal rate is smaller (the rate is larger) than predicted by the equation then there is *antagonism*.

   This classification is widely used for analysing interaction between enzyme inhibitors, but determining the inhibition constants when such interaction occurs requires additional methods. It follows from the top line of equation 5.22 that a plot of $v_0/v_{1,2\ldots n}$ against any inhibitor concentration i_1 at constant concentrations of substrate and all other inhibitors is a straight line with a slope that is independent of all of the other inhibitor concentrations (though not of the substrate concentration):

$$\frac{v_0}{v_{1,2\ldots n}} = 1 + \frac{i_1(K_m/K_{ic1} + a/K_{iu1})}{K_m + a} + \frac{i_2(K_m/K_{ic2} + a/K_{iu2})}{K_m + a} + \ldots \tag{5.23}$$

When plots of this sort are made at different fixed concentrations of the other inhibitors the lines should be parallel if the inhibitors bind exclusively, and the inhibition constants can be determined from the dependences of the slopes on the substrate concentration. When Yonetani and Theorell (1964) applied this method to multiple inhibition of horse-liver alcohol dehydrogenase they found that the lines were indeed parallel when the inhibitors were similar in structure, like ADP and ADP-ribose. However, they observed an intersecting set of lines with inhibitors that were quite different in structure, like ADP and o-phenanthroline, and they concluded that these last two inhibitors bound non-exclusively; in the terminology of Chou and Talalay their effect was synergistic. More generally, it follows from equation 5.23 that a plot of $v_0/v_{1,2\ldots n}$ against *any* linear combinations of inhibitor concentrations is a straight line, and this is the basis of an earlier type of analysis proposed by Yagi and Ozawa (1960). However, use of their method places extra constraints on the choice of inhibitor concentrations to be used experimentally, as they must be varied in constant ratio, and in practice it has been largely supplanted by the approach of Yonetani and Theorell.

As discussed by Martínez-Irujo and co-workers (1998), these methods all embody some mechanistic assumptions, and if they are taken beyond their range of applicability they are likely to lead to meaningless results. For example, various inhibitors of the proteinase from human immunosuppressor virus lead to parallel lines in the Yonetani–Theorell plots (Balzarini and co-workers, 1992). However, as these inhibitors act as chain terminators and not by direct action on the proteinase, the behaviour cannot be interpreted in terms of equation 5.12, though it can be taken to indicate that therapeutic use of combinations of these inhibitors is unlikely to offer any advantage over using them individually.

5.5   Relationship between inhibition constants and the concentration for half-inhibition

In biochemistry inhibitors are usually characterized in terms of the type of inhibition and the inhibition constants, as has been done to this point in this chapter. From the mechanistic point of view this is appropriate, as it is linked in a clear way to the mechanism of inhibition. Nonetheless, enzyme inhibition is of crucial importance also in pharmacology, and for obvious reasons pharmacologists tend to be more interested in the effects of inhibition than in its mechanism. As a result, it is common in the pharmacology literature to characterize an inhibitor by the concentration $i_{0.5}$ that under assay conditions produces a rate half that observed in the absence of the inhibitor. This is also often symbolized as i_{50} (for the concentration that produces 50% inhibition) or in some other similar ways.

This concentration $i_{0.5}$ is not equal to either of the usual inhibition constants except in pure non-competitive inhibition, which occurs too rarely in practice (see Section 5.2.2) to constitute a useful example: in general it is safer to say that $i_{0.5}$ is not an inhibition constant, and that one needs knowledge of the type of inhibition to convert it into one. Cheng and Prusoff (1973) and Brandt, Laux and Yates (1987) have given some detailed analyses of the relationships, which follow from the usual equation for mixed inhibition, equation 5.12. Cortés and co-workers (2001) rearranged this to show the value of i at which $1/v$ (in a Dixon plot) or a/v (in a plot of a/v against i) is zero:

$$i = -\frac{K_m + a}{\dfrac{K_m}{K_{ic}} + \dfrac{a}{K_{iu}}} \tag{5.24}$$

This result is not in itself of immediate importance, but if one ignores the minus sign and substitutes the resulting positive value of i into equation 5.12 the result is a rate equal to half the rate v_0 in the absence of inhibitor:

$$v = \frac{Va}{2(K_m + a)} = 0.5v_0 \tag{5.25}$$

It follows that the abscissa intercept on either a Dixon plot or a plot of a/v against i is equal to $-i_{0.5}$, and that dropping the minus sign from equation 5.24 shows the exact dependence of $i_{0.5}$ on the substrate concentration and other parameters:

$$i_{0.5} = \frac{K_m + a}{\dfrac{K_m}{K_{ic}} + \dfrac{a}{K_{iu}}} \tag{5.26}$$

This result may be obtained more simply by substituting $v_i = 0.5v_0$ into equation 5.12.

5.6 Inhibition by a competing substrate

5.6.1 Competition between substrates

The equations for two reactions occurring at the same site were given in Section 2.4, as equations 2.21–22, for introducing the idea of enzyme specificity. As noted there, they are of exactly the same form as equation 5.1: a competing substrate behaves experimentally like a competitive inhibitor if one measures the rate with respect to the other substrate, but the parameter that corresponds to the competitive inhibition constant is simply the Michaelis constant of the competing substrate. It follows, therefore, that one can use the methods of Section 5.3 to estimate a Michaelis constant by treating it as if it were an inhibition constant, and the result should be the same as if one measured it directly. In the example of fumarase discussed in Section 2.4, the column of Table 2.1 (page 39) labelled "K_i" shows the values that were measured in that way, and it may be seen that they agreed with the direct measurements.

5.6.2   Testing if two reactions occur at the same site

From equations 2.21–22, the combined rate $v_{tot} = v + v'$ for two competing reactions may be expressed as follows:

$$v_{tot} = v + v' = \frac{(V/K_m)a + (V'/K'_m)a'}{1 + \dfrac{a}{K_m} + \dfrac{a'}{K'_m}} \tag{5.27}$$

Suppose now that reference concentrations $a = a_0$ and $a' = a'_0$ are found (experimentally) such that

$$\frac{Va_0}{K_m + a_0} = \frac{V'a'_0}{K'_m + a'_0} = v_0 \tag{5.28}$$

They are thus two concentrations that lead to the same rate (at the same enzyme concentration) for each substrate when the other substrate is absent. If a series of mixtures are prepared in which the concentrations are linearly interpolated between zero and these reference concentrations,

$$\left. \begin{array}{l} a = (1 - r)a_0 \\ a' = ra'_0 \end{array} \right\} \tag{5.29}$$

then equation 5.27 takes the following form:

$$v_{tot} = \frac{(V/K_m)(1 - r)a_0 + (V'/K'_m)ra'_0}{1 + \dfrac{(1 - r)a_0}{K_m} + \dfrac{ra'_0}{K'_m}}$$

$$= \frac{(V/K_m)(1 - r)a_0 + (V'/K'_m)ra'_0}{(1 - r)\left(1 + \dfrac{a_0}{K_m}\right) + r\left(1 + \dfrac{a'_0}{K'_m}\right)}$$

$$= \frac{v_0 \left[(1 - r) \left(1 + \dfrac{a_0}{K_m} \right) + r \left(1 + \dfrac{a'_0}{K'_m} \right) \right]}{(1 - r) \left(1 + \dfrac{a_0}{K_m} \right) + r \left(1 + \dfrac{a'_0}{K'_m} \right)} = v_0 \qquad (5.30)$$

The third line of this equation follows from the second by using equation 5.28 to substitute $V a_0 / K_m$ with $v_0 (1 + a_0 / K_m)$, and $V' a'_0 / K'_m$ similarly. The conclusion is that if equations 2.21–22 are valid the total rate for mixtures prepared according to equation 5.29 is independent of the proportions of the substrates in the mixture. A plot of v_{tot} against r, known as a *competition plot*, therefore provides a test of whether two substrates compete for the same active site of an enzyme. Examination of a broader range of possible models than equations 2.21–22 shows that there are three principal possibilities (Chevillard, Cárdenas and Cornish-Bowden, 1993):

1. *Competition for one site.* The plot shows no dependence of v_{tot} on r, so it gives a horizontal straight line, as illustrated in Figure 5.7a for competition between glucose and fructose for the active site of yeast hexokinase.
2. *Reaction at separate sites.* If the two reactions are completely separate with no interaction between either enzyme and the substrate of the other enzyme, the plot shows a curve with a maximum. In reality this is rather an extreme assumption, because if the two substrates have some similarity with one another each should be capable of inhibiting the enzyme for the other. However, provided that each substrate binds more tightly to its own enzyme than to the other the behaviour is qualitatively the same, a curve with a maximum, as illustrated in Figure 5.7b for phosphorylation of glucose and galactose by a mixture of hexokinase and galactokinase.
3. *Antagonistic reactions.* If the two substrates react at different sites but each interacts with the wrong site more strongly than with the right one, the plot shows a curve with a minimum. Nitrate reductase provides an example of pronounced antagonistic effects (Giordani and co-workers, 1997).

If a spectrophotometric method is used for following the reaction one can ensure that the total rate of reaction in the presence of both substrates is the sum of the individual reactions by following the reaction at an isosbestic point, a wavelength where both reactions have the same specific change of absorbance. However, this is not actually necessary, and the analysis still gives correct results if the rates are measured in units that are essentially arbitrary, such as change of absorbance with time at a wavelength where the two reactions have different spectroscopic properties (Cárdenas, 2001).

A similar method was proposed by Keleti and co-workers (1987), and used by them to show that serine and threonine react at the same site on threonine

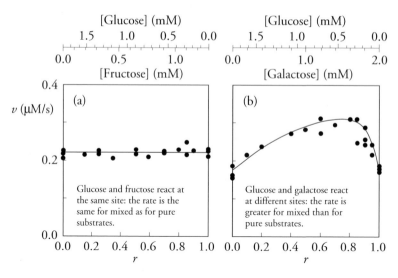

Figure 5.7. Competition plots. Data of Chevillard, Cárdenas and Cornish-Bowden (1993) for phosphorylation of glucose/fructose mixtures by yeast hexokinase (a), and of glucose/galactose by a mixture of yeast hexokinase and galactokinase (b). In both experiments the mixtures are constructed so that the measured rate is the same for the pure substrates (see text for details). If the two substrates compete for the same site the rate is independent of the proportions of substrates, as in (a); when the reactions occur at different sites with little or no cross-inhibition the rate for the intermediate mixtures should be greater than for the pure substrates, as in (b).

dehydratase. Conceptually it is similar to the method described here, but experimentally more demanding, as it requires knowledge of the two Michaelis constants, which are needed for calculating the concentrations of the substrates, so as to vary them in such a way that $a/K_m + a'/K'_m$ is constant.

5.7   Enzyme activation

5.7.1   Miscellaneous uses of the term activation

Discussion of the activation of enzyme-catalysed reactions is complicated by the use of the term in enzymology with diverse meanings. In this book I use it for the converse of reversible inhibition: an activator is a species that combines with an enzyme to increase its activity, without itself undergoing a net change in the reaction. Other processes that are sometimes called activation are the following:

1. Several enzymes, mainly extracellular catabolic enzymes, are secreted as inactive precursors or *zymogens*, which are subsequently converted into active enzymes by partial proteolysis, pepsin being secreted as pepsinogen,

for example. The name "zymogen activation" for this process remains widespread (see for example Khan and James, 1998).

2. Several enzymes important in metabolic regulation exist in the cell in active and inactive states, the two often differing by the presence or absence of a phosphate group (see Section 12.9.2). For example, phosphorylase a is the active form of phosphorylase, and phosphorylase b is the inactive form. Interconversion of the two states requires two separate reactions, transfer of a phosphate group from ATP in one direction and removal of the phosphate group by hydrolysis in the other: these processes do not correspond to activation and inhibition in the dynamic sense used in this book.

3. Many reactions are said to be "activated" by metal ions when the truth is that a metal ion forms part of the substrate. For example, many ATP-dependent kinases are said to be activated by Mg^{2+}, not because of the effect of Mg^{2+} on the enzyme itself but because the true substrate is the ion $MgATP^{2-}$, not ATP^{4-}, the predominant metal-free state at physiological pH. For example, although rat-liver hexokinase D uses $MgATP^{2-}$ as substrate, free Mg^{2+} is actually an inhibitor, not an activator, as indeed ATP^{4-} is also an inhibitor, not a substrate (Storer and Cornish-Bowden, 1977). Although this sort of confusion is understandable if one regards ATP as the substrate of reactions that actually involve $MgATP^{2-}$, it is best to avoid it by expressing results in terms of the concentrations of the actual species involved and restricting the term "activation" to effects on the enzyme, especially as such effects do sometimes occur with enzymes that have $MgATP^{2-}$ as a reactant, such as pyruvate kinase (MacFarlane and Ainsworth, 1972). Because of the great importance of $MgATP^{2-}$ in metabolic reactions, Section 3.4 of this book is devoted to discussion of methods of controlling its concentration.

5.7.2   Essential activation

The simplest kind of true activation is *essential activation* (sometimes known as *specific activation* or *compulsory activation*) in which the free enzyme without activator bound to it has no activity and does not bind substrate. This may be represented by Figure 5.8, in which the activator is represented as X, and the rate constants are shown with primes to emphasize that they refer to the enzyme with activator bound to it. This scheme is similar to that for competitive inhibition, and it generates a rate equation of similar form:

$$v = \frac{V'a}{K'_m(1 + K_x/x) + a} \tag{5.31}$$

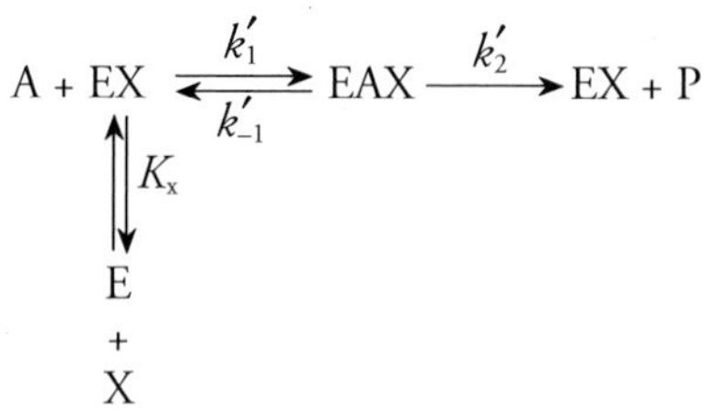

$$A + EX \underset{k'_{-1}}{\overset{k'_1}{\rightleftharpoons}} EAX \xrightarrow{k'_2} EX + P$$

$$\Big\updownarrow K_x$$

$$E$$
$$+$$
$$X$$

Figure 5.8. Mechanism producing essential activation.

in which $V' = k'_2 e_0$ and $K'_m = (k'_{-1} + k'_2)/k'_1$ are the limiting rate and Michaelis constant respectively for the activated enzyme EX, and x is the concentration of X.

This equation differs from that for competitive inhibition, equation 5.1, by having i/K_{ic} replaced by K_x/x. Thus the rate is zero in the absence of activator, as one would expect from the mechanism. Despite the formal similarity between competitive inhibition and essential activation, however, there is an important difference in plausibility. Because a competitive inhibitor is usually assumed to bind at the same site on the enzyme as the substrate, it is easy to imagine that they cannot bind simultaneously, and competitive inhibition is thus common. However, it is less easy to visualize an enzyme that cannot bind substrate at all in the absence of activator. Essential activation is in practice much less frequently observed than its counterpart in inhibition, though it occurs, for example, in the activation of cathepsin C by chloride ions (Cigić and Pain, 1999). In general it is useful mainly as a simple introduction to the more complicated kinds of activation, which unfortunately require correspondingly more complicated rate equations.

An important point to note from this discussion is that there is nothing in the model of Figure 5.8 to suggest any idea of competition, and so regardless of the similarity in the algebra any such name as "competitive activation" is unjustified and likely to be confusing in many contexts. Unfortunately, however, expressions that preserve the algebraic relationship while avoiding mechanistic absurdity, like "the activation analogue of competitive inhibition", tend to be so cumbersome that the temptation to use a shorter form is likely to be irresistible. Nonetheless, if the word "competitive" is applied to activation for this sort of reason it should always be placed in quotation marks, and the first use of it in any context should be accompanied by an explanation of what is actually meant.

The simplest realistic type of activation is the counterpart of mixed inhibition, as shown in Figure 5.9. In this mechanism the activator is not required for substrate binding, but only for catalysis, so one must logically

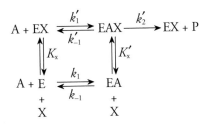

Figure 5.9. Mechanism producing mixed activation.

include steps for binding of substrate both to E and to EX. This means that activator-binding steps are not dead-end reactions, though for simplicity we shall treat them as equilibria nonetheless. The rate equation is then analogous to that for mixed inhibition:

$$v = \frac{V'a}{K'_m(1 + K_x/x) + a(1 + K'_x/x)} \tag{5.32}$$

and provides analogous expressions for the apparent parameters (compare equations 5.6–8):

$$V^{app} = \frac{V'}{1 + K'_x/x} \tag{5.33}$$

$$K^{app}_m = \frac{K'_m(1 + K_x/x)}{1 + K'_x/x} \tag{5.34}$$

$$V^{app}/K^{app}_m = \frac{V'/K'_m}{1 + K_x/x} \tag{5.35}$$

Essentially the same plots and methods can be used for investigating the type of activation as are used in linear inhibition (Section 5.3), replacing i throughout by $1/x$, K_{ic} by $1/K_x$ and K_{iu} by $1/K'_x$. For example, K_x may be determined by a plot analogous to a Dixon plot in which $1/v$ is plotted against $1/x$ at two or more values of a; the abscissa coordinate of the point of intersection of the resulting straight lines then gives $-1/K_x$.

5.7.3 Hyperbolic activation and inhibition

In practice, activators often behave in a more complicated way than suggested by equation 5.32 because there may be some catalytic activity in the absence of the activator, with different rate constants for the two forms of the enzyme. The mechanism may then be represented by Figure 5.10.

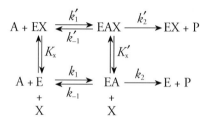

Figure 5.10. General modifier mechanism. This mechanism can produce hyperbolic inhibition or activation, or a combination of the two. All of the simple mechanisms for inhibition and activation (apart from product inhibition) are special cases of this mechanism.

There are several important points to note about this mechanism. First, the activator-dissociation steps are not dead-end reactions and so they are not necessarily at equilibrium. However, if this is taken into account the kinetic behaviour becomes rather complicated, and for discussing it here I shall assume that the rate equation has the form it would have if the activator- and substrate-binding steps were all equilibria. Second, this is not necessarily a mechanism for activation at all: it could equally well result in inhibition if the EX complex is less reactive than the free enzyme. Indeed, it is more complicated than that, because the conditions that decide whether X is an activator or an inhibitor may be different in different ranges of substrate concentration: if $k_2 > k_2'$ then X inhibits at high substrate concentrations; if $k_1 k_2 / (k_{-1} + k_2) > k_2' / (k_{-1}' + k_2')$ then X inhibits at low substrate concentrations; only if the two inequalities are consistent is the behaviour consistent over the whole range. As Figure 5.10 takes account of this possible duality of behaviour, it is a *general modifier mechanism* (Botts and Morales, 1954), where *modifier* is a term that embraces both inhibitors and activators.

When the mechanism is fully analysed, it turns out that plots of $1/V^{app}$ or K_m^{app}/V^{app} against inhibitor or reciprocal activator concentration are not straight lines but rectangular hyperbolas, and for this reason the behaviour is called *hyperbolic activation* or *hyperbolic inhibition**. All of this may seem too complicated for an elementary account (even without worrying that the binding steps should not strictly be treated as equilibria). However, hyperbolic effects ought not to be forgotten about completely, because Figure 5.10 represents a plausible mechanism that one should expect to apply to real enzymes quite often. The fact that few examples (especially of hyperbolic inhibition) have been reported, therefore, is more likely to reflect a failure to

---

*The terms *non-essential activation* and *partial inhibition* are also sometimes used instead, to emphasize that some activity remains in the complete absence of activator or at saturation with inhibitor.

recognize them than a genuinely rare kind of behaviour. Hyperbolic effects are not difficult to recognize experimentally, but the symptoms are often dismissed as just an unwelcome complication. It is important to use a wide enough range of inhibitor and activator concentrations to know whether or not the rate tends to zero at very high inhibitor or very low activator concentration. One should particularly note whether the expected "linear" plots are actually straight lines or not; any systematic curvature should be checked and if confirmed it is likely to indicate hyperbolic effects.

These points are illustrated by a study of yeast alcohol dehydrogenase by Dickenson and Dickinson (1978): they found that saturating concentrations of ethanol decreased the apparent limiting rate for the reaction with acetaldehyde by a factor of about four at pH 7, but to see this behaviour clearly they needed to examine ethanol concentrations as high as 1 M, and to notice that the lines plotted at much lower ethanol concentrations were not strictly straight.

5.8   Design of inhibition experiments

Before embarking on a discussion of the design of inhibition experiments, I should perhaps comment on why this section comes after a discussion of how such experiments are analysed (Sections 5.2–3). As design should obviously precede analysis in any well-planned experiment, it might seem more logical to discuss the two topics in that order. However, it is my experience that the most effective way of learning the importance of design is to suffer the difficulties that arise when one tries to analyse the results of a badly designed experiment, one that has been carried out without any thought being given to the information it is intended to supply. Moreover, one needs some knowledge of analytical methods before one can appreciate the principles of design.

There are two primary aims in inhibition experiments: to identify the type of inhibition and to estimate the values of the inhibition constants. If the experiment is carefully designed it is usually possible to satisfy both of these aims simultaneously. I shall initially assume that the inhibition is linear and that the relations given in Table 5.1 (page 120) apply, because this is likely to be adequate as a first approximation, and it is useful to characterize the linear behaviour before trying to understand any complications that may be present.

It is evident from Table 5.1 that any competitive component in the inhibition will be most pronounced at low concentrations of substrate, because competitive inhibitors decrease the apparent value of V/K_m, which characterizes the kinetics at low concentrations of substrate; conversely,

Table 5.2. Design of inhibition experiments. The Table illustrates the choice of substrate concentrations a required to determine the inhibition constants K_{ic} and K_{iu} in an experiment where the approximate values are known to be $K_m = 1$, $K_{ic} = 2$ and $K_{iu} = 10$, in arbitrary units (though the same units for K_{ic} and K_{iu}). The suggested a values are in the same units as K_m, and are designed to extend from about $0.2K_m^{app}$ to about $10K_m^{app}$ at each value of the inhibitor concentration i.

i	$1 + i/K_{ic}$	$1 + i/K_{iu}$	K_m^{app}	a values						
0	1.0	1.0	1.0	0.2	0.5	1	2	5	10	
1	1.5	1.1	1.4	0.2	0.5	1	2	5	10	
2	2.0	1.2	1.7		0.5	1	2	5	10	20
5	3.5	1.5	2.3		0.5	1	2	5	10	20
10	6.0	2.0	3.0		0.5	1	2	5	10	20
20	11.0	3.0	3.7			1	2	5	10	20	50

any uncompetitive component will be most noticeable at high substrate concentrations. Obviously, therefore, the inhibition can be fully characterized only if it is investigated at both high and low substrate concentrations. Conditions that are ideal for assaying an enzyme may well therefore be unsatisfactory for investigating its response to inhibitors. For example, a simple calculation shows that a competitive inhibitor at a concentration equal to its inhibition constant decreases the measured rate by less than 10% if $a = 10K_m$; although an effect of this size ought to be easily detected in any careful experiment, it might still be dismissed as of little consequence if it was not realized that the effect at low substrate concentration would be much bigger.

Just as in a simple experiment without inhibitors it is prudent to include substrate concentrations from about $0.2K_m$ to as high as conveniently possible (Section 3.3.1), so in an inhibition experiment the inhibitor concentrations should extend from about $0.2K_{ic}$ or $0.2K_{iu}$ (whichever is the smaller) to as high as possible without making the rate too small to measure accurately (and, at least for the lower substrate concentrations, observations without inhibitor should also be included). At each i value the a values should be chosen as in Section 3.3.1, but relative to K_m^{app}, not to K_m. This is because it is the *apparent* values of the kinetic constants that characterize the Michaelis–Menten behaviour at each inhibitor concentration, not the true values. A simple example of a possible experimental design is given in Table 5.2. Note that there is no requirement to use the same set of a values at each i value, or the same set of i values at each a value and doing so means devoting experimental effort to combinations that provide little information. Nonetheless, for plotting the results in any of the ways discussed in Section 5.3 it is convenient if several of the same a and i values are used in each experiment, to produce a reasonable number of points on each line whether one uses a as abscissa (for example, in determining apparent Michaelis–Menten parameters at each i value) or i as abscissa (for example, in plots of $1/v$ or a/v against i). Accordingly, Table 5.2 includes some combinations that would probably not be included if each line of the Table were considered in isolation.

If these recommendations are followed (as a guide, of course, not rigidly, because the particular numbers used in Table 5.2 are unlikely to apply exactly to any specific example), any hyperbolic character in the inhibition ought to be obvious without the need for further experiments. Consequently there is no need to add appreciably to the remarks at the end of Section 5.7.3. The main consideration is to include i values that are high enough and numerous enough to indicate whether the rate approaches zero or not when the inhibitor approaches saturation.

5.9   Inhibitory effects of substrates

5.9.1   Non-productive binding

Much of the information that exists about the general properties of enzymes has come from the study of a small group of enzymes, the extracellular hydrolytic enzymes, which include pepsin, lysozyme, ribonuclease and, most notably, α-chymotrypsin. Although this is less true now than it was 30 years ago, one still encounters α-chymotrypsin in the literature, by implication as a "typical enzyme", far more often than any biological importance it may have could possibly justify. These enzymes share various properties that make them eminently suitable for detailed study: they are effectively one-substrate enzymes, as the second substrate is always water, and they are abundant, easily crystallized, stable, monomeric and unspecific. However, all of these properties disqualify them from being regarded as "typical enzymes", which catalyse reactions of more than one substrate (not counting water), and are present in low amounts, difficult to purify and crystallize, unstable, oligomeric and highly specific.

All of these extracellular hydrolytic enzymes share another unusual characteristic, one that is a definite disadvantage: their natural substrates are all ill-defined polymers, and as a result they are nearly always studied with unnatural substrates that are much less bulky than the real ones. However, an enzyme that can bind a macromolecule is likely to be able to bind a small molecule in many different ways. Thus, instead of a single enzyme–substrate complex that breaks down to products, there may be numerous *non-productive complexes* in addition that do not form products. This is

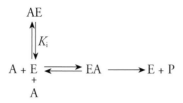

Figure 5.11. Non-productive binding. A is the substrate of the reaction, but as well as its "correct" mode of binding to produce the complex EA that can undergo reaction it can also bind incorrectly to give a complex AE that does not react further.

illustrated in Figure 5.11, in which AE represents all such non-productive complexes. This scheme is the same as that for linear competitive inhibition (Figure 5.2, page 115) with the inhibitor replaced by substrate, and the rate equation is likewise analogous to equation 5.1:

$$v = \frac{k_2 e_0 a}{\left(\dfrac{k_{-1} + k_2}{k_1}\right)\left(1 + \dfrac{a}{K_i}\right) + a} \tag{5.36}$$

If the *expected* values of the kinetic parameters are defined as the values they would have if no non-productive complexes were formed, so $\tilde{V} = k_2 e_0$ and $\tilde{K}_m = (k_{-1} + k_2)/k_1$ (compare "pH-independent" constants, Section 9.4), then equation 5.36 can be rearranged to give the Michaelis–Menten equation:

$$v = \frac{Va}{K_m + a} \tag{5.37}$$

with parameters defined as follows:

$$V = \frac{\tilde{V}}{1 + \tilde{K}_m/K_i} \tag{5.38}$$

$$K_m = \frac{\tilde{K}_m}{1 + \tilde{K}_m/K_i} \tag{5.39}$$

$$V/K_m = \tilde{V}/\tilde{K}_m \tag{5.40}$$

Thus the Michaelis–Menten equation is obeyed exactly for this mechanism and so the observed kinetics do not indicate whether non-productive binding is significant or not. Unfortunately, it is often the expected values that are of interest in an experiment, as they refer to the productive pathway. Hence the measured values V and K_m may be less, by an unknown and unmeasurable

Table 5.3. Non-productive complexes in chymotrypsin catalysis. The Table shows data of Ingles and Knowles (1967) for the chymotrypsin-catalysed hydrolysis of the p-nitrophenyl esters of various acetylamino acids. For these substrates the hydrolysis of the corresponding acetylaminoacylchymotrypsins is rate-limiting, and so the measured values of the catalytic constant k_{cat} are actually first-order rate constants for this hydrolysis reaction. The values are compared with the second-order rate constants k_{OH^-} for base-catalysed hydrolysis of the p-nitrophenylesters of the corresponding benzyloxycarbonyl amino acids.

Acyl group	k_{cat} (s^{-1})	k_{OH^-} (M^{-1} s^{-1})	k_{cat}/k_{OH^-} (M)
Acetyl-L-tryptophanyl-	52	0.16	330
Acetyl-L-phenylalanyl-	95	0.54	150
Acetyl-L-leucyl-	5.0	0.35	14
Acetylglycyl-	0.30	0.51	0.58
Acetyl-D-leucyl-	0.034	0.35	0.097
Acetyl-D-phenylalanyl-	0.015	0.54	0.027
Acetyl-D-tryptophanyl-	0.0028	0.16	0.018

amount, than the quantities of interest. Only V/K_m gives a correct measure of the catalytic properties of the enzyme.

For highly specific enzymes, plausibility arguments can be used to justify excluding non-productive binding from consideration, but for unspecific enzymes, such as chymotrypsin, comparison of results with different substrates may sometimes provide evidence of the phenomenon. For example, Ingles and Knowles (1967) measured the rates of hydrolysis of a series of acylchymotrypsins, by measuring k_{cat} values for the chymotrypsin-catalysed hydrolysis of the corresponding p-nitrophenyl esters, in which the "deacylation" step, the hydrolysis of the acylchymotrypsin intermediate, was known to be rate-limiting. The results (Table 5.3) were somewhat complicated by the unequal reactivity of the various acyl groups towards nucleophiles. Ingles and Knowles therefore measured the corresponding rate constants for hydrolysis catalysed by hydroxide ion. Dividing the rate constants for chymotrypsin catalysis by those for base catalysis produced an interesting pattern: the order of the reactivity of the specific L substrates was exactly reversed with the poor D substrates, so Ac-L-Trp > Ac-L-Phe >> Ac-L-Leu >> Ac-Gly >> Ac-D-Leu >> Ac-D-Phe > Ac-D-Trp. The simplest explanation is in terms of non-productive binding: for acyl groups with the correct L configuration, the large hydrophobic side chains permit tight and rigid binding in the correct mode, largely ruling out non-productive complexes, but for acyl groups with the D configuration, the same side chains favour tight and rigid binding in non-productive modes.

Non-productive binding is not usually considered in the context of inhibition; indeed, it is usually not considered at all, but it follows exactly

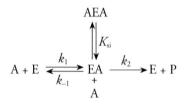

Figure 5.12. Mechanism that produces substrate inhibition, or inhibition by excess substrate.

the same mechanism as that most often considered for competitive inhibition, and it is important to take account of it as a possibility when comparing results with different substrates of an unspecific enzyme. The term *substrate inhibition* is usually reserved for the uncompetitive analogue of non-productive binding, which is considered next.

5.9.2 Substrate inhibition

For some enzymes it is possible for a second substrate molecule to bind to the enzyme–substrate complex EA to produce an inactive complex AEA, as shown in Figure 5.12. The mechanism is analogous to that usually considered for uncompetitive inhibition (Section 5.2.3), and the rate equation is as follows:

$$v = \frac{k_2 e_0 a}{\dfrac{k_{-1} + k_2}{k_1} + a(1 + a/K_{si})} = \frac{V'a}{K'_m + a + a^2/K_{si}} \qquad (5.41)$$

where V' and K'_m are defined as $k_2 e_0$ and $(k_{-1} + k_2)/k_1$ respectively, so that they satisfy the usual definitions of the Michaelis–Menten parameters in the simple two-step model (Section 2.3); however, they are written with primes because they are *not* Michaelis–Menten parameters, as equation 5.41 is not equivalent to the Michaelis–Menten equation: the effect of the term in a^2 is to make the rate approach zero rather than V' when a is large, and K'_m is not equal to the value of a when $v = V'/2$. The curve of v as a function of a is plotted in Figure 5.13, together with the corresponding plots of a/v against a and $1/v$ against $1/a$; these last two are not straight lines but a parabola and hyperbola respectively, but if K_{si} is large compared with K'_m (as it usually is), they are straight enough at low values of a for V' and K'_m to be estimated from them in the usual way.

Substrate inhibition is not usually important if substrate concentrations are kept at or below their likely physiological values (though there are some important exceptions, such as phosphofructokinase), but it can become so

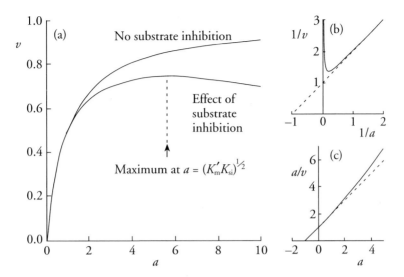

Figure 5.13. Substrate inhibition. The continuous lines were calculated from equation 5.41 with $K'_m = 1$, $V' = 1$, $K_{si} = 30$. The broken lines were calculated from the Michaelis–Menten equation with $K_m = 1$, $V = 1$, with no substrate inhibition. (a) In the direct plot of v against a the maximum occurs when $a^2 = K'_m K_{si}$. The smaller panels show the appearance of the corresponding plots of (b) $1/v$ against $1/a$ and (c) a/v against a. The impression that the plot in (b) shows the inhibition more clearly than that in (c) is an illusion caused by the continuation of the curve in (b) to $1/a = 0.0167$, that is to say to $a = 60$, whereas in (c) it stops at $a = 5$. In general the likelihood of recognizing whether substrate inhibition occurs is determined more by the range of data plotted than by the type of plot.

at high substrate concentrations, and provides a useful diagnostic tool for distinguishing between possible reaction pathways, as discussed in Section 7.5.

Problems

5.1 Initial studies of an esterase using a racemic mixture as substrate revealed that the L enantiomer was the true substrate, as it was completely converted into product whereas the D enantiomer could be recovered unchanged at the end of the reaction. On the basis of this result the kinetics of the reaction were analysed assuming that the D enantiomer had no effect on the enzyme, and a Michaelis constant for the L enantiomer was estimated to be 2 mM. Subsequent work made it clear that it would have been more reasonable to assume that the D enantiomer was a competitive inhibitor with K_{ic} equal to the K_m value of the L enantiomer. How should the original K_m estimate be revised to take account of this information?

5.2 The following data show the initial rates (in arbitrary units) measured for an enzyme-catalysed reaction at various concentrations i and a of inhibitor and substrate respectively. What information can be deduced about the type of inhibition? How could the design of the experiment be improved to reveal this more clearly?

i (mM)	$a = 1$ mM	$a = 2$ mM	$a = 3$ mM
0	2.36	3.90	5.30
1	1.99	3.35	4.40
2	1.75	2.96	3.98
3	1.60	2.66	3.58
4	1.37	2.35	3.33

5.3 The symbols K_{is} and K_{ii} are sometimes used for the inhibition constants symbolized in this chapter as K_{ic} and K_{iu} respectively. The second subscripts s and i stand for *slope* and *intercept* respectively, because if the slopes and intercepts of one of the plots described in Section 2.6 are replotted against the inhibitor concentration they provide values of the two inhibition constants.

(a) Which primary plot is referred to?

(b) Which intercept (ordinate or abscissa) is replotted?

(c) How do the inhibition constants appear in the secondary plot?

5.4 The Japanese traditional soup stock Katsuobushi owes its character-istic flavour to 5′-inosinic acid, a degradation product from RNA that results from fungal fermentation of fish. Studies of a nuclease from *Aspergillus* (Ito, Matsuura and Minamiura, 1994) yielded the following kinetic parameters for the hydrolysis of various dinucleoside monophos-phates:

	C–A	U–A	G–A	A–A	A–G	A–C	A–U
K_m (mM)	1.00	1.03	1.10	0.803	0.437	0.495	2.61
V (μM min^{-1})	11.5	10.4	8.78	13.2	9.81	11.6	11.8

Assuming that the behaviour of the bonds in the dinucleoside monophos-phates is representative of the same kinds of bond in RNA, which kind of bond (of the seven shown) would be expected to be hydrolysed most rapidly during digestion of RNA by the enzyme?

5.5 The following data of Doumeng and Maroux (1979) refer to the hydrolysis of various tripeptides into their N-terminal amino acids and C-terminal dipeptides, catalysed by intestinal aminotripeptidase at pH 7.0 and 37 °C:

Substrate	k_{cat} (s^{-1})	K_m (mM)
L-Pro–Gly–Gly	385	1.3
L-Leu–Gly–Gly	190	0.55
L-Ala–Gly–Gly	365	1.4
L-Ala–L-Ala–L-Ala	298	0.52

(a) Which substrate would be hydrolysed most rapidly in the early stages of reaction if a sample of enzyme was added to a mixture of all four substrates in equimolar concentrations?

(b) When L-Ala–Gly–Gly was studied as an inhibitor of the hydrolysis of L-Pro–Gly–Gly the competitive inhibition constant was found to be 1.4 mM. Is this value consistent with the view that the enzyme has a single active site at which both substrates are hydrolysed?

5.6 At any given ratio of inhibitor concentration to the appropriate inhibition constant, a competitive inhibitor decreases the rate more than an uncompetitive inhibitor does if the substrate concentration is less than K_m; the reverse is true if the substrate concentration is greater than K_m. Prove this relationship algebraically, and explain it conceptually, without reference to algebra.

5.7 If two inhibitors I and J both act as ordinary competitive inhibitors with inhibition constants K_i and K_j respectively when examined one at a time, their combined effect when present together may be expressed by an equation of the following form:

$$v = \frac{Va}{K_m\left[1 + \dfrac{i}{K_i}\left(1 + \dfrac{j}{K_j'}\right) + \dfrac{j}{K_j}\right] + a}$$

in which K_j' is the dissociation constant for release of J from a hypothetical EIJ complex. If plots of $1/v$ against i are made at various values of j, what is the abscissa coordinate of the common intersection point?

In a study of hexokinase D in which I was N-acetylglucosamine, Vandercammen and Van Schaftingen (1991) obtained parallel lines in such plots when J was glucosamine, and lines intersecting on the abscissa axis when J was a specific regulator protein. What do these results imply about the values of K_j' in the two cases, and hence about the binding sites of the three inhibitors? How would the two kinds of behaviour be categorized in the terminology of Chou and Talalay?

5.8 Protein-tyrosine kinases catalyse the phosphorylation of particular tyrosine residues in proteins. Songyang and co-workers (1995) studied their selectivity for particular sequences around the target residue by means of a library of peptides Met-Ala-Xaa$_4$-Tyr-Xaa$_4$-Ala-Lys$_3$, in which Xaa represents any amino acid apart from Trp, Cys or Tyr. Taking the peptides in the library to have a mean molecular mass of 1.6 kDa and the sample as 1 mg in a final volume of 300 μl, calculate (a) the number of different kinds of peptide in the sample; (b) the average number of molecules of each; (c) the average concentration of each. Comparing this concentration with a "typical" K_m of 5 μM for a protein-tyrosine kinase acting on a good substrate, discuss whether the experiment provided a useful indication of the specificity of the enzymes.

Chapter 6

Tight-binding and Irreversible Inhibitors

6.1 Tight-binding inhibitors

Tight-binding inhibitors fall at the boundary between reversible and irreversible inhibitors, because although the inhibition is reversible (for example by thorough dialysis to remove the inhibitor) the dissociation may be slow enough to give the impression of an irreversible effect. In some ways it would be better to call them slowly dissociating inhibitors. Tightness of binding and slowness of dissociation may appear to be two distinct properties, but they are closely linked, because there is a limit to the rate at which binding can occur, known as the *diffusion limit*: molecules cannot bind to one another any faster than they can diffuse to one another through the solution. *On* rate constants cannot therefore exceed a limit of about $7 \times 10^9 \, M^{-1} \, s^{-1}$ (Fersht, 1999), and in practice they are often much smaller, because it is not sufficient for a small molecule to encounter a protein molecule; it must do so at an appropriate location if any reaction is to ensue. In the data tabulated by Fersht for binding of substrates to enzymes, for example, most of the *on* rate constants are smaller than $10^8 \, M^{-1} \, s^{-1}$. Consider now the implications of this limit for the dissociation constant of an inhibitor. Let us regard the dissociation as slow if it occurs with a rate constant of less than $1 \, s^{-1}$. Given that an inhibition constant K_i can be written as the ratio k_{-i}/k_i of the *off* and *on* rate constants, this gives a limit of $K_i = 1/10^8 \, M = 10^{-8} \, M$, or 10 nM. Many pharmacologically important inhibitors bind more tightly than this, and if they cannot achieve lower K_i values by increasing k_i they must do so by decreasing k_{-i}, that is to say by dissociating more slowly.

Independent of the slowness of dissociation, tight-binding inhibitors present another difficulty for analysis of the inhibition by the methods described in the previous chapter, because an inhibition constant of the order of 10 nM or less is likely to be less than the concentration of the enzyme. This means that to vary the degree of inhibition one needs to vary the inhibitor concentration in a range lower than the enzyme concentration, and one cannot make the usual assumption that the inhibitor concentration is unaffected by binding to the enzyme and that therefore the free and total inhibitor concentrations are the same. If we analyse a simple binding process $I + E = EI$ with dissociation constant K_i, without assuming that either I or E is in excess, the definition of K_i is

$$K_i = (e_0 - x)(i_0 - x)/x \tag{6.1}$$

where x is the concentration of complex EI. This can be rearranged into a quadratic equation in x:

$$x^2 - (e_0 + i_0 + K_i)x + e_0 i_0 = 0 \tag{6.2}$$

which may be solved by the usual methods for quadratic equations to yield

$$x = 0.5\{e_0 + i_0 + K_i - [(e_0 + i_0 + K_i)^2 - 4e_0 i_0]^{0.5}\} \tag{6.3}$$

Although mathematically a quadratic equation has two solutions, one may readily show that the other solution, with the first minus sign in equation 6.3 replaced by a plus sign, leads to physically impossible results; for example, putting $e_0 = 0$ or $i_0 = 0$ leads to a positive value of x, even though there obviously can be no EI formed in the absence of E or I.

The full rate equation for a reaction inhibited by a tight-binding inhibitor takes the following form:

$$\frac{v_i}{v_0} = 1 - \frac{e_0 + i_0 + i_{0.5}^{\mathrm{app}} - [(e_0 + i_0 + i_{0.5}^{\mathrm{app}})^2 - 4e_0 i_0]^{0.5}}{2e_0} \tag{6.4}$$

where $i_{0.5}^{\mathrm{app}} = (K_m + a)/(K_m/K_{ic} + a/K_{iu})$. The non-linear character of this equation is not a problem for modern computer fitting, but before this became true various graphical methods for analysing tight binding were proposed, for example by Dixon (1972) and Henderson (1973).

Derivation of equation 6.4 is rather complicated (Morrison, 1969; Henderson, 1972), though the result can be recognized as having equation 6.3 as a basis. It is also not obvious how to simplify it to yield the appropriate result when e_0 is negligible compared with $i_{0.5}^{\mathrm{app}}$ and i_0: one cannot just set e_0 to zero as this leads to an indefinite result. Instead, defining the contents of the square brackets in equation 6.4 as g^2 it is simple to rearrange its expression to read as follows:

$$g^2 = (i_0 + i_{0.5}^{\mathrm{app}})^2 \left[1 + \frac{e_0^2 + 2e_0 i_{0.5}^{\mathrm{app}} - 2e_0 i_0}{(i_0 + i_{0.5}^{\mathrm{app}})^2} \right] \tag{6.5}$$

Here e_0^2 is negligible when e_0 is small, and the square root of the sum in the square brackets can be approximated by the first two terms of the binomial expansion, using the general relationship $(1+x)^{0.5} \approx 1 + 0.5x$ when x is small, so

$$g \approx i_0 + i_{0.5}^{\text{app}} + \frac{e_0(i_{0.5}^{\text{app}} - i_0)}{i_0 + i_{0.5}^{\text{app}}} \tag{6.6}$$

One can then substitute this expression for g into equation 6.4, and simplify and rearrange the result to obtain the following equation:

$$\frac{v_i}{v_0} = \frac{1}{1 + i_0/i_{0.5}^{\text{app}}} \tag{6.7}$$

which it is then straightforward to rearrange further as follows:

$$\frac{i_0 v_i}{v_0 - v_i} = i_{0.5}^{\text{app}} \tag{6.8}$$

The relationship to equation 5.20 is now made obvious by recognizing that $i_{0.5}^{\text{app}}$ is the quantity defined in equation 5.26 as the inhibitor concentration for half-inhibition. This definition is still valid for tight-binding inhibition, but it provides the *free* value of this concentration, whereas the total value is of greater practical interest, and for a tight-binding inhibitor it is not the same as the free value. Substituting $v_i/v_0 = 0.5$ into equation 6.4, we can solve it for i_0 to obtain the total inhibitor concentration for half-inhibition:

$$i_{0.5} = i_{0.5}^{\text{app}} + 0.5e_0 \tag{6.9}$$

This relationship was derived by Myers (1952) from a more general one given by Goldstein (1944). It can be rationalized by realizing that at half-inhibition exactly half of the enzyme is complexed with inhibitor, and so $0.5e_0$ is the difference between the free and total inhibitor concentrations.

The practical behaviour of equation 6.4 was examined in detail by Morrison (1969). An important characteristic is that although it predicts deviations from the ordinary inhibition equations described in Chapter 5, the deviations are small enough be overlooked if examined over an inadequate range of inhibitor and substrate concentrations. However, the slopes and intercepts of the apparently straight lines do not behave in the ways expected from the simpler analysis, and, in particular, competitive inhibitors can easily be mistaken for mixed inhibitors if inhibitor depletion is ignored. For example, Turner, Lerea and Kull (1983) re-examined some supposedly pure non-competitive inhibitors of ribonuclease with very small inhibition constants and found that they were in reality competitive inhibitors.

6.2   Irreversible inhibitors

6.2.1   Non-specific irreversible inhibition

We now consider true *irreversible inhibitors* or *catalytic poisons*, which have effects that cannot be reversed by removing the inhibitor by dialysis or dilution. In most examples the inhibitor reacts with an essential group in the enzyme to bring about a covalent change, but occasionally, especially with heavy-metal ions, there may not be any covalent change but simply a binding that cannot be reversed on any reasonable time scale. In fact, many enzymes are poisoned by traces of heavy-metal ions, and for this reason it is common practice to carry out kinetic studies in the presence of complexing agents, such as ethylene-diaminetetraacetate. This is particularly important in the purification of enzymes: in crude preparations, the total protein concentration is high, and the many protein impurities present sequester almost all the metal ions that may be present, but the purer an enzyme becomes the less it is protected by other proteins and the more important it is to add alternative sequestering agents. Knowledge of which metals poison particular enzymes can be used as a clue to the groups necessary for enzyme activity; for example, poisoning by mercury(II) compounds has often been used to implicate sulphydryl groups in the catalytic activity of enzymes.

Various organic inhibitors act in the same sort of way, reacting with particular functional groups in proteins and inactivating enzymes if the groups in question are necessary for activity. For example, other inhibitors that react with sulphydryl groups include 5,5′-dithiobis-(2-nitrobenzoate) and iodoacetate, of which the latter also reacts more slowly with other kinds of group, such as the hydroxyl groups of tyrosine residues; acetic anhydride reacts with groups capable of being acetylated, principally amino groups but also hydroxyl and sulphydryl groups. All of these inhibitors are rather unspecific and appear to react in a single step without prior formation of an enzyme–inhibitor complex. Kinetically, therefore, we observe an irreversible second-order reaction, as discussed in Section 1.2.

6.2.2   Specific irreversible inhibition

A more specific type of irreversible inhibition occurs when the inactivating reaction follows an initial binding, similar to the formation of an enzyme–substrate complex, produced by affinity between the inhibitor and a particular site on the enzyme, normally close to or part of the active site. The scheme to represent this resembles the ordinary Briggs–Haldane scheme (equation 2.9), with the important difference that the free enzyme is not regenerated in the second step; instead the initial complex EI is transformed into a second complex E′ that does not undergo any further reaction:

$$\underset{e_{\text{act}} - x}{\text{E}} + \underset{i}{\text{I}} \underset{k_{-1}}{\overset{k_1}{\rightleftharpoons}} \underset{x}{\text{EI}} \overset{k_2}{\longrightarrow} \text{E}' \text{ (inactive)} \tag{6.10}$$

Here e_{act} represents the total concentration of *active* enzyme, including E and EI but not E′. The kinetics for this scheme depend on the relative magnitudes of k_{-1} and k_2. If $k_2 \ll k_{-1}$ then a treatment developed by Kitz and Wilson (1962) is applicable. They were concerned with the effects of various compounds on acetylcholinesterase and made essentially the same assumptions as those of Michaelis and Menten (1913) for the catalytic process: they assumed that the inactivation of an enzyme E by an inhibitor I would proceed through an intermediate EI that was in equilibrium with E and I throughout the process. With these assumptions the concentration of EI at any time is equal to $(e_{\text{act}} - x)i/K_i$ and the rate of inactivation v is given by

$$-\frac{de_{\text{act}}}{dt} = k_2 x = \frac{k_2 e_{\text{act}} i}{K_i + i} \tag{6.11}$$

Note that although $k_2 e_{\text{act}}$ is the limiting rate of inactivation at saturating concentrations of inhibitor, and thus corresponds to V in the Michaelis–Menten equation, it is not a constant, because e_{act} decreases as the inactivation proceeds. If the inhibitor is in sufficient excess for i to be essentially constant the loss of activity is a pseudo-first-order process, and analysis by the methods of Section 1.5 would give an apparent first-order rate constant $k_{\text{app}} = k_2 i/(K_i + i)$. As this has the form of the Michaelis–Menten equation one can use any of the methods of Chapter 2 to estimate k_2 and K_i from measurements of k^{app} at different values of i.

If k_2 is not small enough to allow formation of EI to be treated as an equilibrium, but is still smaller than k_{-1}, the treatment of Kitz and Wilson can be modified as proposed by Malcolm and Radda (1970), with K_i now defined as $(k_{-1} + k_2)/k_1$, essentially the same as the definition of the Michaelis constant in Briggs–Haldane conditions (Section 2.3). With this modification the analysis is essentially the same as before, and describes the observed behaviour well. This simple treatment breaks down, however, if k_2 exceeds k_{-1}, because then it is no longer possible to ignore the accumulation of E′, so e_{act} as defined above is no longer a useful quantity to consider. In these conditions, therefore, the loss of activity does not follow simple first-order kinetics: there is no exact analytical solution, but the kinetics may still be analysed by numerical methods.

6.3   Substrate protection experiments

The substrate of an enzyme often protects it against inactivation by an irreversible inhibitor; in other words, the substrate acts as an inhibitor of the inactivation reaction. If the substrate is itself able to undergo its ordinary reaction (either because it is a one-substrate reaction or because the other necessary components, such as water, are present also), the reaction scheme,

$$E + I \rightleftarrows EI \rightarrow E'(\text{inactive}); E + A \rightleftarrows EA \rightarrow E + P \qquad (6.12)$$

can be regarded as a combination of the scheme of competing substrates, Figure 2.3 (page 38), with the Kitz–Wilson scheme for enzyme inactivation, equation 6.10, and the corresponding rate equation is likewise a combination of equations 6.11 and 2.21:

$$v = \frac{Vi}{K_i(1 + a/K_m) + i} \qquad (6.13)$$

The similarity to equation 5.1 is obvious, though to avoid confusion one must emphasize that the substrate and inhibitor have reversed their usual roles, and v represents the rate of loss of activity, not the rate of the ordinary reaction catalysed by the enzyme. It follows that one can measure K_m by the methods one would use to measure K_{ic} in ordinary competitive inhibition, but one must remember that V is not a constant, as it decreases as the enzyme becomes inactivated: thus one must first obtain apparent inhibition constants K_m^{app} by the method described in Section 6.2.2 and then obtain K_m from the linear dependence on a, $K_m^{app} = K_i(1 + a/K_m)$.

As a method of determining K_m this may seem to have no obvious advantage over more conventional ones. It is useful, however, because it can be applied when A is the first substrate of a reaction that cannot proceed in the absence of other substrates (compare Section 7.4.1). Binding of substrate in the inactivation experiment is then an equilibrium, and so the analysis yields the true dissociation constant K_s instead of K_m. This approach requires only catalytic amounts of enzyme and has accordingly been applied to a number of enzymes (for example, Malcolm and Radda, 1970; Anderton and Rabin, 1970) that were not available in the quantities needed for equilibrium dialysis or other methods of studying binding at equilibrium.

6.4 Chemical modification as a means of identifying essential groups

It is common practice to deduce the nature of groups required for enzyme catalysis by observing whether activity is lost when certain residues are chemically modified. Unfortunately, however, a particular residue may be essential for catalytic activity without necessarily playing any part in the catalytic process; it may, for example, be essential for maintaining the active structure of the enzyme. Nonetheless, identification of the essential groups in an enzyme is an important step in characterizing the mechanism, and many reagents are now available for modifying specific types of residue. The logic used for interpreting such experiments is often loose, though proper analytical approaches have been available for a long time. Of these, the method of Ray and Koshland (1961) is appropriate when the loss of activity can readily be measured as a function of time, and the method of Tsou (1962) is convenient when there are no rate data but there is information on the amount of activity remaining at various degrees of chemical modification. These two approaches will now be discussed.

6.4.1 Kinetic analysis of chemical modification

Consider an enzyme that has two groups g_1 and g_2 that are both essential for catalytic activity, in the sense that if either of them is lost then all catalytic activity is lost, as illustrated in Figure 6.1a. If g_1 is converted to an inactive form with first-order rate constant k_1 and g_2 is converted to a (different) inactive form with first-order rate constant k_2 then the activity A remaining after time t is given by

$$A = A_0 e^{-k_1 t} e^{-k_2 t} = A_0 e^{-(k_1 + k_2)t} = A_0 e^{-k_{\text{inactivation}} t} \tag{6.14}$$

where A_0 is the value of A at zero time, and $k_{\text{inactivation}}$, the observed first-order rate constant for inactivation, is the sum $k_1 + k_2$ of the rate constants for the separate reactions. Note that the inactive forms produced in the first steps may, and probably do, react further with the same or different rate constants, as indicated in the Figure, but no knowledge of these later processes or their rate constants is required, because all activity has been assumed to be lost in the first steps.

Although for simplicity Figure 6.1a has been drawn with just two essential groups, and equation 6.14 has been written accordingly, it is obvious from inspection that the same treatment can be generalized to any number of

(a) Two groups that react   (b) Two groups that react
    at different rates, both   equally fast, one of them
    essential   essential

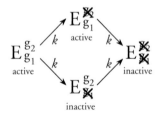

Figure 6.1. Two simple models for loss of activity of an enzyme during chemical modification.

essential groups, and the observed rate constant is then the sum of the rate constants for all the individual processes. In practice, unless all the groups are of a very similar kind we should expect considerable variations in the values of the rate constants, and the slowest processes will then make little contribution to the total, which will be dominated by a small number of relatively rapid processes.

Returning now to the case of two groups, suppose now that they are both of the same chemical type (two cysteine residues, for example) and that they both react with the modifying reagent with the same rate constant k, but only one of them is essential for catalytic activity: this may happen, for example, if one of them is in the active site of the enzyme and the other is on the other side of the molecule. The model is illustrated in Figure 6.1b, and the observed rate constant is now the same as the rate constant k, the equation for loss of activity being as follows:

$$A = A_0 e^{-kt} = A_0 e^{-k_{inactivation} t} \tag{6.15}$$

This result is again easy to generalize for the more realistic assumption that there are many more than two groups sensitive to the modifying agent and more than one of them is essential for activity: with two essential groups, for example, the observed rate constant is equal to $2k$, and so on.

Ray and Koshland (1961) extended the analysis to more complicated models in which modification of one group results in only partial loss of activity (or no loss of activity at all), so that two or more groups need to be modified for complete loss of activity. Such examples may be recognized experimentally because the time course for loss of activity does not follow first-order kinetics; not surprisingly, they are more complicated to analyse than those illustrated in Figure 6.1.

6.4.2 Remaining activity as a function of degree of modification

The simplest example to consider in the method of Tsou (1962) is one in which n groups on each monomeric enzyme molecule react equally fast with the modifying agent, and $n_{essential}$ of these are essential for catalytic activity. After modification of an average of $n_{modified}$ groups on each molecule, the probability that any particular group has been modified is $n_{modified}/n$ and the probability that it remains unmodified is $1 - n_{modified}/n$. For the enzyme molecule to retain activity, *all* of its $n_{essential}$ essential groups must remain unmodified, for which the probability is $(1 - n_{modified}/n)^{n_{essential}}$. Thus the fraction $f = A/A_0$ of activity remaining after modification of $n_{modified}$ groups per molecule (with A and A_0 defined as in Section 6.4.1) must be

$$f = (1 - n_{modified}/n)^{n_{essential}} \tag{6.16}$$

Hence

$$f^{1/n_{essential}} = 1 - n_{modified}/n \tag{6.17}$$

and a plot of $f^{1/n_{essential}}$ against $n_{essential}$ should be a straight line. Initially, of course, one does not know what value of $n_{modified}$ to use in the plot, so one plots $f, f^{1/2}, f^{1/3}$ and so on in turn against $n_{essential}$ to decide which value gives the straightest line.

One objection to this treatment is that not all of the modification reactions may be equally fast, and they may anyway not be independent, as modification of one group may alter the rates at which neighbouring groups are modified. Tsou's paper should be consulted for a full discussion of these and other complications, but two additional classes of group can be accommodated without losing the essential simplicity of the method. If there are n_{fast} non-essential groups that react rapidly compared with the essential groups, these will produce an initial region of the plot (regardless of $n_{essential}$) in which there is no decrease in $f^{1/n_{essential}}$ as $n_{modified}$ increases. Further groups, whether essential or not, that react slowly compared with the fastest-reacting essential groups will not become appreciably modified until most of the activity has been lost; they must therefore be difficult or impossible to detect. In practice, therefore, a Tsou plot is likely to resemble the one shown in Figure 6.2, which was used by Paterson and Knowles (1972) as evidence that at least two carboxyl groups are essential to the activity of pepsin, that three non-essential groups were modified rapidly in their experiments, and that the two essential groups belong to a class of ten that were modified at similar rates.

Pepsin is a monomeric enzyme, but Tsou's analysis can be extended without difficulty to oligomeric enzymes provided that one can assume that the

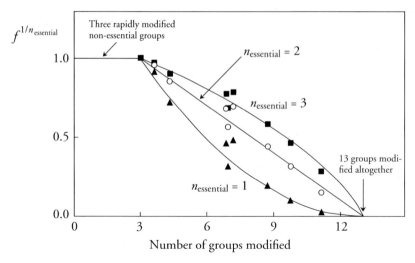

Figure 6.2. Tsou plot for determining the number of essential groups in an enzyme. The plot shows data of Paterson and Knowles (1972) for the inactivation of pepsin by trimethyloxonium fluoroborate, a reagent that reacts specifically with carboxyl groups. The variable f represents the fraction of activity remaining after modification of the number of groups shown, and is raised to the power $1/n_{essential}$, where $n_{essential} = 1, 2$ or 3. The straight line observed with $n_{essential} = 2$ indicates that at least two carboxyl groups are essential to the activity of pepsin, out of a total of 13 modified. The horizontal line at the beginning of the plot indicates that there are three rapidly reacting groups that are not needed for catalytic activity.

subunits react independently with the modifying agent and that inactivation of one subunit does not affect the activity of the others. With these assumptions, oligomeric enzymes can be treated in the same way as monomeric enzymes, except that $n_{essential}$ is now the number of essential groups per subunit, even though n, $n_{modified}$ and n_{fast} are still defined per molecule (Norris and Brocklehurst, 1976).

On the other hand, if inactivation of one subunit does cause inactivation of the others the whole oligomer must be treated as one molecule. For example, Li and co-workers (1998) studied the effect of chaperonin 60 on the reactivation of the tetrameric enzyme glyceraldehyde 3-phosphate dehydrogenase from rabbit muscle after it had been denatured with guanidine hydrochloride. They found that two groups were essential for reassembly of the tetramer, and interpreted this to mean that both of the dimers formed in the denaturation needed to be in a free and unbound state.

In the past, putative identification of a modified group was often followed by partial hydrolysis and sequencing of the peptide fragment containing the modified amino acid residue, in the hope of learning about the structure of the catalytic site. Modern genetic engineering techniques have largely supplanted this approach, but it remains profitable to complement them with kinetic analysis of the properties of the artificial mutants that can be produced. With a mutant of known structure one can predict exactly how a Tsou plot ought to be altered by a particular mutation, and experimental verification that it is in fact so altered is a way of confirming the correctness of the interpretation.

Problems

6.1 Consider a mixture in which the total enzyme concentration is $1\,nM$ ($10^{-9}\,M$), and the substrate concentration is equal to the K_m value of $1\,mM$ ($10^{-3}\,M$), and a competitive inhibitor is present at a concentration equal to its inhibition constant. What is the rate relative to the uninhibited rate if the inhibition constant is (a) $1\,mM$; (b) $1\,nM$?

6.2 The fastest inactivation measured in the experiments of Kitz and Wilson (1962) was estimated to have $k_2 = 5 \times 10^{-3}\,s^{-1}$, $K_i = 0.1\,mM$, for the mechanism shown as equation 6.10. Assuming that K_i can be expressed as $(k_{-1} + k_2)/k_1$, how big would k_1 have to be for this expression to be significantly different from the equilibrium ratio k_{-1}/k_1 assumed by Kitz and Wilson? Childs and Bardsley (1975) argued that Kitz and Wilson were wrong to assume a pre-equilibrium: what light do the typical values of $10^6\,M^{-1}\,s^{-1}$ or greater for second-order rate constants for specific binding of small molecules to proteins shed on this criticism?

6.3 The following table shows data of Norris and Brocklehurst (1976) for the effect on the activity of urease of modification with 2,2′-dipyridyl disulphide, a compound that reacts specifically with thiol groups. The number of groups modified per molecule and the activity relative to the untreated enzyme are shown as γ and α respectively. Assuming that urease has six subunits per molecule that act independently both in the catalytic and in the modification reactions, estimate (a) the number of essential thiol groups per subunit, and (b) the number of inessential thiol groups that are modified rapidly in comparison with the essential groups.

γ	α	γ	α	γ	α
0.0	1.000	23.0	0.957	27.0	0.547
2.0	1.000	24.0	0.896	27.5	0.442
4.0	1.000	25.0	0.853	28.0	0.353
18.0	1.000	25.5	0.799	29.0	0.198
20.0	1.000	26.0	0.694	29.5	0.104
22.0	0.982	26.5	0.597	30.0	0.011

Chapter 7

Reactions of More than One Substrate

7.1 Introduction

Much of the earlier part of this book has been concerned with reactions of a single substrate and a single product. Such reactions are actually rather rare in biochemistry, being confined to a few isomerizations, such as the interconversion of glucose 1-phosphate and glucose 6-phosphate, catalysed by phosphoglucomutase. Nonetheless, the development of enzyme kinetics was greatly simplified by two facts: first, many hydrolytic enzymes can be treated as single-substrate enzymes, because the second substrate, water, is always present in such large excess that its concentration can be treated as a constant; second, most enzymes behave much like single-substrate enzymes if only one substrate concentration is varied, as will be clear from the rate equations to be introduced in this chapter (especially Section 7.4.2).

There are three principal steady-state kinetic methods for elucidating the order of addition of substrates and release of products: measurement of initial rates in the absence of product; testing the nature of product inhibition; and tracer studies with radioactively labelled substrates. The last of these will be discussed in Chapter 8, the others in this one, using a general reaction with two substrates and two products as an example:

$$A + B \rightleftarrows P + Q \tag{7.1}$$

This type of reaction is by far the most common in biochemistry: in compilations of all the known enzyme-catalysed reactions, such as that in *Enzyme Nomenclature* (International Union of Biochemistry and Molecular Biology, 1992), some 60% of reactions are in the first three classes (oxidoreductases, group-transfer reactions and hydrolases), all of which satisfy equation 7.1. More complicated reactions also occur, with as many as four or five substrates, but these can for the most part be studied by generalizing the principles developed for the study of two-substrate two-product reactions. Even the reaction in equation 7.1 can proceed in many different ways, but I shall confine discussion to a small number of important mechanisms, rather than attempt to treat all of them. An encyclopaedic approach, such as that of Segel (1975), is largely self-defeating, because nature can be relied upon to provide examples that fall outside any "exhaustive" treatment; more important, readers who understand the methods used to discriminate between the simple mechanisms are well equipped to adapt them to special experimental circumstances and to understand the more detailed discussions found elsewhere.

A point that is perhaps worth emphasizing here, as it is sometimes mis-understood, is that the distinction between substrates and coenzymes, useful though it may be in physiological studies, has no meaning in relation to enzyme mechanisms. Discussions of the metabolic role of alcohol dehydrogenase, for example, often distinguish between ethanol, the "substrate", and oxidized NAD, the "coenzyme". So far as studies of the enzyme are concerned they are both substrates; neither is more fundamental than the other, and both are required for the reaction to take place. In Chapter 12 I shall consider why a different view may be justified in studies of metabolism.

7.2   Classification of mechanisms

7.2.1   Ternary-complex mechanisms

Almost all two-substrate two-product reactions are formally *group-transfer* reactions [though *Enzyme Nomenclature* (International Union of Biochem-istry and Molecular Biology, 1992) reserves the name *transferase* for enzymes in Class 2, group-transfer enzymes that are neither oxidoreductases nor hydro-lases], and for discussing possible mechanisms it is helpful to show this more explicitly than was done in equation 7.1:

$$GX + Y \rightleftarrows X + GY \qquad (7.2)$$

writing A as GX and B as Y, so that one can see that the reaction consists of transferring a group G from a donor molecule GX to an acceptor molecule Y. The need to satisfy valence requirements normally implies that some other group is transferred simultaneously in the opposite direction: for example, when a phosphoryl group is transferred from ATP to glucose by hexokinase a proton is transferred in the other direction and released into solution; however, it is not usually necessary to show this reverse transfer explicitly. The type of symbolism used in equation 7.2 was introduced by Wong and

Hanes (1962) in a paper that laid the foundations of the modern classification of kinetic mechanisms. Although it is convenient for discussing the mechanisms themselves it becomes cumbersome for writing rate equations, and so I shall return to the use of single letters later in this chapter.

Wong and Hanes showed how most reasonable mechanisms of group transfer could be regarded as special cases of a general mechanism. Perhaps fortunately, enzymes that require the complete mechanism seem to be rare, and some supposed examples, such as pyruvate carboxylase, may well have been misinterpreted (see Warren and Tipton, 1974ab). Accordingly, I shall discuss the three simplest group-transfer mechanisms as separate cases.

The main division is between mechanisms that proceed through a *ternary complex*, EGXY, so-called because it contains the enzyme and both substrates in a single complex, and those that proceed through a *substituted enzyme*, EG, which contains the enzyme and the transferred group but neither of the complete substrates. Early investigators, such as Woolf (1929, 1931) and Haldane (1930), assumed that the reaction would pass through a ternary complex, and that either of the two binary complexes EGX and EY could serve as an intermediate in its formation. In other words, the substrates could bind to the enzyme in *random order*, as illustrated in Figure 7.1. The rigorous steady-state equation for this mechanism is complicated, and includes terms in $[GX]^2$ and $[Y]^2$, but, as Gulbinsky and Cleland (1968) showed, the contribution of these terms to the rate may often be so slight that the experimental rate equation is indistinguishable from one derived on the assumption that all steps apart from interconversion of the ternary complexes EXG•Y and EX•GY are at equilibrium. If this assumption is made there are no squared terms in the rate equation, and for simplicity I shall make the rapid-equilibrium assumption in

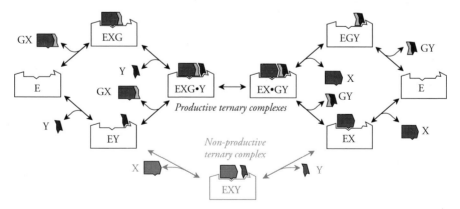

Figure 7.1. Random-order ternary-complex mechanism. The part of the Figure shown in grey at the bottom represents a non-productive pathway that is not a necessary component of the mechanism but can often be expected to occur as there is nothing to prevent it.

discussing this mechanism. However, the frequent failure of the experimental observations to disprove the rapid-equilibrium assumption does *not* show that the assumption is usually incorrect. The step that interconverts EXG•Y and EX•GY cannot be detected by steady-state measurements, but it is logical to show it explicitly in the random-order mechanism because it is treated as rate-determining in deriving the rate equation.

The non-productive complex EXY is not a necessary participant in the random-order mechanism, but it can normally be expected to exist, because if both EY and EX are significant intermediates there is no reason to exclude EXY. Another non-productive complex (not included in Figure 7.1) is possible if the transferred group G is not too bulky: EXG•GY can result from binding of GY to EGX or of GX to EGY. This is less likely than the formation of EXY, however.

It is now generally recognized that many enzymes cannot be regarded as rigid templates, as implied by Figure 7.1. Instead, it is likely that the conformations of both enzyme and substrate are altered on binding, in accordance with the "induced-fit" hypothesis of Koshland (1958, 1959ab; see also Section 11.4). It may well happen therefore that no binding site exists on the enzyme for one of the two substrates until the other has bound. There is then a *compulsory order* of binding, as illustrated in Figure 7.2. If both substrates and products are taken into consideration, four different orders are possible, but the induced-fit explanation of compulsory-order mechanisms leads us to expect that the reverse reaction should be structurally analogous to the forward reaction, so that the second product ought to be the structural analogue of the first substrate; thus only two of the four possibilities are very likely.

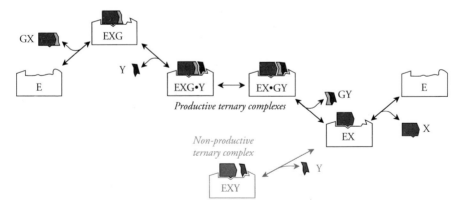

Figure 7.2. Compulsory-order ternary-complex mechanism. Formally this is the same as the random-order versions shown in Figure 7.1 with some steps omitted. To explain why these steps do not occur, that is to say why the substrates must bind in a particular order, one may suppose that binding of the first substrate GX induces a conformational change that allows the binding site for the second substrate Y to become recognizable.

Daniel Edward Koshland (1920–)

Daniel Koshland was born in New York, but spent much of his early life in the area of San Francisco. After wartime service on the Manhattan Project, he carried out doctoral and post-doctoral work at Chicago, and subsequently worked at the Brookhaven National Laboratory, where he became the foremost influence on the way biochemists think about enzyme catalysis. His many fundamental ideas included the relationship of stereochemistry to enzyme mechanism, the role of the enzyme in bringing reacting molecules into correctly oriented proximity, and, most important, enzyme flexibility and induced fit. After he returned to California he continued at Berkeley to be a fertile source of original and stimulating ideas. Some of these, such as his analysis of the role of bacterial memory in chemotaxis, are not very close to the theme of this book, but others, such as the development of our understanding of cooperativity, are fundamental.

In NAD-dependent dehydrogenases, for example, the coenzymes are often found to be first substrate and second product. Nonetheless, the less plausible possibilities should not be dismissed too lightly: a tentative suggestion that hexokinase D might release its products in an unexpected order (Storer and Cornish-Bowden, 1977), received support much later from experiments with substrate analogues (Monasterio and Cárdenas, 2003).

7.2.2 Substituted-enzyme mechanisms

Early in the development of multiple-substrate kinetics, Doudoroff, Barker and Hassid (1947) showed by isotope-exchange studies that the reaction catalysed by sucrose glucosyltransferase proceeded through a substituted-enzyme intermediate rather than a ternary complex. Since then, studies with numerous and diverse enzymes, including α-chymotrypsin, transaminases and flavoenzymes, have shown that the substituted-enzyme mechanism, illustrated in Figure 7.3, is common and important. In the ordinary form of this mechanism, occurrence of a ternary complex is structurally impossible because the binding sites for X and Y are either the same or overlapping. For example, aspartate transaminase catalyses a reaction involving four dicarboxylate anions of similar size:

$$\text{glutamate} + \text{E–pyridoxal} \rightleftarrows \ldots \rightleftarrows \text{2-oxoglutarate} + \text{E–pyridoxamine}$$

$$\text{oxaloacetate} + \text{E–pyridoxamine} \rightleftarrows \ldots \rightleftarrows \text{aspartate} + \text{E–pyridoxal} \qquad (7.3)$$

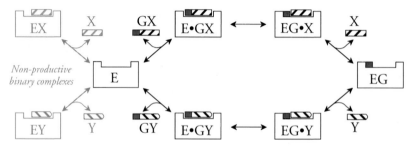

Figure 7.3. Substituted-enzyme mechanism. As in Figure 7.1, the non-productive pathway shown in grey at the left is not a necessary component of the mechanism.

Here E–pyridoxal represents the enzyme with its coenzyme as pyridoxal phosphate (strictly it is not bound as an aldehyde but as an internal aldimine formed by condensation with a lysine residue, but that is not important for the present purpose), and E–pyridoxamine represents the form with pyridoxamine phosphate. As all four reactants are structurally similar, it is reasonable to expect the binding sites for 2-oxoglutarate and oxaloacetate (X and Y in the general case) to be virtually identical and the second half of the reaction to be essentially the reverse of the first half.

In this mechanism, it is usually possible for the substrates to be able to bind to the "wrong" form of the enzyme, so that, for example, in equation 7.3 it is difficult to imagine a structure for E–pyridoxal that would allow it to bind glutamate but not 2-oxoglutarate; one thus expects to see substrate inhibition at high substrate concentrations (Section 7.5.4). This is almost always true of E, X and Y, as drawn in Figure 7.3, but binding of GX or GY to the wrong form EG may be prevented by steric interference between the two G groups.

The substituted-enzyme mechanism shown in Figure 7.3 is a compulsory-order mechanism, but this is less noteworthy than with ternary-complex mechanisms because there is only one mechanistically reasonable order, and no random-order alternative: even if X and Y bind to E, there is no reasonable way for the resulting complexes to break down to give GX or GY. The kinetic properties of the random-order substituted-enzyme mechanism have nonetheless been analysed, together with numerous other mechanisms that are difficult to visualize in chemical terms (Fisher and Hoagland, 1968; Sweeny and Fisher, 1968). The method of King and Altman (1956) can be applied to unreasonable mechanisms as easily as to reasonable ones, and if one regards kinetics as a branch of algebra, largely unrelated to chemistry, one risks having to deal with a bewildering array of possibilities. For this reason one should always regard algebra as the servant of enzyme kinetics and not its master.

7.2.3   Comparison between chemical and kinetic classifications

In a substituted-enzyme mechanism the group G is transferred twice, first from the substrate GX to the free enzyme E, then from the substituted enzyme EG to the second substrate Y. For this reason Koshland (1954) introduced the term *double-displacement reaction* for this type of mechanism. Conversely, ternary-complex mechanisms, in which G is transferred only once, are *single-displacement reactions*. This terminology is still sometimes used, especially in non-kinetic contexts. It leads naturally to consideration of the stereo-chemistry of group-transfer reactions, which Koshland discussed in detail and which forms the subject of Problem 7.1 at the end of this chapter.

The essential idea, illustrated in Figure 7.4, is that each substitution at a chiral centre (usually a carbon or phosphorus atom) results in inversion of configuration at that centre. Thus a ternary-complex mechanism should

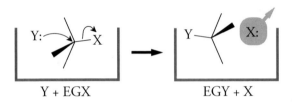

Y + EGX   EGY + X

(a) Single-displacement reaction:
one inversion of configuration

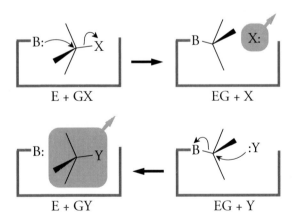

E + GX   EG + X

E + GY   EG + Y

(b) Double-displacement reaction:
two inversions of configuration

Figure 7.4. Stereochemical aspects of group transfer. Normally each substitution results in inversion of configuration, so a double displacement results in restoration of the original configuration: experimentally this is observed as retention of configuration.

result in inversion of configuration, but a substituted-enzyme mechanism should result in retention of configuration (as two inversions regenerate the original configuration). For example, studies of liver hexokinase D by nuclear magnetic resonance (Pollard-Knight and co-workers, 1982) showed that it catalysed phosphoryl transfer with inversion of configuration at phosphorus, in agreement with kinetic evidence that the reaction follows a ternary-complex mechanism. Similar stereochemistry has been found with other kinases believed to follow ternary-complex mechanisms, but nucleoside diphosphate kinase provides a significant exception: not only does the reaction proceed with retention of configuration (Sheu, Richard and Frey, 1979), but it also shows the typical kinetics of a substituted-enzyme mechanism (Garcés and Cleland, 1969).

However, a ternary-complex mechanism is not the only way to explain inversion: as Spector (1982) pointed out, one is not the only odd number, and all of the stereochemical evidence usually advanced in support of single-displacement mechanisms is just as consistent with mechanisms involving three (or five, seven, and so on) displacements. For acetate kinase the triple-displacement mechanism has been proved beyond reasonable doubt (Spector, 1980).

At one time it seemed possible to express experimental results in terms of some broad generalizations, for example, that kinases followed random-order ternary-complex mechanisms, NAD-dependent dehydrogenases followed compulsory-order ternary-complex mechanisms (with oxidized NAD as first substrate and reduced NAD as second product), and transaminases followed substituted-enzyme mechanisms. This sort of classification is not wholly wrong, and it can give a useful guide to what to expect when studying a new enzyme, but it is certainly oversimplified. For example, alcohol dehydrogenase from horse liver was once regarded as an archetypal example of an enzyme obeying a compulsory-order ternary-complex mechanism, but is now believed to bind its substrates in a random order but to release the products in a compulsory order (Hanes and co-workers, 1972). In the strictest sense compulsory-order ternary-complex mechanisms may not occur at all, but they remain useful as a basis for discussion.

It is important to realize that the sequence of events suggested by kinetic studies, as discussed later in this chapter, does not have to agree with the chemical mechanism, the actual sequence of events in the catalytic site. Kinetically speaking, a ternary-complex mechanism means that both substrates must remain associated with the enzyme before the first product is released, but it is perfectly possible for the first product to be released only after the second substrate arrives even if the reaction proceeds through a substituted enzyme. Although this does not seem likely for a transaminase type of reaction as shown in equation 7.3, where the four reactants are so similar that one expects them all to bind at the same site, it is not easy to exclude it when the transferred group is large and the four reactants are quite different from one another.

Although the operation of a substituted-enzyme mechanism can often be shown on the basis of *positive* evidence, such as isolation of the putative substituted-enzyme intermediate (something easily done for many transaminases), operation of a ternary-complex mechanism is usually deduced on the basis of negative evidence, such as failure to isolate a substituted enzyme. Spector (1982) argued that this is always true (not just usually). As noted above, he pointed out that stereochemical evidence for a ternary-complex mechanism is just as consistent with a triple-displacement mechanism, and he argued that the claim that one displacement is "simpler" than three or five is naive. Many tasks in everyday life are facilitated by breaking them up into several smaller tasks instead of attempting a once-for-all solution, and there is no reason why this should not also be true for enzyme chemistry (it is certainly true of metabolic pathways, in which difficult tasks, such as coupling the oxidation of palmitate to the synthesis of ATP, are always handled as sequences of simpler ones).

Conceptually, and chemically, the problem with a ternary-complex mechanism is that the enzyme is little more than a bystander. In Figure 7.2 it at least has some conformational work to do, but in Figure 7.1 it does nothing more than give the reactants some moral encouragement. By contrast, in a substituted-enzyme mechanism (Figure 7.3), the enzyme is an essential participant in every step and without the enzyme there would be no reaction. It would be misleading to suggest that Spector's thesis has been generally accepted; in reality it has been met mainly with indifference or hostility, but there are no strong reasons for rejecting it.

7.2.4 Schematic representation of mechanisms

As noted already, the system of writing the substrates in a group-transfer reaction as GX and Y, and the products as X and GY, is convenient for making the transferred group explicit and for discussing chemical aspects of mechanisms. However, it is less convenient for writing rate equations, and in the remainder of this chapter we shall therefore return to the more common system of representing the substrates as A and B and the products as P and Q.

It is also appropriate to mention at this point a schematic way of representing mechanisms for reactions of two or more substrates that was introduced by Cleland (1963). In this system the various forms of the enzyme are

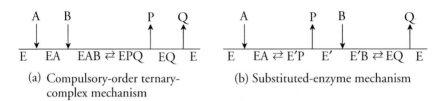

(a) Compulsory-order ternary-   (b) Substituted-enzyme mechanism
    complex mechanism

Figure 7.5. Compulsory-order mechanisms represented with the type of diagrams introduced by Cleland (1963).

written below a horizontal line, and vertical arrows are used to represent addition of substrates and release of products. The two compulsory-order mechanisms for two-substrate reactions (Figures 7.2 and 7.3) are thus drawn as shown in Figure 7.5. This system provides a tidy and clear way of showing compulsory-order mechanisms (even for more complicated examples than those shown here, as illustrated below in Figure 7.12, on page 187), and is widely used for this purpose. However, it becomes cumbersome if one needs to incorporate branches or other complications, and does not readily lend itself to inclusion of rate constants in the diagram in an unambiguous way. It is also somewhat unsatisfactory for considering reverse reactions, because then each arrow needs to be read backwards, a downward-pointing arrow representing an upward-proceeding step, and so on.

7.3   Rate equations

7.3.1   Compulsory-order ternary-complex mechanism

Steady-state kinetic measurements have proved to be of great value for distinguishing between the various reaction mechanisms for group-transfer reactions. The development of these methods was a considerable task, on account of the large number of possibilities and the relatively small kinetic differences between them. Segal, Kachmar and Boyer (1952) were among the first to recognize the need for a systematic approach, and derived the equations for several mechanisms. Subsequently, Alberty (1953, 1958) and Dalziel (1957) made major advances in the understanding of group-transfer reactions, and introduced most of the methods described in this chapter.

As all steady-state methods for distinguishing between mechanisms depend on differences between the complete rate equations, it is appropriate to give a brief account of these equations before discussing methods. The equation for the compulsory-order ternary-complex mechanism will be given first, as it was derived in Section 4.3 as an illustration of the method of

King and Altman, and it had the following form (equation 4.12):

$$v = \frac{e_0(N_1ab - N_{-1}pq)}{D_0 + D_1a + D_2b + D_3p + D_4q + D_5ab + D_6ap + D_7bq + D_8pq + D_9abp + D_{10}bpq}$$

(7.4)

This equation contains 13 coefficients, but these were defined in terms of only eight rate constants, so there must be relationships between the coefficients that are not explicit in the equation. Moreover, the coefficients lack obvious mechanistic meaning. Numerous systems have been used for re-writing rate equations in more meaningful terms; the one used in this book follows the recommendations of the International Union of Biochemistry and Molecular Biology (International Union of Biochemistry, 1982), and derives ultimately from the classification of Cleland (1963) of the constants into three types: *limiting rates, Michaelis constants* and *inhibition constants*. In general a Michaelis constant K_{mA} for a substrate A corresponds to K_m in a single-substrate reaction, and an inhibition constant K_{iA} is related to the K_{ic} and K_{iu} values obtained when the substrate A is used as a product inhibitor of the reverse reaction (but is not necessarily identical to either of them). In some circumstances the inhibition constants are true substrate-dissociation constants, and when this is true one can emphasize it by using symbols such as K_{sA} rather than K_{iA}, and so on. If actual enzyme concentrations are known the limiting rates can be converted into *catalytic constants*, and the definition of *specificity constants* follows likewise in a natural way from the definition for single-substrate reactions (Section 2.3.4).

In this system equation 7.4 may be rewritten as follows:

$$v = \frac{\dfrac{V_+ab}{K_{iA}K_{mB}} - \dfrac{V_-pq}{K_{mP}K_{iQ}}}{1 + \dfrac{a}{K_{iA}} + \dfrac{K_{mA}b}{K_{iA}K_{mB}} + \dfrac{K_{mQ}p}{K_{mP}K_{iQ}} + \dfrac{q}{K_{iQ}} + \dfrac{ab}{K_{iA}K_{mB}} + \dfrac{K_{mQ}ap}{K_{iA}K_{mP}K_{iQ}} + \dfrac{K_{mA}bq}{K_{iA}K_{mB}K_{iQ}} + \dfrac{pq}{K_{mP}K_{iQ}} + \dfrac{abp}{K_{iA}K_{mB}K_{iP}} + \dfrac{bpq}{K_{iB}K_{mP}K_{iQ}}}$$

(7.5)

and comparison with the definitions given after equation 4.12 shows that the kinetic parameters must have the values given in Table 7.1. Although this equation may appear complicated, it contains more regularities than are obvious at first sight. The terms that contain q are in general similar to those that contain a, whereas those that contain p are similar to those that contain b. The whole expression does, of course, satisfy the usual dimensional

Table 7.1. Definitions of kinetic parameters for compulsory-order mechanisms. The Table shows the relationships between the parameters that appear in equations 7.5 and 7.7 and the rate constants for the individual steps of the two principal compulsory-order mechanisms for group-transfer reactions. Although equation 7.7 does not contain K_{iB} it can be written so that it does by means of the identity $K_{iA}K_{mB}/K_{iP}K_{mQ} = K_{mA}K_{iB}/K_{mP}K_{iQ}$.

	Ternary-complex mechanism	Substituted-enzyme mechanism
	$E \underset{k_{-1}}{\overset{k_1 a}{\rightleftarrows}} EA$ $k_4 \big\Vert k_{-4}q \quad k_{-2}\big\Vert k_2 b$ $EQ \underset{k_{-3}p}{\overset{k_3}{\rightleftarrows}} \begin{bmatrix} EAB \\ EPQ \end{bmatrix}$	$E \underset{k_{-1}}{\overset{k_1 a}{\rightleftarrows}} \begin{bmatrix} EA \\ E'P \end{bmatrix}$ $k_4 \big\Vert k_{-4}q \quad k_{-2}p\big\Vert k_2$ $\begin{bmatrix} E'B \\ EQ \end{bmatrix} \underset{k_3 b}{\overset{k_{-3}}{\rightleftarrows}} E'$
V_+	$\dfrac{k_3 k_4 e_0}{k_3 + k_4}$	$\dfrac{k_2 k_4 e_0}{k_2 + k_4}$
V_-	$\dfrac{k_{-1}k_{-2}e_0}{k_{-1} + k_{-2}}$	$\dfrac{k_{-1}k_{-3}e_0}{k_{-1} + k_{-3}}$
K_{mA}	$\dfrac{k_3 k_4}{k_1(k_3 + k_4)}$	$\dfrac{(k_{-1} + k_2)k_4}{k_1(k_2 + k_4)}$
K_{mB}	$\dfrac{(k_{-2} + k_3)k_4}{k_2(k_3 + k_4)}$	$\dfrac{k_2(k_{-3} + k_4)}{(k_2 + k_4)k_3}$
K_{mP}	$\dfrac{k_{-1}(k_{-2} + k_3)}{(k_{-1} + k_{-2})k_{-3}}$	$\dfrac{(k_{-1} + k_2)k_{-3}}{(k_{-1} + k_{-3})k_{-2}}$
K_{mQ}	$\dfrac{k_{-1}k_{-2}}{(k_{-1} + k_{-2})k_{-4}}$	$\dfrac{k_{-1}(k_{-3} + k_4)}{(k_{-1} + k_{-3})k_{-4}}$
K_{iA}	k_{-1}/k_1	k_{-1}/k_1
K_{iB}	$(k_{-1} + k_{-2})/k_2$	k_{-3}/k_3
K_{iP}	$(k_{-1} + k_4)/k_1$	k_2/k_{-2}
K_{iQ}	k_4/k_{-4}	k_4/k_{-4}

requirements (Section 1.3), but one can go further than this by ignoring, for the moment, the fact that all the concentrations, Michaelis constants and inhibition constants have the same dimensions: if a, K_{mA} and K_{iA} have special A dimensions, and b, K_{mB} and K_{iB} have (different) B dimensions, then even by this more restricted definition all the terms in the denominator are dimensionless; whatever reactants appear in the numerator of any term, whether as concentrations or as subscripts, also appear in the denominator.

7.3.2 Random-order ternary-complex mechanism

The equation for the rapid-equilibrium random-order ternary-complex mechanism is as follows:

$$v = \frac{\dfrac{V_+ab}{K_{iA}K_{mB}} - \dfrac{V_-pq}{K_{mP}K_{iQ}}}{1 + \dfrac{a}{K_{iA}} + \dfrac{b}{K_{iB}} + \dfrac{p}{K_{iP}} + \dfrac{q}{K_{iQ}} + \dfrac{ab}{K_{iA}K_{mB}} + \dfrac{pq}{K_{mP}K_{iQ}}} \tag{7.6}$$

It is perhaps surprising that the simpler equation should refer to the more complicated mechanism; the explanation is that equation 7.6, unlike equation 7.5, was derived with the assumption that all steps apart from interconversion of the ternary complexes EAB and EPQ are at equilibrium, an assumption that causes many terms, including all terms in squared concentrations, to vanish from the equation. With this assumption, K_{iA}, K_{iB}, K_{iP} and K_{iQ} are the dissociation constants of EA, EB, EP and EQ respectively; K_{mA} and K_{mB} are the dissociation constants for release of A and B respectively from EAB, and K_{mP} and K_{mQ} are the dissociation constants for release of P and Q respectively from EPQ. (Although K_{mA} and K_{mQ} do not appear explicitly in equation 7.6 they can be introduced because in this mechanism A and B are interchangeable, so $K_{mA}K_{iB}$ is the same as $K_{iA}K_{mB}$, and $K_{iP}K_{mQ}$ is similarly the same as $K_{mP}K_{iQ}$. These substitutions cannot be made in equation 7.5 because the compulsory-order mechanism is not symmetrical in A and B or in P and Q, an additional reason why equation 7.5 is more complicated.) Equation 7.6 may apply within experimental error whether the equilibrium assumption is correct or not, however, and the Michaelis and inhibition constants cannot in general be safely interpreted as true dissociation constants.

7.3.3 Substituted-enzyme mechanism

The steady-state rate equation for the substituted-enzyme mechanism is as follows:

$$v = \frac{\dfrac{V_+ab}{K_{iA}K_{mB}} - \dfrac{V_-pq}{K_{iP}K_{mQ}}}{\dfrac{a}{K_{iA}} + \dfrac{K_{mA}b}{K_{iA}K_{mB}} + \dfrac{p}{K_{iP}} + \dfrac{K_{mP}q}{K_{iP}K_{mQ}} + \dfrac{ab}{K_{iA}K_{mB}} + \dfrac{ap}{K_{iA}K_{iP}} + \dfrac{K_{mA}bq}{K_{iA}K_{mB}K_{iQ}} + \dfrac{pq}{K_{iP}K_{mQ}}} \tag{7.7}$$

where the kinetic parameters are again defined in Table 7.1. In coefficient form this equation is the same as equation 7.5 without the constant 1 and the terms in *abp* and *bpq* in the denominator, but the relationships between the parameters are different, and equation 7.7 has $K_{iP}K_{mQ}$ wherever $K_{mP}K_{iQ}$ might be expected by analogy with equation 7.5. See Section 4.7.1 for an illustration based on this equation.

7.3.4   Calculation of rate constants from kinetic parameters

Although Table 7.1 shows how the kinetic parameters of the two compulsory-order mechanisms can be expressed in terms of rate constants, it does not give the inverse relationships. For the substituted-enzyme mechanism no unique inverse relationship exists, so it is not possible to calculate the rate constants from measurements of kinetic parameters, because there are infinitely many different sets of rate constants capable of generating the same parameters. For the random-order ternary-complex mechanism a unique relationship does exist (Cleland, 1963), and is shown in Table 7.2. It should be used with caution, however, because it assumes that the simplest form of the mechanism applies; in other words it makes no allowance for the possibility that the mechanism may contain more than the minimum number of steps. If any of the binary or ternary complexes isomerize, the form of the steady-state rate equation is unaffected, but the interpretation of the parameters is different and some or all of the expressions in Table 7.2 become invalid.

For the random-order ternary-complex mechanism it is obvious that none of the rate constants apart from those for the rate-limiting interconversion of ternary complexes can be determined from steady-state measurements if the rapid-equilibrium assumption is correct (because as soon as one makes a rapid-equilibrium assumption one foreswears the possibility of getting

Table 7.2. Calculation of rate constants from kinetic parameters. The Table is obtained by rearranging the definitions given in Table 7.1 for the kinetic parameters of the compulsory-order ternary-complex mechanism. Analogous rearrangement of the definitions for the substituted-enzyme mechanism is not possible.

Rate constant	Expression	Rate constant	Expression
k_1	$V_+/K_{mA}e_0$	k_{-1}	$V_+K_{iA}/K_{mA}e_0$
k_2	$\dfrac{V_+(k_{-2}+k_3)}{k_3K_{mB}e_0}$	k_{-2}	$\dfrac{V_+V_-K_{iA}}{e_0(V_+K_{iA}-V_-K_{mA})}$
k_3	$\dfrac{V_+V_-K_{iQ}}{e_0(V_-K_{iQ}-V_+K_{mQ})}$	k_{-3}	$\dfrac{V_-(k_{-2}+k_3)}{k_{-2}K_{mP}e_0}$
k_4	$V_-K_{iQ}/K_{mQ}e_0$	k_{-4}	$V_-/K_{mQ}e_0$

any rate information about the steps assumed to be at equilibrium). However, if the rapid-equilibrium assumption is not made, and there are detectable deviations from equation 7.6, it may be possible to use curve-fitting techniques to deduce information about the rate constants (Cornish-Bowden and Wong, 1978).

7.4 Initial-rate measurements in the absence of products

7.4.1 Meanings of the parameters

If no products are included in the reaction mixture, the equation for the initial rate for a reaction following the compulsory-order ternary-complex mechanism is obtained from equation 7.5 by omitting all terms containing p or q and multiplying all the others by $K_{iA}K_{mB}$:

$$v = \frac{Vab}{K_{iA}K_{mB} + K_{mB}a + K_{mA}b + ab} \tag{7.8}$$

The meanings of the parameters, and their relation to the parameters of the one-substrate Michaelis–Menten equation (Section 2.3.2), become apparent if the equation is examined at extreme values of a and b. If both a and b are so large that any terms that do not contain both of them are negligible, the equation simplifies to $v = V$. Thus V has exactly the meaning it had for the single-substrate reaction: it defines the limiting rate when the enzyme is saturated, as long as one takes "saturated" to mean saturated by both substrates. The catalytic constant $k_{cat} = V/e_0$ likewise has the same meaning as for a single-substrate reaction.

The meanings of the Michaelis constants come from consideration of how the equation simplifies if only one substrate concentration is large whereas the other remains moderate. For example, if b is large enough for terms that do not contain it to be negligible, then equation 7.8 simplifies to the Michaelis–Menten equation in terms of A, with K_{mA} as the Michaelis constant:

$$v = \frac{Va}{K_{mA} + a} \tag{7.9}$$

Thus K_{mA} is defined as the limiting Michaelis constant for A when B is saturating. A parallel argument shows that K_{mB} is the Michaelis constant for B when A is saturating. More generally, for a mechanism with an arbitrary number of substrates, the Michaelis constant for any substrate is defined as the limiting Michaelis constant for that substrate when all other substrates are at saturation. K_{iA} is *not* the same as K_{mA}, and its meaning can be seen by considering the effect on equation 7.8 of making b very small (but not zero), so that although the numerator remains non-zero all terms in b in the denominator are negligible:

$$v = \frac{\left(\dfrac{Vb}{K_{mB}}\right)a}{K_{iA} + a} = \frac{(k_B b)e_0 a}{K_{iA} + a} \tag{7.10}$$

In this equation $k_B = k_{cat}/K_{mB}$ is the *specificity constant* for B, and the specificity constant for A is defined similarly as $k_A = k_{cat}/K_{mA}$. It follows that K_{iA} is the true equilibrium dissociation constant of EA, because when b approaches zero the rate of reaction of B with EA must also approach zero; nothing then prevents establishment of equilibrium in the binding of A to E, so the Michaelis–Menten assumption of equilibrium binding is valid in this instance. K_{iB} does not appear in equation 7.8, because B does not bind to the free enzyme. It does, however, occur in the equation for the complete reversible reaction, equation 7.5, and its magnitude affects the behaviour of B as an inhibitor of the reverse reaction. Although equation 7.8 is not symmetrical in A and B, because $K_{iA}K_{mB}$ is not the same as $K_{mA}K_{iB}$, it is symmetrical in *form*; measurement of initial rates in the absence of products does not therefore distinguish A from B, and does not allow a conclusion as to which substrate binds first.

It is interesting to examine the definitions of the specificity constants in the simplest form of the substituted-enzyme mechanism. It follows from the definitions in Table 7.1 that the specificity constant for A may be expressed in terms of rate constants as follows:

$$k_A = \frac{k_1 k_2}{k_{-1} + k_2} \tag{7.11}$$

Notice that this definition only includes rate constants from the half-reaction that involves A; the rate constants for the steps involving B are absent. The specificity constant for B is likewise independent of the steps that involve A. In principle, therefore, we should expect that if B is replaced by an analogue B′ whose corresponding reaction with A is catalysed by the same enzyme the value of k_A should be unchanged. The appropriate tests have been done for a few enzymes, but when they have been done the results have usually *not* been in accordance with this prediction. Aspartate transaminase (equation 7.3), one of the most thoroughly investigated of the enzymes that follow a substituted-enzyme mechanism, and sometimes regarded as the archetypal example, does behave in the expected way (Katz and Westley, 1979), but several enzymes studied by the same group do not (Jarabak and Westley, 1974; Katz and Westley, 1979, 1980). With other enzymes, variation of k_A with the identity of B has often been taken as evidence that a substituted-enzyme mechanism cannot apply (for example, Morpeth and Massey, 1982).

How can we explain this discrepancy? Does it imply that the relationships in Table 7.1 have been incorrectly derived, or that there is an error in the assumptions made for analysing them? Westley and co-workers argue that the error lies in the implied assumption that exactly the same substituted enzyme, with the same kinetic properties, is produced regardless of the half-reaction that has generated it. They consider that the enzyme retains some conformational "memory" of the reaction it has undergone, sufficiently long-lived to affect its kinetic properties in the next reaction that it undergoes. We do not need to analyse this idea thoroughly here; the important point is that the models of enzyme behaviour given in textbooks commonly embody the simplest possible assumptions, but real enzymes often behave in more complicated ways. This does not mean that simple models are useless, but only that they should be regarded as starting points for analysing enzyme mechanisms and not as complete descriptions of them.

Comparison of k_{cat} values for different substrates of an enzyme that follows a substituted-enzyme mechanism may also provide valuable information. For example, Zerner, Bond and Bender (1964) found that the ethyl, methyl and p-nitrophenyl esters of N-acetyltryptophan all had similar k_{cat} values in chymotrypsin-catalysed hydrolysis, despite having K_m values varying over about a fiftyfold range, whereas the corresponding amide substrate, N-acetyltryptophanamide, had a much smaller k_{cat} value. They interpreted these results (and similar ones for derivatives of phenylalanine) to mean that the constancy of k_{cat} for the esters meant that a step identical for the three substrates, occurring therefore after loss of the alcohols that made them different from one another, was sufficiently slow that it largely accounted for the overall kinetics. This step was presumably the hydrolysis of the acetyltryptophanyl enzyme intermediate (commonly called "deacylation" in studies of chymotrypsin and similar enzymes). The different value for the amides was accounted for by supposing that with amide substrates the initial formation of the acetyltryptophanyl enzyme intermediate ("acylation") was much slower than with ester substrates.

7.4.2 Apparent Michaelis–Menten parameters

If the concentration of one substrate is varied at a constant (but not necessarily very high or very low) concentration of the other, equation 7.8 still has the form of the Michaelis–Menten equation with respect to the varied substrate. For example, if a is varied at constant b, terms that do not contain a are constant, and equation 7.8 can be rearranged into the following form:

$$v = \frac{\left(\dfrac{Vb}{K_{mB} + b}\right)a}{\left(\dfrac{K_{iA}K_{mB} + K_{mA}b}{K_{mB} + b}\right) + a} = \frac{V^{app}a}{K_m^{app} + a} \tag{7.12}$$

The apparent values of the Michaelis–Menten parameters are functions of b:

$$V^{app} = \frac{Vb}{K_{mB} + b} \tag{7.13}$$

$$K_m^{app} = \frac{K_{iA}K_{mB} + K_{mA}b}{K_{mB} + b} \tag{7.14}$$

$$V^{app}/K_m^{app} = \frac{(V/K_{mA})b}{(K_{iA}K_{mB}/K_{mA}) + b} \tag{7.15}$$

Note that the expressions for V^{app} and V^{app}/K_m^{app} are themselves of Michaelis–Menten form with respect to b: this will be used later for constructing secondary plots (Section 7.4.4).

This reduction of equation 7.8 to the Michaelis–Menten equation with apparent parameters is a particular example of a more general kind of behaviour with wide implications for enzymology. It means that even if a reaction actually has two or more substrates it can be treated as a single-substrate reaction if only one substrate concentration is varied at a time. This explains why the analysis of single-substrate reactions continued to be an essential component of all steady-state kinetic analysis even after it was realized that comparatively few enzymes actually have just one substrate.

7.4.3 Primary plots for ternary-complex mechanisms

A typical experiment to characterize a reaction that follows equation 7.8 involves several sub-experiments, each at a different value of b, and each treated as a single-substrate experiment in which apparent Michaelis–Menten parameters are determined by measuring the rate as a function of a. Any of the plots discussed in Section 2.6 can be used for this purpose, the only difference from the single-substrate case being that they yield apparent rather than real parameters, and such a plot is called a *primary plot*, to distinguish it from the secondary plots discussed in Section 7.4.4. In the text I shall refer only to the appearance of plots of a/v against a, but the other plots are also illustrated

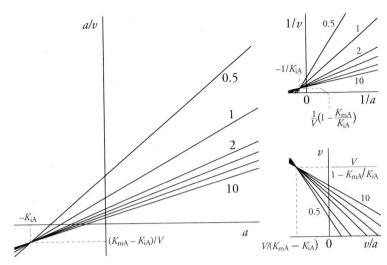

Figure 7.6. Primary plots for ternary-complex mechanisms (ignoring substrate inhibition). The lines are labelled with the values of b/K_{mB} (some values are omitted to avoid crowding the Figure, the complete series being 0.5, 1, 2, 3, 5, 10). The appearance of each of the three commonly used straight-line plots is shown. The primary plots are qualitatively the same (with a and A replaced by b and B throughout) when b is varied at various values of a instead of vice versa.

in the figures, for comparison and for the use of readers who prefer these plots.

Figure 7.6 shows a typical set of primary plots for an enzyme that obeys equation 7.8. The lines intersect at a unique point at which $a = -K_{iA}$ and $a/v = (K_{mA} - K_{iA})/V$, which must occur at the left of the a axis, but may be above or below the a/v axis depending on the relative magnitudes of K_{iA} and K_{mA} (contrast the plot with the substituted-enzyme mechanism, Figure 7.8 in Section 7.4.5). These coordinates give the value of K_{iA} directly, but determination of the other parameters requires additional plots of the apparent parameters, as discussed in Section 7.4.4.

For the rapid-equilibrium random-order ternary-complex mechanism, the complete rate equation (equation 7.6) also simplifies to equation 7.8 if terms in p and q are omitted. Thus the primary plots are the same as for the compulsory-order mechanism, and it is impossible to tell from measurements of the initial rate in the absence of products whether there is a compulsory order of binding of substrates.

7.4.4 Secondary plots

As equations 7.13 and 7.15 have the form of the Michaelis–Menten equation, it follows that plots of V^{app} or V^{app}/K_m^{app} against b describe rectangular hyperbolas through the origin and that they can be analysed by the usual plots, which are now called *secondary plots* or *replots* as they represent further processing of the apparent parameters obtained from the primary plots. For example, the slopes of the primary plots of a/v against a (or the ordinate intercepts of the plots of $1/v$ against $1/a$) provide values of $1/V^{app}$, whose expression can be found by writing equation 7.13 as follows:

$$\frac{1}{V^{app}} = \frac{1}{V} + \frac{K_{mB}}{V} \cdot \frac{1}{b} \qquad (7.16)$$

so a plot of $1/V^{app}$ against $1/b$ is a straight line of slope K_{mB}/V and intercept $1/V$ on the $1/V^{app}$ axis.

A different secondary plot may be made with ordinate variable K_m^{app}/V^{app}, the ordinate intercepts of the plots of a/v against a (or slopes of the plots of $1/v$ against $1/a$), whose expression can be found by writing equation 7.15 as follows:

$$\frac{K_m^{app}}{V^{app}} = \frac{K_{mA}}{V} + \frac{K_{iA}K_{mB}}{V} \cdot \frac{1}{b} \qquad (7.17)$$

Thus a secondary plot of K_m^{app}/V^{app} against $1/b$ is a straight line of slope $K_{iA}K_{mB}/V$ and intercept K_{mA}/V on the K_m^{app}/V^{app} axis. All four parameters of equation 7.8, V, K_{iA}, K_{mA} and K_{mB}, can readily be calculated from these plots, which are illustrated in Figure 7.7.

The equation for K_m^{app}, equation 7.14, also describes a rectangular hyperbola, but the curve does not pass through the origin. Instead K_m^{app} approaches K_{iA} as b approaches zero (and it approaches K_{mA} as b becomes very large, as already discussed). It is thus a three-parameter hyperbola, and cannot be redrawn as a straight line. As in other contexts, K_m^{app} is a less convenient parameter to examine than K_m^{app}/V^{app}.

One can equally well treat B as the variable substrate instead of A, making primary plots of b/v against b at the different values of a; indeed, until it is known which substrate binds first it is arbitrary which is designated A and which B. The analysis is the same, and so there is no need to describe it again.

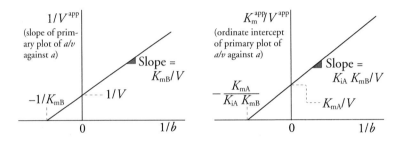

Figure 7.7. Secondary plots for ternary-complex mechanisms.

The only important difference is that K_{iB} does not occur in equation 7.8, and $K_{iA}K_{mB}/K_{mA}$ occurs wherever K_{iB} might be expected from simple interchange of A and B.

7.4.5   Plots for the substituted-enzyme mechanism

For the substituted-enzyme mechanism, the initial rate in the absence of products is as follows:

$$v = \frac{Vab}{K_{mB}a + K_{mA}b + ab} \tag{7.18}$$

The most striking feature of this equation is the absence of a constant from the denominator. (Problem 4.5 at the end of Chapter 4 explores why this should be so.) It causes behaviour recognizably different from that seen with ternary-complex mechanisms when either substrate concentration is varied: for example, if a is varied at a constant value of b, the apparent values of the Michaelis–Menten parameters are as follows:

$$V^{app} = \frac{Vb}{K_{mB} + b} \tag{7.19}$$

$$K_m^{app} = \frac{K_{mA}b}{K_{mB} + b} \tag{7.20}$$

$$V^{app}/K_m^{app} = V/K_{mA} \tag{7.21}$$

Although equation 7.19 is identical to equation 7.13, equation 7.20 is simpler than equation 7.14, and equation 7.21, unlike equation 7.15, shows no dependence on b. What this means is that only V^{app} behaves in the same way as in ternary-complex mechanisms, and the important characteristic is that V^{app}/K_m^{app} is independent of b, with a constant value of V/K_{mA}. It is also constant if b is varied at different values of a; its value is then V/K_{mB}. Primary plots of a/v against a or of b/v against b form series of straight lines intersecting on the a/v or b/v axes, as shown in Figure 7.8. The pattern is easily distinguishable from that given by ternary-complex mechanisms (Figure 7.6) unless K_{iA} is much smaller than K_{mA}.

The secondary plots for the substituted-enzyme mechanism are illustrated in Figure 7.9. The plot of $1/V^{app}$ against $1/b$ has the same slope and intercepts as the corresponding plot for the ternary-complex mechanisms (Figure 7.7). The plot of K_m^{app}/V^{app} against $1/b$ is not needed for parameter estimation, as the only parameter that it provides, K_{mA}/V, is already known from the

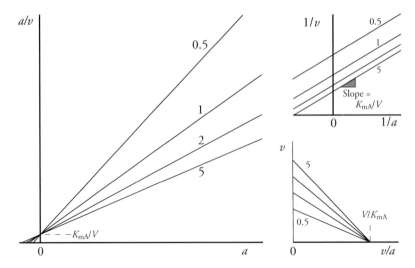

Figure 7.8. Primary plots for substituted-enzyme mechanisms (ignoring substrate inhibition). The lines are labelled with the values of b/K_{mB}. The appearance of each of the three commonly used straight-line plots is shown. The primary plots are qualitatively the same (with a and A replaced by b and B throughout) when b is varied at various values of a instead of vice versa.

primary plots. However, it still has some use for illustrative purposes, to confirm that K_m^{app}/V^{app} is indeed independent of b.

7.5   Substrate inhibition

7.5.1   Why substrate inhibition occurs

The analysis given in Section 7.4 is strictly valid only at low substrate concentrations because, in all reasonable mechanisms, at least one of the four reactants can bind to the wrong form of the enzyme. In the substituted-enzyme mechanism, the substrate and product that lack the transferred group (Y and X respectively in the symbolism used in Section 7.2) can be expected to bind to the wrong form of the free enzyme; in the random-order ternary-complex

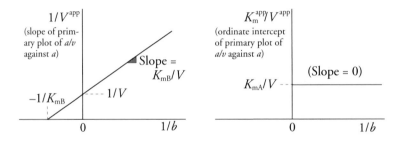

Figure 7.9. Secondary plots for the substituted-enzyme mechanism.

mechanism, the same pair may bind to the wrong binary complexes; and, in the compulsory-order ternary-complex mechanism, either the second substrate or the first product may bind to the wrong binary complex. In this last case, substrate inhibition can occur in either the forward or the reverse reaction, but not in both, because only one of the two binary complexes is available. For convenience, I shall take B as the reactant that displays substrate inhibition for each mechanism, but the results can readily be transposed for other reactants if required.

7.5.2 Compulsory-order ternary-complex mechanism

The non-productive complex EBQ in the compulsory-order ternary-complex mechanism was considered in Section 4.7.3. It can be allowed for in the rate equation by multiplying every denominator term that refers to EQ by $(1 + k_5 b/k_{-5})$, where k_{-5}/k_5 is the dissociation constant of EBQ. Equation 7.8 then becomes:

$$v = \frac{Vab}{K_{iA}K_{mB} + K_{mB}a + K_{mA}b + ab(1 + b/K_{siB})} \tag{7.22}$$

where K_{siB} is a constant that defines the strength of the inhibition. It is *not* the same as the dissociation constant k_{-5}/k_5, because the coefficient of ab is derived not only from EQ but also from the ternary complex (EAB + EPQ), as should be clear from the derivation of equation 7.4 (as equation 4.12) in Section 4.4. According to the relative amounts of these two complexes in the steady state, K_{siB} may approximate to k_{-5}/k_5, or it may be much greater. Thus substrate inhibition in this mechanism is not necessarily detectable at any attainable concentration of B.

Substrate inhibition according to equation 7.22 is effective only at high concentrations of A, so it is potentiated by A and thus resembles uncompetitive inhibition. Primary plots of b/v against b are parabolic, with a common intersection point at $b = -K_{iA}K_{mB}/K_{mA}$. Primary plots of a/v against a are linear, but have no common intersection point. These plots are illustrated in Figure 7.10.

7.5.3 Random-order ternary-complex mechanism

In the random-order ternary-complex mechanism, the concentration of EQ is zero in the absence of added Q if the rapid-equilibrium assumption holds. As B cannot bind to a species that is not present, substrate inhibition does not occur with this mechanism unless Q is added. If the rapid-equilibrium assumption does not hold, there is no reason why substrate inhibition should

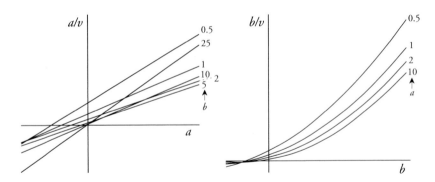

Figure 7.10. Effect of substrate inhibition by B (with $K_{siB} = 10K_{mB}$) on primary plots for ternary-complex mechanisms (compare Figure 7.6).

not occur, but its nature cannot easily be predicted, because the rate equation is too complicated. In this mechanism, EBQ is *not* a dead-end complex, because it can be formed from either EB or EQ, and need not therefore be in equilibrium with either.

7.5.4 Substituted-enzyme mechanism

In the substituted-enzyme mechanism, the non-productive complex EB results from binding of B to E (or binding of Y to E in the symbolism of Section 7.2). It is a dead-end complex, and so it can be allowed for by multiplying terms that refer to E in the denominator of the rate equation by $(1 + b/K_{siB})$, where K_{siB} is the dissociation constant of EB. Equation 7.18 therefore takes the following form:

$$v = \frac{Vab}{K_{mB}a + K_{mA}b(1 + b/K_{siB}) + ab} \tag{7.23}$$

Inhibition according to this equation is most effective when a is small, and thus it resembles competitive inhibition. Primary plots of b/v against b are parabolic and intersect at a common point on the b/v axis, in other words at $b = 0$. Primary plots of a/v against a are linear, with no common intersection point, but every pair of lines intersects at a positive value of a, to the right of the a/v axis. These plots are illustrated in Figure 7.11.

7.5.5 Diagnostic value of substrate inhibition

Substrate inhibition may at first sight seem a tiresome complication in the analysis of kinetic data. Actually, it is usefully informative, because it accentuates the difference in behaviour predicted for ternary-complex and substituted-enzyme mechanisms, and is usually straightforward to interpret.

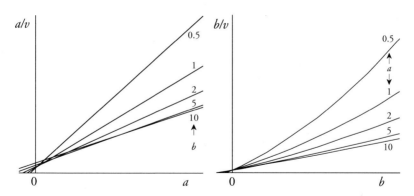

Figure 7.11. Effect of substrate inhibition by B (with $K_{siB} = 10K_{mB}$) on primary plots for substituted-enzyme mechanisms (compare Figure 7.8).

As a substrate normally binds more tightly to the right enzyme species than to the wrong one, substrate inhibition is rarely severe enough to interfere with the analysis described in Section 7.4. Substrate inhibition by one substrate at low concentrations of the other provides strong *positive* evidence that the substituted-enzyme mechanism applies. In contrast, the observation of a V^{app}/K_m^{app} value independent of the concentration of the other substrate (equation 7.21) is only negative evidence for a substituted-enzyme mechanism because it can also be explained as a special case of a ternary-complex mechanism in which the expected variation in V^{app}/K_m^{app} has not been detected.

In a compulsory-order ternary-complex mechanism, substrate inhibition allows the substrate that binds second to be identified without product-inhibition studies.

7.6 Product inhibition

Product-inhibition studies are among the most useful of methods for elucidating the order of binding of substrates and release of products, as they are both informative and simple to understand. Provided that only one product is added to a reaction mixture, the term in the numerator that refers to the reverse reaction must be zero (except in one-product reactions, which are not common). The only effect of adding product, therefore, is to increase the denominator of the rate equation, and thus to inhibit the forward reaction.

The question of whether a particular product acts as a competitive, un-competitive or mixed inhibitor does not have an absolute answer, because it depends on which substrate concentration is considered to be variable. Once this has been decided, however, the answer is straightforward: the denominator of the rate equation can be separated into *variable* and *constant* terms according to whether they contain the variable substrate concentration or not; the expression for V^{app} depends on the variable terms, whereas the expression for $V^{\mathrm{app}}/K_{\mathrm{m}}^{\mathrm{app}}$ depends on the constant terms, as in Section 7.4. As described in Section 5.2 and summarized in Table 5.1, the various kinds of inhibition are classified according to whether they affect $V^{\mathrm{app}}/K_{\mathrm{m}}^{\mathrm{app}}$ (competitive inhibition), V^{app} (uncompetitive inhibition) or both (mixed inhibition). So a product is a competitive inhibitor if its concentration appears only in constant terms, an uncompetitive inhibitor if it appears only in variable terms, and a mixed inhibitor if it appears in both. If the product can combine with only one form of the enzyme, only linear terms in its concentration are possible, and so the inhibition is linear, but more complicated inhibition becomes possible if the product can also bind to "wrong" enzyme forms to give dead-end complexes.

These principles can be illustrated with reference to the compulsory-order ternary-complex mechanism under conditions where P is added to the reaction mixture but Q is not, and A is the variable substrate. The complete rate equation is equation 7.5, but it can be simplified by omitting terms that contain q, and rearranging the denominator D so that terms that contain a are separated from terms that do not. When this is done the denominator may be written as follows:

$$D = 1 + \underbrace{\frac{K_{\mathrm{mA}}b}{K_{\mathrm{iA}}K_{\mathrm{mB}}} + \frac{K_{\mathrm{mQ}}p}{K_{\mathrm{mP}}K_{\mathrm{iQ}}}}_{\text{constant}} + \underbrace{\frac{a}{K_{\mathrm{iA}}}\left(1 + \frac{b}{K_{\mathrm{mB}}} + \frac{K_{\mathrm{mQ}}p}{K_{\mathrm{mP}}K_{\mathrm{iQ}}} + \frac{bp}{K_{\mathrm{mB}}K_{\mathrm{iP}}}\right)}_{\text{variable}} \qquad (7.24)$$

As both the constant and variable parts of this expression contain p it follows that P is a mixed inhibitor when A is the variable substrate. Similar analysis shows that when B is the variable substrate both P and Q behave as mixed inhibitors. However, when one considers inhibition by Q with A as the variable substrate, the results are different, because the denominator of equation 7.5 contains no terms in which a and q are multiplied together, though q does occur in terms that do not contain a:

$$D = 1 + \underbrace{\frac{K_{\mathrm{mA}}b}{K_{\mathrm{iA}}K_{\mathrm{mB}}} + \frac{q}{K_{\mathrm{iQ}}} + \frac{K_{\mathrm{mA}}bq}{K_{\mathrm{iA}}K_{\mathrm{mB}}K_{\mathrm{iQ}}}}_{\text{constant}} + \underbrace{\frac{a}{K_{\mathrm{iA}}}\left(1 + \frac{b}{K_{\mathrm{mB}}}\right)}_{\text{variable}} \qquad (7.25)$$

Thus Q is a competitive inhibitor with respect to A. These results, together with the corresponding ones for the substituted-enzyme mechanism, are

Table 7.3. Product inhibition in the two principal compulsory-order mechanisms. The Table shows the type of inhibition expected for each combination of product and variable substrate. The arrows show the tendency to modify the type of inhibition at saturating concentrations of the constant substrate.

Product	Variable substrate	Type of inhibition	
		Ternary-complex mechanism	Substituted-enzyme mechanism
P	A	Mixed → uncompetitive	Mixed → no inhibition
P	B	Mixed	Competitive
Q	A	Competitive	Competitive
Q	B	Mixed → no inhibition	Mixed → no inhibition

summarized in Table 7.3. The types of inhibition expected for the random-order ternary-complex mechanism are considered in Problem 7.5 at the end of this chapter.

At very high concentrations of the constant substrate, the types of product inhibition become modified, because terms in the rate equation that do not contain the constant substrate concentration become negligible. The constant part of equation 7.24, for example, contains no term in bp, and so becomes independent of p if b is large; the variable part, on the other hand, does contain a term in bp and therefore remains dependent on p when b is large: this implies that when A is the variable substrate product inhibition by P becomes linear uncompetitive when B approaches saturation. The same analysis applied to equation 7.25 shows that the constant term continues to depend on q when b is large; so Q continues to be competitive with respect to A when B approaches saturation. These results, as well as the corresponding ones for other substrate-product combinations, are included in Table 7.3.

It is easy to predict the product-inhibition characteristics of any mechanism. The most reliable method is to study the form of the complete rate equation, but one can usually arrive at the same result by inspecting the mechanism in the light of the method of King and Altman, as described in Section 4.7.1. For any combination of product and variable substrate, one must search for a King–Altman pattern that gives rise to a term containing the product concentration but not the variable substrate concentration; if one is successful the product concentration must appear in the constant part of the denominator and there must be a competitive component in the inhibition. One must then search for a King–Altman pattern that gives rise to a term containing both the product concentration and the variable substrate concentration; if one is successful the product concentration must appear in the variable part of the denominator and there must be an uncompetitive component in the inhibition. With this information it is a simple matter to decide on the type of inhibition. In searching for suitable King–Altman patterns, one should remember that product-release steps are irreversible if the product in question is not present in the reaction mixture.

In two-product reactions, uncompetitive inhibition is largely confined to the case mentioned, inhibition by the first product in a compulsory-order ternary-complex mechanism when the second substrate is saturating. It becomes more common in reactions with three or more products, and occurs with at least one product in all compulsory-order mechanisms for such reactions (Section 7.8).

7.7   Design of experiments

The design of an experiment to study an uninhibited two-substrate reaction rests on principles similar to those for studying linear inhibition (Section 5.8). The values of the Michaelis and inhibition constants for the various substrates will not of course be known in advance, and some trial experiments must be done in ignorance. However, a few experiments in which one substrate concentration is varied at each of two concentrations of the other, one as high and the other as low as practically convenient, should reveal the likely range of apparent K_m values for the first substrate. This range can then be used to select the concentrations of this substrate to be used in a more thorough study. The concentrations of the other substrate can be selected similarly on the basis of a converse trial experiment. At each concentration of constant substrate the variable substrate concentrations should extend from about $0.2 K_m^{app}$ to about $10 K_m^{app}$ or as high as conveniently possible, as in the imaginary inhibition experiment outlined in Table 5.2 (Section 5.8). It is not necessary to have exactly the same set of varied concentrations of one substrate at each constant concentration of the other. It is, however, useful to have sets based loosely on a grid (as in Table 5.2), because this allows the same experiment to be plotted both ways, with each substrate designated "variable" in turn. Note that the labels "variable" and "constant" are experimentally arbitrary, and are convenient only for analysing the results, and especially for defining what we mean by "competitive", "uncompetitive", and so on, in reactions with more than one substrate.

The design of product-inhibition experiments for multiple-substrate experiments requires no special discussion beyond that given in Section 5.8 for linear inhibition studies. It should be sufficient to emphasize that the experiment should be done in such a way as to reveal whether significant competitive and uncompetitive components are present.

7.8 Reactions with three or more substrates

The methods for studying reactions with three or more substrates are a logical extension of those described earlier in this chapter; they do not therefore need as much detailed discussion as two-substrate reactions. Nonetheless, they are not uncommon or unimportant in biochemistry — they include, for example, the important group of reactions catalysed by the aminoacyl-tRNA synthetases — and in this section I shall outline some of the main points, with particular attention to characteristics that are not well exemplified by two-substrate kinetics.

Three-substrate reactions do not necessarily have three products — indeed, reactions with three substrates and two products are common — but to keep the discussion within manageable limits I shall consider only a reaction with three substrates, A, B and C, and three products, P, Q and R. If the mechanism is branched (so that some steps can occur in random order), the complete rate equation contains terms in the squares and possibly higher powers in the reactant concentrations, but if no such higher-order dependence is observed the most general equation for the initial rate in the absence of products is as follows:

$$v = \frac{Vabc}{K_{ABC} + K_{BC}a + K_{AC}b + K_{AB}c + K_{mC}ab + K_{mB}ac + K_{mA}bc + abc} \qquad (7.26)$$

in which V is the limiting rate when all three substrate concentrations are extrapolated to saturation; K_{mA}, K_{mB} and K_{mC} are the Michaelis constants for the three substrates when the other two substrate concentrations are extrapolated to saturation; and K_{ABC}, K_{BC}, K_{AC} and K_{AB} are products of Michaelis and other constants with specific meanings that depend on the particular mechanism considered, but which are analogous to the product $K_{iA}K_{mB}$ that occurs in equation 7.8.

Equation 7.26 applies in full if the reaction proceeds through a quaternary complex EABC that exists in the steady state in equilibrium with the free enzyme E and all possible binary and ternary complexes, EA, EB, EC, EAB, EAC and EBC. In addition to this fully random-order rapid-equilibrium mechanism, a range of other quaternary-complex mechanisms are possible, in which the order of binding is fully or partly compulsory. The extreme case is the fully compulsory-order mechanism, in which there is only one binary complex, say EA, and only one ternary complex, say EAB, possible between E and EABC. Plausible intermediate cases are ones in which there are two binary complexes and one ternary complex, say EA, EB and EAB, or one binary complex and two ternary complexes, say EA, EAB and EAC.

The classification of two-substrate mechanisms into ternary-complex and substituted-enzyme mechanisms also has its parallel for three-substrate mechanisms, but again the range of possibilities is considerably greater. The extreme type of substituted-enzyme mechanism is one in which only binary complexes occur and each substrate-binding step is followed by a product-release step. Alternatively, a three-substrate three-product reaction may combine aspects of both of these kinds of mechanism, so that two substrate molecules may bind to form a ternary complex, with release of the first product before the third substrate binds. For example, in most aminoacyl-tRNA synthetases the amino acid and ATP react in the active site of the enzyme to release pyrophosphate and form a complex composed of enzyme and aminoacyl-AMP; this then reacts with the appropriate tRNA to complete the reaction, releasing aminoacyl-tRNA and AMP from the enzyme. With threonyl-tRNA synthetase, for example, initial isolation of the enzyme–threonyl-AMP complex (J. E. Allende and co-workers, 1964) was followed by kinetic analysis to confirm the initial interpretation (C. C. Allende and co-workers, 1970). Many similar studies have since been done with the other enzymes of the same group.

It will be clear that the number of conceivable mechanisms is extremely large, and even if chemically implausible ones are excluded (something that is not always done) there are still about 18 reasonable three-substrate three-product mechanisms (listed, for example, by Wong and Hanes, 1969), without considering such complications as non-productive complexes and isomerizations. It is thus especially important to take account of chemical plausibility in studying the kinetics of three-substrate reactions. Moreover, provided the rate appears to obey the Michaelis–Menten equation for each substrate considered separately, it is usual practice to use rate equations derived on the assumption that random-order portions of the mechanism are at equilibrium, whereas compulsory-order portions are in a steady state. This, of course, prevents the appearance of higher-order terms in the rate equation, and provides much scope for the use of Cha's method (Section 4.6).

Kinetically, the various mechanisms differ in that they generate equations similar to equation 7.26 with some of the denominator terms missing, as first noted by Frieden (1959). For example, with the mechanism illustrated in Figure 7.12 it is evident from inspection that the constant and the terms in a and b are missing from the denominator of the rate equation (because one cannot find any King–Altman patterns that contain no concentrations, or a only, or b only: compare Section 4.7). Thus instead of equation 7.26 the

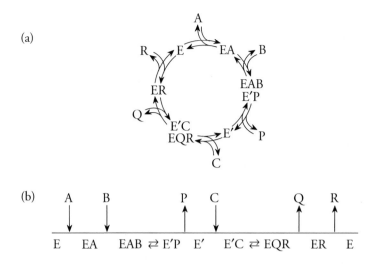

Figure 7.12. Example of a three-substrate three-product mechanism, shown both (a) as an explicit sequence of steps, and (b) in Cleland's symbolism.

rate equation for this mechanism is as follows:

$$v = \frac{Vabc}{K_{ABC} + K_{mC}ab + K_{mB}ac + K_{mA}bc + abc} \tag{7.27}$$

For any one substrate concentration varied at constant values of the other two this equation is of the form of the Michaelis–Menten equation, with apparent constants as listed in Table 7.4. As usual, the behaviour of K_m^{app} is too complicated to be directly useful, but the other parameters are informative. I shall here only discuss the behaviour of V^{app}/K_m^{app}, but it is also instructive to examine the expressions for V^{app} and compare them with those from equations 7.8 and 7.18. With a variable, V^{app}/K_m^{app} increases with b but is independent of c; with b variable, V^{app}/K_m^{app} increases with a but is independent of c; with c variable, V^{app}/K_m^{app} is constant, independent of both a and b. This immediately distinguishes A and B from C, but not from each other.

A and B can be distinguished by considering the effect of adding a single product. Although the rate equation contains no term in a alone, it does contain a term in ap if P is added to the reaction mixture, as one may readily confirm by inspection. Terms in aq or ar cannot, however, be generated by addition of Q or R, and none of the three products alone can generate a term in bp, bq or br. Treating p as a constant, we can, in the terminology of Wong and Hanes (1969), say that addition of P *recalls* the missing term in a to the rate equation. On the other hand, Q and R cannot recall the term in a and none of the three products can recall the term in b. The practical consequence of this is that if P is present in the reaction mixture, V^{app}/K_m^{app} for variable

Table 7.4. *Apparent constants for an example of a three-substrate mechanism. The Table gives expressions for the apparent values of the parameters of the Michaelis–Menten equation for a three-substrate reaction that obeys equation 7.27.*

Variable substrate	V^{app}	V^{app}/K_m^{app}	K_m^{app}
A	$\dfrac{Vbc}{K_{mC}b + K_{mB}c + bc}$	$\dfrac{Vb}{K_{AB} + K_{mA}b}$	$\dfrac{Vbc}{K_{mC}b + K_{mB}c + bc}$
B	$\dfrac{Vac}{K_{mC}a + K_{mA}c + ac}$	$\dfrac{Va}{K_{AB} + K_{mB}a}$	$\dfrac{(K_{AB} + K_{mB}a)c}{K_{mC}a + K_{mA}c + ac}$
C	$\dfrac{Vab}{K_{AB} + K_{mB}a + K_{mA}b + ab}$	$\dfrac{V}{K_{mC}}$	$\dfrac{K_{mC}ab}{K_{AB} + K_{mB}a + K_{mA}b + ab}$

b becomes dependent on c, but V^{app}/K_m^{app} for variable a remains independent of c regardless of which product is present.

Product inhibition in three-substrate three-product reactions obeys principles similar to those outlined in Section 7.6, with the additional feature that uncompetitive inhibition becomes relatively common: it occurs with at least one substrate–product pair in all compulsory-order mechanisms. For the mechanism we have been discussing, for example, Q must be uncompetitive with respect to both A and B, because, in the absence of both P and R, all King–Altman patterns giving a dependence on q also include ab. Similarly, R must be uncompetitive with respect to C.

This brief discussion of some salient points of three-substrate mechanisms, with emphasis on a single example, cannot be more than an introduction to a large subject. For more information, see Wong and Hanes (1969) and Dalziel (1969). Dixon and Webb (1979) discuss the application of isotope exchange (Sections 8.2–5) to three-substrate reactions, though the comments of Dalziel (1969) about possibly misleading results should be noted.

The analysis of four-substrate reactions has been outlined by Elliott and Tipton (1974). It follows principles similar to those for three-substrate reactions.

Problems

7.1 The progressive hydrolysis of the $\alpha(1 \to 4)$ glucosidic bonds of amylose is catalysed both by α-amylase and by β-amylase. With α-amylase the newly formed reducing group has the same α-configuration (before mutarotation) as the corresponding linkage in the polymer, whereas with β-amylase it has the β-configuration. Suggest reasonable mechanisms for the two group-transfer reactions that would account for these observations.

7.2 Petersen and Degn (1978) reported that when laccase from *Rhus vernicifera* catalyses the oxidation of hydroquinone by molecular oxygen, the rate increases indefinitely as the concentrations of both substrates are increased in constant ratio, with no evidence of saturation. They account for these observations in terms of a substituted-enzyme mechanism in which the initial oxidation of enzyme by oxygen occurs in a single step, followed by a second step in which the enzyme is regenerated as a result of reduction of the oxidized enzyme by hydroquinone. Explain why this mechanism accounts for the inability of the substrates to saturate the enzyme.

7.3 Derive an equation for the initial rate in the absence of added products of a reaction obeying a compulsory-order ternary-complex mechanism, with A binding first and B binding second, assuming that both substrate-binding steps are at equilibrium. How does the equation differ in form from the ordinary steady-state equation for this mechanism? What would be the appearance of primary plots of b/v against b?

7.4 The rate of an enzyme-catalysed reaction with two substrates is measured with the two concentrations a and b varied at a constant value of a/b. What would be the expected shape of a plot of a/v against a if the reaction followed (a) a ternary-complex mechanism? (b) a substituted-enzyme mechanism?

7.5 What set of product-inhibition patterns would be expected for an enzyme that obeyed a rapid-equilibrium random-order ternary-complex mechanism?

7.6 Consider a reaction with substrates A and B that follows a substituted-enzyme mechanism. Without deriving a complete rate equation, determine the type of inhibition expected for an inhibitor that binds in a dead-end reaction to the form of the free enzyme that binds B, but has no effect on the other form of the free enzyme. (Assume that no products are present.)

7.7 The symbolism of Dalziel (1957) is sometimes found in the literature. In this system, equation 7.8 would be written as follows:

$$\frac{e_0}{v} = \phi_0 + \frac{\phi_1}{S_1} + \frac{\phi_2}{S_2} + \frac{\phi_{12}}{S_1 S_2}$$

in which e_0 is the total enzyme concentration, v is the initial rate, S_1 and S_2 represent a and b respectively, and ϕ_0, ϕ_1, ϕ_2 and ϕ_{12} are constants, sometimes known as *Dalziel coefficients**. What are the values of these constants in terms of the symbols used in equation 7.8? At what point (expressed in terms of Dalziel coefficients) do the straight lines obtained by plotting S_1/v against S_1 at different values of S_2 intersect?

7.8 Consider a three-substrate three-product reaction $A+B+C \rightleftarrows P+Q+R$ that proceeds by a quaternary-complex mechanism in which the substrates bind and the products are released in the order shown in the equation. The initial rate in the absence of products may be obtained from equation 7.26 by deleting one term. By inspecting a mechanistic scheme (without deriving a complete rate equation), answer the following questions:

(a) Which term of equation 7.26 needs to be deleted?

(b) Which product, if any, can "recall" this term to the rate equation?

(c) Which product behaves as an uncompetitive inhibitor (in the absence of the other two products) regardless of which substrate concentration is varied?

(d) Which product behaves as a competitive inhibitor when A is the variable substrate?

---

*One hears the name Dalziel pronounced in a variety of ways. Dalziel himself pronounced it in the traditional Scottish way as dee-ell, with only slightly more stress on the second syllable than on the first (essentially the same as the way the prefix in DL-lactic acid is pronounced).

Chapter 8

Use of Isotopes for Studying Enzyme Mechanisms

8.1 Isotope exchange and isotope effects

It is little exaggeration to say that the availability of isotopes ranks with that of the spectrophotometer in importance for the development of classical biochemistry. Hardly any of the metabolic pathways currently known could have been elucidated without the use of isotopes. In enzyme kinetics the balance of importance is much more strongly in favour of spectrophotometry than it is in biochemistry as a whole, but there are still some major uses of isotopes that need to be considered. Before discussing these, it is important to emphasize that the two major classes of application are not merely different from one another, but diametrically opposite from one another.

Isotope exchange, and other uses of isotopes as labels, depend on the assumption that isotopic substitution of an atom has no effect on its chemical properties and that any effects on its kinetic properties are small enough to be neglected in the analysis: the isotopic label is used just as a way of identifying molecules that would otherwise be lost in a sea of identical ones. Analysis of isotope effects, on the other hand, assumes that there are measurable differences in kinetic or equilibrium properties of isotopically substituted molecules. The two sets of assumptions cannot, of course, be true simultaneously, but in practice this presents little problem, as one can normally set up conditions where the appropriate assumptions are valid. In fact, the experimental differences between the two sorts of application are considerable, as listed in Table 8.1. Nature has been generous in providing two heavy isotopes of hydrogen, the most important element for studying isotope effects, because these depend on relative differences in atomic mass (Section 8.6), which are at their greatest for the lightest elements. Radioactive isotopes of three of the most important biological elements are also available, but there are none

Table 8.1. Kinetic[1] uses of isotopes.

Condition	Isotopic label	Isotope effects
Location of substitution	Remote from reaction site	Reacting bond: primary isotope effect
		One bond away from reacting bond: secondary isotope effect
		Solvent: solvent isotope effect
Extent of substitution	Trace	100% (ideally[2])
Isotopic property	Radioactive	Different atomic mass
Typical isotopes	^3H, ^{14}C, ^{32}P	^2H, ^3H

[1]Non-kinetic uses of isotopes do not necessarily agree with the generalizations given. For example, heavy isotopes of oxygen (^{17}O and ^{18}O) have been used with success to study the stereochemistry of substitution reactions at phosphorus atoms, but as these isotopes are not radioactive they have to be measured by other techniques, such as nuclear magnetic resonance (Jarvest, Lowe and Potter, 1981) or mass spectrometry (Abbott and co-workers, 1979), which are not sensitive enough to be used with only trace amounts of isotopic label.

[2]Substitution with ^3H is only possible at trace levels. This means that experiments that require saturation of the enzyme with substituted molecules are not possible with ^3H.

for nitrogen or oxygen: although this has serious consequences for the study of metabolism, it is less important for isotope exchange as a kinetic probe of enzyme mechanisms, because with nearly all molecules it is possible to label a suitable carbon or hydrogen atom remote from the site of reaction.

8.2   Principles of isotope exchange

Study of the initial rates of multiple-substrate reactions in both forward and reverse directions, and in the presence and absence of products, will usually eliminate many possible reaction pathways and give a good idea of the gross features of the mechanism, but it will not usually reveal the existence of any minor alternative pathways if these contribute negligibly to the total rate. Further information is therefore required to provide a definitive picture. Even if a clear mechanism does emerge from initial-rate and product-inhibition experiments, it is valuable to be able to confirm its validity independently. The important technique of isotope exchange can often satisfy these requirements. It was introduced to enzyme kinetics by Boyer (1959), though the fundamental point was noted earlier by McKay (1938): even if a chemical reaction is at equilibrium, when its net rate is zero by definition, the unidirectional rates through steps or groups of steps can be measured by means of isotopic tracers.

Use of non-isotopic tracers is older still, dating back to the investigation of fatty acid oxidation by Knoop (1904), but although chemical tracers such as the phenyl group could yield important qualitative information, kinetic use of tracers had to wait until isotopes became available.

Two important assumptions are normally needed for analysing isotope exchange experiments kinetically. They are usually true and are often merely implied, but it is as well to state them clearly to avoid misunderstanding. The first assumption is that a reaction that involves radioactive substrates follows the same mechanism as the normal reaction, with the same rate constants. In other words, isotope effects (Section 8.6) are assumed to be negligible. This assumption is usually true, provided that ^{3}H (tritium) is not used as a radioactive label. Even then, isotope effects are likely to be negligible if the ^{3}H atom is not directly involved in the reaction or in binding the substrate to the enzyme. The second assumption is that the concentrations of all radioactive species are so low that they have no perceptible effect on the concentrations of the unlabelled species. This assumption can usually be made to be true, and it is important, because it allows labelled species to be ignored in calculating the concentrations of unlabelled species, thereby simplifying the analysis considerably.

Isotope exchange is most easily understood in relation to an example, such as the one shown in Figure 8.1, which represents the transfer of a radioactive atom (represented by an asterisk) from A* to P* in the compulsory-order ternary-complex mechanism. As this exchange requires A* to bind to E, it can occur only if there is a significant concentration of E. The exchange reaction must therefore be inhibited by high concentrations of either A or Q, as they compete with A* for E. The effects of B and P are more subtle: on the one hand, the exchange reaction includes the binding of B to EA*, and so a finite concentration of B is required. On the other hand, if B and P are present at very high concentrations, the enzyme will exist largely as a ternary complex, (EAB + EPQ), and so there will be no E for A* to bind to. One would therefore expect high concentrations of B and P to inhibit the exchange, and it is not difficult to show that this expectation is correct. The rates of change

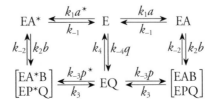

Figure 8.1. Isotope exchange in a compulsory-order ternary-complex mechanism. The Scheme shows the steps needed to transfer label from A* to P*.

of intermediate concentrations can be written in the usual way and set to zero according to the steady-state assumption:

$$\frac{d[EA^*]}{dt} = k_1 a^*[E] - (k_{-1} + k_2 b)[EA^*] + k_{-2}[EA^*B] = 0 \qquad (8.1)$$

$$\frac{d[EA^*B]}{dt} = k_2 b[EA^*] - (k_{-2} + k_3)[EA^*B] + k_{-3} p^*[EQ] = 0 \qquad (8.2)$$

These are a pair of simultaneous equations with $[EA^*]$ and $[EA^*B]$ as unknowns. The solution for $[EA^*B]$, with p^* set to zero, is as follows:

$$[EA^*B] = \frac{k_1 k_2 a^* b[E]}{k_{-1}(k_{-2} + k_3) + k_2 k_3 b} \qquad (8.3)$$

The initial rate of exchange v^* is given by $k_3[EA^*B]$, or

$$v^* = \frac{k_1 k_2 k_3 a^* b[E]}{k_{-1}(k_{-2} + k_3) + k_2 k_3 b} \qquad (8.4)$$

This equation is independent of any assumptions about the effect of the radioactively labelled reactants on the unlabelled reaction, but before it can be used we need an expression for $[E]$ to insert into it. As indicated above, the treatment is greatly simplified by assuming that the concentrations of the unlabelled reactants are unaffected by the presence of trace amounts of labelled species; accordingly the value of $[E]$ is the same as if there were no label. If the experiment is done with the unlabelled reaction in the steady state the expression for $[E]$ derived in Section 4.3 must be used, but this leads to complicated expressions and is fortunately unnecessary. There are two ways of avoiding the complications: either one can study isotope exchange with the unlabelled reaction at equilibrium, as discussed in Section 8.3, or one can consider not the actual rates of exchange but ratios of such rates, discussed in Section 8.5.1. The equilibrium method is by far the better known and more widely used of the two, but both are powerful and useful methods.

8.3 Isotope exchange at equilibrium

If the unlabelled reaction is at equilibrium the concentration of free enzyme in the compulsory-order ternary-complex mechanism is as follows:

$$[E] = \frac{e_0}{1 + \dfrac{k_1 a}{k_{-1}} + \dfrac{k_1 k_2 ab}{k_{-1} k_{-2}} + \dfrac{k_1 k_2 k_3 abp}{k_{-1} k_{-2} k_{-3}}} \qquad (8.5)$$

The four terms in the denominator refer to the four enzyme species E, EA, (EAB + EPQ) and EQ, and each is in the appropriate equilibrium ratio to the preceding one, for example, $[EA]/[E] = k_1 a/k_{-1}$, and so on. The equation does not contain q, because, if equilibrium is to be maintained, only three of the four reactant concentrations can be chosen at will. Any one of a, b and p can be replaced with q by means of the following identity:

$$K_{eq} = \frac{k_1 k_2 k_3 k_4}{k_{-1} k_{-2} k_{-3} k_{-4}} = \frac{pq}{ab} \tag{8.6}$$

For example, if we wanted to examine the effect of increasing b and p in constant ratio at some values of a and q, it would be appropriate to replace p by q using equation 8.6, and if this is done substitution of equation 8.5 into equation 8.4 gives the rate of isotope exchange at chemical equilibrium:

$$v^* = \frac{k_1 k_2 k_3 e_0 a^* b}{\left(1 + \dfrac{k_1 a}{k_{-1}} + \dfrac{k_1 k_2 ab}{k_{-1} k_{-2}} + \dfrac{k_{-4} q}{k_4}\right)[k_{-1}(k_{-2} + k_3) + k_2 k_3 b]} \tag{8.7}$$

As the numerator of this equation is proportional to b whereas the denominator is a quadratic in b it is evident that it has the same form as the equation for linear substrate inhibition, equation 5.41. Thus as b and p are increased from zero to saturation the rate of exchange increases to a maximum and then decreases back to zero.

The equations for any other exchange reaction can be derived similarly. In the compulsory-order ternary-complex mechanism, exchange between B* and P* or Q* is not inhibited by A, because saturating concentrations of A do not remove EA from the system but instead bring its concentration to a maximum. Similar results apply to the reverse reaction (as they logically must for a system at equilibrium, as in such a system the forward rate of any step is the same as the reverse rate): exchange from Q* is inhibited by excess of P, but exchange from P* is not inhibited by excess of Q.

The random-order ternary-complex mechanism differs from the compulsory-order mechanism in that no exchange can be completely inhibited by the substrate that is not involved in the exchange. For example, if B is present in excess, the particular pathway for A* to P* exchange considered above is inhibited because E is removed from the system, but the exchange is not inhibited completely because an alternative pathway is available: at high concentrations of B, A* can enter into exchange reactions by binding to EB to produce EA*B. As radioactive counting can be made very sensitive, it is possible to detect minor pathways by isotope exchange.

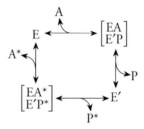

Figure 8.2. Isotope exchange in a substituted-enzyme mechanism. Notice that with this type of mechanism isotope exchange can occur in an incomplete reaction mixture: it is not necessary for the second substrate B or the second product Q to be present.

8.4 Isotope exchange in substituted-enzyme mechanisms

Isotope exchange allows a useful simplification of the substituted-enzyme mechanism, because one can study one half of the reaction at a time (see Figure 8.2). This mechanism has the same form as the complete mechanism, with P* and A* replacing B and Q respectively, but the kinetics are simpler because the rate constants are the same for the two halves of the reaction. This type of exchange represents a major qualitative difference between the substituted-enzyme mechanism and ternary-complex mechanisms, because in ternary-complex mechanisms no exchange can occur unless the system is complete. This method of distinguishing between the two types of mechanism was, in fact, used and discussed (Doudoroff, Barker and Hassid, 1947; Koshland, 1955) well before the introduction of isotope exchange as a kinetic technique.

The possibility of studying only parts of mechanisms in this way is especially valuable with more complicated substituted-enzyme mechanisms with three or more substrates. In such cases, any simplification of the kinetics is obviously to be welcomed, and this approach has been used with some success, for example, by Cedar and Schwartz (1969) in the study of asparagine synthetase.

This capacity of many enzymes to catalyse partial reactions implies that isotope-exchange experiments require more highly purified enzyme than conventional experiments if the results are to be valid. The reason for this requirement is simple. Suppose one is studying aspartate transaminase, which catalyses the following reaction (shown in more detail in equation 7.3):

$$\text{glutamate} + \text{oxaloacetate} \rightleftarrows \text{2-oxoglutarate} + \text{aspartate} \qquad (8.8)$$

Small quantities of contaminating enzymes, for example, other transaminases, are of little importance if one is following the complete reaction, because it is unlikely that any of the contaminants is a catalyst for the complete reaction. Exchange between glutamate and 2-oxoglutarate is another matter, however, especially if one is using a highly sensitive assay, and one must be certain that contaminating transaminases are not present if one wants to get valid information about aspartate transaminase. Although this warning applies especially to enzymes that operate by substituted-enzyme mechanisms, no harm can come of paying extra attention to enzyme purity with other enzymes also, even though these are not expected to catalyse partial reactions; as discussed at the end of Section 7.2.3, there may be many enzymes that react according to substituted-enzyme mechanisms in addition to those that are generally acknowledged to do so.

8.5   Non-equilibrium isotope exchange

8.5.1   Chemiflux ratios

As noted above, the rate equations for isotope exchange become excessively complicated if the unlabelled reaction is not maintained at equilibrium during the exchange. Nonetheless, there is good reason to study enzyme reactions that are not at equilibrium, because at chemical equilibrium it is impossible to disentangle effects of substrate binding from effects of product release, and unless the enzyme is capable of catalysing a half-reaction (Section 8.4) one is forced to consider a whole reaction at a time. Non-equilibrium isotope exchange, by contrast, allows the part of a mechanism responsible for a particular exchange to be isolated for separate examination.

The crucial step in making non-equilibrium isotope exchange a usable technique without becoming enmired in hopeless complications was made by Britton (1966). He realized that measuring and comparing simultaneous exchanges from one reactant to two others allows one to greatly simplify the rate expressions by dividing one by the other. This is because mechanistic features outside the part of the mechanism responsible for the two exchanges affect both rates in the same way and therefore do not affect their ratio.

This idea can be illustrated by reference to a mechanistic fragment capable of converting two substrates into a product P, as follows:

$$
E \underset{v_{-1}}{\overset{A \quad v_1}{\rightleftharpoons}} EA \underset{v_{-2}}{\overset{B \quad v_2}{\rightleftharpoons}} E'P \underset{v_{-3}}{\overset{P \quad v_3}{\rightleftharpoons}} E' \tag{8.9}
$$

This can represent a complete reaction (with E′ identical to E), or it can be a fragment of a larger scheme, of which the rest can be arbitrarily complicated without affecting the analysis, provided that it does not include any alternative way of interconverting A + B with P. Let us suppose now that we have P doubly labelled (P*†) in such a way that we can measure simultaneous conversion into A* and into B†. As we do not need to mention the labels explicitly for deriving the equations they will be omitted: it is sufficient to remember that when we refer to the unidirectional conversion of (say) P into A the existence of a suitable label is implicit. In this context it is conventional to refer to a uni-directional rate of this kind as a *flux* but unfortunately this distinction between rates and fluxes is quite different from (almost opposite to) the way in which the same pair of words is used in metabolic control analysis (Section 12.4). For this reason I shall use the longer term *chemiflux* recommended by the International Union of Pure and Applied Chemistry (1981) for a unidirectional rate, with the symbol $F(X \rightarrow Y)$ to represent the chemiflux from X to Y.

The chemiflux from P to E′P is clearly $v_{-3} = k_{-3}[E′]p$. However, this is not the chemiflux from P to B, and still less is it that from P to A, because not all of the molecules of E′P produced by binding P to E′ continue in the same direction of reaction and release B; some of them return to regenerate E′ and P. Furthermore, some of the EA molecules produced by the release of B from E′P also return to E′ and P (with the uptake of a different molecule of B) instead of releasing A. It is obvious immediately that if the mechanism is as shown in Figure 8.2 the chemiflux from P to B is greater than the chemiflux from P to A, but we can put this on a quantitative basis by considering the probabilities at each point of continuing in the same direction or reversing. For a molecule of E′P produced by binding of P to E′ this probability is $v_{-2}/(v_{-2} + v_3)$, because there are only two possible outcomes, either continuation to EA + B or return to E′ + P, which have rates v_{-2} and v_3 respectively. So the chemiflux from P to B is

$$F(P \rightarrow B) = \frac{F(P \rightarrow EP)F(EP \rightarrow B)}{F(P \rightarrow EP) + F(EP \rightarrow B)} = \frac{v_{-2}v_{-3}}{v_{-2} + v_3} = \frac{k_{-2}k_{-3}[E]p}{k_{-2} + k_3} \qquad (8.10)$$

To calculate the chemiflux from P to A, we first note that $F(P \rightarrow EA)$ is the same as $F(P \rightarrow B)$, just given in equation 8.10. We also need the reverse chemiflux $F(EA \rightarrow P)$, which may be calculated similarly and is as follows:

$$F(EA \rightarrow P) = \frac{v_2v_3}{v_{-2} + v_3} = \frac{k_2k_3[EA]b}{k_{-2} + k_3} \qquad (8.11)$$

Then the chemiflux from P to A is constructed in a manner analogous to that used for equation 8.10: it is the chemiflux from P to EA multiplied by the probability that the molecule of EA, once formed, will release A rather than returning to P:

$$F(P \rightarrow A) = \frac{F(P \rightarrow EA)F(EA \rightarrow A)}{F(EA \rightarrow A) + F(EA \rightarrow P)} = \frac{v_{-1}v_{-2}v_{-3}}{(v_{-2} + v_3)\left(v_{-1} + \dfrac{v_2 v_3}{v_{-2} + v_3}\right)}$$

$$= \frac{k_{-1}k_{-2}k_{-3}[E]p}{(k_{-2} + k_3)\left(k_{-1} + \dfrac{k_2 k_3 b}{k_{-2} + k_3}\right)} \qquad (8.12)$$

At this point one is tempted to despair: if the expression for a chemiflux in a simple three-step mechanistic fragment is as complicated as this, before even the concentration of E has been substituted to produce a usable equation, what hope is there of obtaining manageable equations for mechanisms of real interest? The essential point is that although such equations are indeed hopelessly complicated if each one is examined in isolation, they simplify dramatically when one is divided by another to obtain an expression for a chemiflux ratio.

This will become clear in a moment, but first we must dispose of a second question that is likely to arise in relation to expressions like equation 8.12: would it not be tidier to multiply the numerator and denominator by $k_{-2} + k_3$ and then multiply out all the brackets? This is, however, a temptation that has to be resisted: the potential simplification by this process is actually so slight as to be negligible, and if chemiflux expressions are left in their untidy forms it is easy (with practice) to recognize by inspection the reason for every term in the equation, whereas this logic is obscured by multiplying out. Moreover, equations in their untidy forms are convenient for comparing with one another. Thus, it is evident from inspection that equation 8.12 differs from equation 8.10 only by the factor k_{-1} in the numerator and the second term in parentheses in the denominator. The chemiflux ratio is therefore as follows:

$$\frac{F(P \rightarrow B)}{F(P \rightarrow A)} = 1 + \frac{k_2 k_3 b}{k_{-1}(k_{-2} + k_3)} \qquad (8.13)$$

Noteworthy features of this equation are that a, the concentration of the substrate that binds first in the forward reaction, does not appear in it, and that it expresses a simple straight-line dependence on b, the concentration of the substrate that binds second.

The conclusions embodied in equation 8.13 depend on remarkably few assumptions. The reaction may be proceeding from A + B to P, or from P to A + B, or it may be at equilibrium. The concentrations a, b and p may have any values, either in absolute units or relative to one another. The recycling of E′ to E may be extremely simple (they can even be the same species), or it can be arbitrarily complicated (as long as it does not contain an alternative way of converting A + B into P). This means also, of course, that there is no assumption that the reaction as a whole obeys Michaelis–Menten kinetics or any other simple kinetic equation. If other reactants are involved they may participate in any order and may be intermingled with the steps shown in equation 8.9; for example, instead of the one-step reaction shown the binding of B to EA may be a three-step process in which another substrate C also binds and another product Q is released. Dead-end reactions, whether inside or outside the fragment that converts A + B into P, have no effect on the form of the chemiflux ratio. It follows, therefore, that measurement of chemiflux ratios provides a remarkably powerful method of studying specific questions of binding order in isolation from other mechanistic features.

It is obvious from symmetry considerations that equation 8.13 cannot apply when A and B can bind to the enzyme in random order, as it would not then be possible for the chemiflux ratio to depend on one substrate but not the other. Analysis of the full random-order ternary-complex mechanism shows this to be correct, and that when A and B bind in random order the chemiflux ratio from P shows a hyperbolic dependence on both substrate concentrations: an example of such behaviour may be found in experiments on rabbit muscle phosphofructokinase (Merry and Britton, 1985). More detail about the theory of these experiments may be found elsewhere (Cornish-Bowden, 1989).

Hexokinase D from rat liver provides an example of an enzyme showing the simpler behaviour (Gregoriou, Trayer and Cornish-Bowden, 1981). Although it does not obey Michaelis–Menten kinetics with respect to glucose, and in consequence conventional product-inhibition experiments are difficult to analyse, measurements of the chemifluxes from glucose 6-phosphate to ATP and glucose gave the results shown in Figure 8.3: the results expected from equation 8.13 if glucose is the substrate that binds first.

8.5.2   Isomerase kinetics

One might suppose that the simplest kinds of kinetic experiments would be appropriate with isomerases (including mutases, racemases and epimerases), as these are enzymes that catalyse genuine one-substrate one-product reactions. However, isomerases present a special difficulty that is obvious when it is pointed out but easy to overlook: it is impossible to do conventional product-inhibition experiments with a reaction with only one product, because a reaction cannot be irreversible if the mixture contains both substrate and product simultaneously. Consequently, the negative term in the numerator of the rate expression cannot be neglected, invalidating the crucial assumption

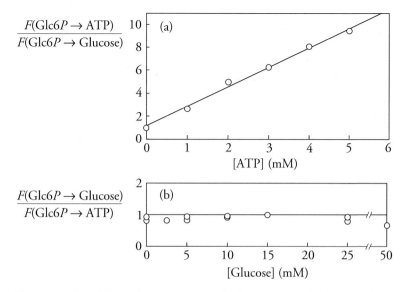

Figure 8.3. Chemiflux-ratio measurements for hexokinase D. The Figure shows data of Gregoriou, Trayer and Cornish-Bowden (1981) for the simultaneous transfer of ^{32}P to ATP and ^{14}C to glucose from glucose 6-phosphate labelled with ^{14}C and ^{32}P. The chemiflux ratio is expected to vary with the concentration of the substrate that binds second in a compulsory-order reaction, but to be independent of the concentration of substrate that binds first, as seen in (a) and (b) respectively.

that allows most kinetic equations to be simplified to the point of allowing conventional graphical and statistical methods.

This limitation is especially serious for isomerases because of a mechanistic feature that is important for them but can more justifiably be neglected for other enzymes. In any mechanism one can postulate that there may be a compulsory enzyme isomerization step, with the form of free enzyme released at the end of the reaction different from the form entering into the first step, so that an enzyme isomerization is needed to complete the cycle. However, in a reaction with multiple substrates and products there is no particular reason to postulate this: if one examines the substituted-enzyme mechanism (equation 7.3), for example, there is no reason to expect the chemical process to be rendered more facile by introducing such a step. By contrast, for an isomerase the advantages of such a step are much clearer. For example, a mutase, an enzyme that catalyses movement of a group from one position to another in the same substrate molecule, can easily be imagined as a protein that exists in two forms, each complementary to one of the two reactants, but which can rapidly switch between its two states. Thus an essential part of a mechanistic study of an isomerase ought to be to determine whether it does operate in this way; Albery and Knowles (1987) go so far as to argue that "no kinetic investigation of an enzyme can be considered complete until experiments have been conducted to determine whether the rate-limiting steps in the saturated region are concerned with substrate handling or with free enzyme interconversion". This view is certainly justified in relation to isomerases, though it may seem exaggerated if applied to other enzymes.

Conventional kinetics cannot answer this question, however. If we apply the method of King and Altman to a mechanism for a one-substrate one-product reaction with enzyme isomerization

$$E + A \xrightleftharpoons[k_{-1}]{k_1} EA \xrightleftharpoons[k_{-2}]{k_2} E' \xrightleftharpoons[k_{-3}]{k_3} E \qquad (8.14)$$

with P produced at the k_2 step,

this is not immediately obvious, because the rate equation contains a term in ap that might appear to be detectable by ordinary techniques:

$$v = \frac{k_1 k_2 k_3 e_0 a - k_{-1} k_{-2} k_{-3} e_0 p}{(k_{-1} + k_2)(k_{-3} + k_3) + k_1(k_2 + k_3)a + (k_{-1} + k_{-3})k_{-2}p + k_1 k_{-2}ap} \qquad (8.15)$$

However, the "ordinary technique" in question is product inhibition, which, as noted above, is not possible under irreversible conditions with a one-product enzyme. Nonetheless, the problem still seems to be accessible to analysis by ordinary techniques if one measures the rate while varying a and p in constant ratio, putting $p = ra$ for example, where r is a constant, because then the rate equation takes a simple form with only one term in the numerator:

$$v = \frac{(k_1 k_2 k_3 - k_{-1} k_{-2} k_{-3}r)e_0 a}{(k_{-1} + k_2)(k_{-3} + k_3) + [k_1(k_2 + k_3) + (k_{-1} + k_{-3})k_{-2}r]a + k_1 k_{-2}ra^2} \qquad (8.16)$$

This equation is of the standard form for substrate inhibition, equation 5.41, and in principle therefore if enzyme isomerization occurs the rate must show a maximum when studied as a function of both reactant concentrations varied in constant ratio. However, whether this maximum is observable within an accessible range of concentrations will depend on the rate constants. One possibility is obvious: if k_{-3} and k_3 are both very large E and E' are not kinetically distinguishable, the mechanism becomes equivalent to the ordinary one-step Michaelis–Menten mechanism and the term in a^2 is undetectable.

Britton (1973), however, pointed out a second possibility that is much less obvious and, indeed, remains surprising and unintuitive even after one has checked the algebra: the term in a^2 also becomes undetectable if k_{-1} and k_2 are both very large, which means that it is undetectable if isomerization of the free enzyme is rate-limiting at high concentrations. Even if it is detectable, Britton (1973, 1994) has shown that any particular set of observed parameters will be consistent with two different sets of rate constants, one giving much greater importance to enzyme isomerization than the other. In summary, therefore, analysis of the sort of kinetics expressed by equations 8.15–16 cannot give an unambiguous indication of the importance of enzyme isomerization.

8.5.3   Tracer perturbation

Tracer perturbation is an isotope-exchange technique that overcomes the sort of interpretational problems discussed in Section 8.5.2. Consider an isomerase reaction that follows the mechanism shown in equation 8.14. In an equilibrium mixture between labelled reactants A* and P*, both E and E′ exist, in such proportions that the chemifluxes from A* to P* and from P* to A* are equal. If a large excess of unlabelled A is added to such an equilibrium mixture, the ensuing reaction from A to P will perturb the equilibrium between E and E′, in such a way that in the limit E disappears from the reaction mixture, leaving no enzyme form capable of reacting with A*, though E′ remains available to react with P*. Thus the only direction possible for the labelled reaction is from P* to A*. It follows, therefore, that if enzyme isomerization occurs as in equation 8.14 swamping the system with unlabelled reactant will perturb the labelled reaction away from equilibrium in the opposite direction from that of the unlabelled reaction. Similar analysis of a two-step Michaelis–Menten mechanism without enzyme isomerization shows no such effect, because A and P react with the same form of free enzyme and there is no equilibrium of enzyme forms to be perturbed. In this case, therefore, addition of unlabelled reactants has no effect on the labelled equilibrium.

With some mechanisms enzyme isomerization can cause perturbation of the labelled reaction in the same direction as the unlabelled reaction. We have tacitly assumed in discussing isomerase reactions that the product P results from rearranging the atoms of the substrate A into a different structure. This does not have to be so, however, and it is not necessarily even likely. In a mutase, for example, E and E′ may be different phosphoenzymes, so that the enzyme–substrate complex is a bisphosphate, with the phospho group that appears in P derived from E and not from A. Transfer of a labelled phospho group from A then requires two catalytic cycles, from A to E′ in the first and from E to P in the second. As the second cycle requires participation of unlabelled A, the whole process can only proceed in the same direction as the unlabelled reaction, albeit very slowly because of the lack of E to react with A*. It follows, therefore, that if a tracer-perturbation experiment is done using ^{32}P as the label there will be a small perturbation in the direction of the unlabelled reaction, whereas if the same system is studied using ^{14}C or ^{3}H as the label there will be a larger perturbation in the opposite direction.

Rabbit-muscle phosphoglucomutase behaved exactly in accordance with these predictions (Britton and Clarke, 1968).

More recently the tracer-perturbation method has formed part of detailed studies of the kinetics of proline racemase (Fisher, Albery and Knowles, 1986) and triose phosphate isomerase (Raines and Knowles, 1987), again with results consistent with kinetically significant enzyme isomerization.

8.6   Theory of kinetic isotope effects

8.6.1   Primary isotope effects

For most purposes one can assume that the different isotopes of an element have the same thermodynamic and kinetic properties. Indeed, the whole use of radioactive isotopes as tracers, on which a large part of the whole edifice of modern biochemistry is built, depends on just this assumption. Nonetheless, it is not precisely correct, and, in reactions where the rate-limiting step involves breaking a bond between hydrogen and a heavier atom, the difference in properties between the isotopes of hydrogen becomes large enough to cause serious errors if they are not taken into account. This has both negative and positive consequences. The negative consequence is that when using tritium (^{3}H) as a radioactive tracer one must be careful to ensure that it is remote from the site of any reaction. In practice this is usually easy to achieve, because "remote" just means separated by at least two atoms; however, it should not be forgotten.

The positive consequence of kinetic isotope effects is that they can be used to obtain mechanistic information that would be difficult to obtain in other ways. Because of this useful aspect, I shall give a brief account of their origin in this section. It will inevitably be oversimplified, but more rigorous accounts are given by Jencks (1969) and Bell (1973).

The energy of a C—H bond as a function of the distance separating the two atoms depends only on the electron clouds surrounding the two atoms and is the same for all isotopes of carbon and hydrogen. However, as the bond vibrates, the particular energies available are quantized, and the specific quantum levels available to a bond depend on the masses of the vibrating atoms, so they are different for different isotopes. This would have no importance if the temperature were high enough for the bonds in a sample of matter to have vibrational energies randomly distributed among several states, but at ordinary temperatures virtually every C—H bond is in its vibrational ground state. This does not correspond to the actual minimum on the potential energy curve, but to the *zero point energy* of the bond, which reflects the persistence of vibration

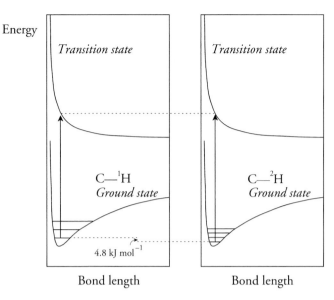

Energy

Transition state   *Transition state*

C—^{1}H
Ground state   C—^{2}H
                          Ground state

4.8 kJ mol^{-1}

Bond length   Bond length

Figure 8.4. Elementary interpretation of primary kinetic isotope effects. As essentially all molecules are in the lowest vibrational energy state at ordinary temperatures, and as this state is lower by about 4.8 kJ mol^{-1} in a C—^{2}H bond than in a C—^{1}H bond, the C—^{2}H bond requires input of about 4.8 kJ mol^{-1} more energy to reach the same transition state.

even at absolute zero. It turns out that the ground state for a C—^{2}H bond is typically about 4.8 kJ mol^{-1} lower (deeper in the potential well) than that of a C—^{1}H bond.

To a first approximation, breaking a C—H bond in the rate-limiting step of a reaction involves passing through a transition state of the molecule in which the bond to be broken is stretched to the point where it has lost one degree of freedom for vibration, but all other bonds are in their ground states. This implies that the transition state is the same for C—^{2}H as for C—^{1}H, but as the ground state is lower for C—^{2}H it requires 4.8 kJ mol^{-1} more enthalpy of activation to reach it, as illustrated in Figure 8.4. Putting this value into equation 1.46, we have

$$\frac{k_{^1H}}{k_{^2H}} = \frac{e^{-\Delta H^{\ddagger}/RT}}{e^{-(\Delta H^{\ddagger}+4.8\,\text{kJ mol}^{-1})/RT}} = e^{4.8\,\text{kJ mol}^{-1}/RT} \tag{8.17}$$

This ratio is about 6.9 at 298 K (25 °C). To the extent that this simple model applies, therefore, we expect bond-breaking reactions to be about 7-fold slower for C—^{2}H than for C—^{1}H bonds.

The type of isotope effect that we have considered is called a *primary isotope effect*, and in the terminology commonly used the calculation leads us to expect that the primary deuterium isotope effect for a C—H bond should be about 7. Similar calculations indicate that the corresponding tritium isotope effect should be about 16.5, and that all other primary isotope effects (for ^{17}O compared with ^{16}O, for example) should be much smaller. These calculations are subject to considerable error, because the full theory of the origin of kinetic isotope effects is much more complicated than that outlined above. However, the deuterium and tritium isotope effects often deviate from the expected values in a consistent way, so that they are related by the following equation (Swain and co-workers, 1958):

$$k_{1H}/k_{3H} = (k_{1H}/k_{2H})^{1.442} \qquad (8.18)$$

This equation is expected to be obeyed if two conditions are fulfilled: first, that the reaction is "kinetically simple", in other words that the isotope effects are due to a single step in the mechanism, and second that hydrogen tunnelling is negligible. This is a quantum mechanical property whereby small particles, most notably electrons but to a much less extent particles as massive as protons, have a finite probability of being found on the opposite side of an energy barrier from where they are expected to be. This property can be quite significant for ^{1}H, smaller for ^{2}H, and much smaller or negligible for ^{3}H and heavier atoms. Tunnelling cause the exponent in equation 8.18 to be larger than 1.442. By contrast, kinetic complications, or effects due to two or more steps in the reaction, cause some averaging out of the effects on the individual steps, resulting in an observed exponent smaller than 1.442. For example, Karsten, Hwang and Cook (1999), who used a different (but equivalent) form of equation 8.18 in which the expected exponent was 3.26 rather than 1.442 (see Problem 8.2 at the end of this chapter), found an observed exponent of 2.2 in a study of a malate dehydrogenase from the worm *Ascaris suum*.

8.6.2   Secondary isotope effects

In discussing the reason for primary isotope effects it was assumed that conversion of a ground state into a transition state affected only the bond that was stretched in the reaction, but this is too simple. In reality the whole molecule is affected, but by amounts that diminish rapidly as one moves away from the reaction site. For bonds adjacent to the one that is broken the vibrational energy levels are a little closer together for different isotopes than they are in the ground state. Thus although the isotopic differences almost cancel, there is a small residual difference in enthalpy of activation between ^{1}H and ^{2}H. This leads to a correspondingly small difference in rate, a *secondary isotope effect*. Such effects are typically of the order of 1.3, so reactions with ^{2}H attached to one of the reacting atoms are typically about 30% slower than the corresponding reactions with ^{1}H.

The smallness of secondary isotope effects makes them more difficult to measure than primary effects, and, probably for that reason, they have been less used in enzymology. They can, however, provide useful mechanistic information not easily available from other kinds of measurement. Many mechanisms proposed both for ordinary chemical reactions and for enzyme-catalysed reactions involve a change of carbon coordination (for example, from tetrahedral to trigonal) when the transition state is formed, and this change can be detected as a secondary isotope effect by isotopic substitution of a hydrogen atom bound to the carbon atom involved in the reaction, even though the C—H bond responsible for the isotope effect is unchanged by the reaction. Sinnott and Souchard (1973), for example, used this type of approach to study β-galactosidase.

8.6.3   Equilibrium isotope effects

Essentially the same arguments apply to equilibrium isotope effects as to secondary isotope effects: the effects of isotopic substitution on energy levels operate in opposite directions on the two sides of the equilibrium, and although they may not exactly cancel the net effect is normally small, and so equilibrium isotope effects are typically close to 1.

8.7   Primary isotope effects in enzyme kinetics

Isotope effects have found increasing use in enzyme studies as a mechanistic probe. The essential theory is given in an article by Northrop (1977), and other articles in the same volume, and a thorough application to triose phosphate isomerase is described by Albery and Knowles (1976).

Although the detailed analysis can become complicated, the basic idea is simple enough to be summarized in an elementary text. It can be presented by reference to the three-step Michaelis–Menten mechanism (equation 2.47) discussed in Section 2.7, for which the definitions of the catalytic and specificity constants were given in equation 2.48 as:

$$k_{\text{cat}} = \frac{k_2 k_3}{k_{-2} + k_2 + k_3} \tag{8.19}$$

$$k_A = \frac{k_1 k_2 k_3}{k_{-1} k_{-2} + k_{-1} k_3 + k_2 k_3} \tag{8.20}$$

with similar expressions for the reverse reaction given in equation 2.49. As steady-state experiments only allow the four Michaelis–Menten parameters to be determined, although there are six rate constants in the mechanism, it might seem impossible to determine the relative contributions of the different rate constants in equations 8.19–20 from such experiments. However, the possibility of making measurements with isotopically labelled reactants permits a useful extension. If we make the reasonable assumption that isotopic substitution in a bond broken in the reaction is likely to generate a primary isotope effect on the chemical step, but little or no isotope effects on the binding steps, we can expect substantial isotope effects on k_2 and k_{-2} but negligible effects on the other rate constants. This implies that the ratios of catalytic and specificity constants for reactants with ^1H and ^2H can be written as follows, where the superscript D (deuterium) refers to ^2H and no superscript is written for ^1H:

$$\frac{k_{cat}}{k_{cat}^D} = \frac{k_2(k_{-2}^D + k_2^D + k_3)}{k_2^D(k_{-2} + k_2 + k_3)} \tag{8.21}$$

$$\frac{k_A}{k_A^D} = \frac{k_2(k_{-1}k_{-2}^D + k_{-1}k_3 + k_2^D k_3)}{k_2^D(k_{-1}k_{-2} + k_{-1}k_3 + k_2 k_3)} \tag{8.22}$$

If it is true that there are no isotope effects on the rate constants for binding, the ratio of isotope effects on k_2 and k_{-2} must be equal to the equilibrium isotope effect for the complete reaction. For simplicity I shall take the equilibrium isotope effect to be exactly unity, but as it can anyway be measured independently the analysis that follows can readily be corrected for equilibrium effects if necessary. Assuming, then, that $k_2/k_2^D = k_{-2}/k_{-2}^D$, equation 8.21 can be rearranged to give an expression for the ratio of $k_3/(k_{-2} + k_2)$ in terms of the measured isotope effect on k_{cat} and the unknown isotope effects on k_2 and k_{-2}:

$$\frac{k_3}{k_{-2} + k_2} = \frac{(k_{cat}/k_{cat}^D) - 1}{(k_2/k_2^D) - (k_{cat}/k_{cat}^D)} \tag{8.23}$$

As k_2/k_2^D is unknown, this result may seem of questionable value. However, even if the exact value is unknown, the theory outlined in Section 8.6.1, together with a large body of experimental information about isotope effects in chemical systems, gives us a good idea of what sort of values are likely. Thus, if we assume a value of 7, it is clear that a measured isotope effect of the order of 7 (or more) on k_{cat} provides strong evidence that the chemical step is slow enough to make an appreciable contribution to the observed kinetics at saturation, whereas a value of the order of 1 indicates that product release is rate-limiting.

Similar arguments can be applied to equation 8.22, but the magnitude of the isotope effect on the specificity constant k_A then provides a measure of the relative importance of the chemical step in comparison not with k_3 alone but with k_{-1} and k_3 together. Although this analysis is somewhat more complicated than for k_{cat} it has the advantage that it can be applied to ^3H as well as to ^2H, whereas ^3H isotope effects on k_{cat} cannot be measured, because an enzyme cannot be saturated with a species that is present only in trace quantities.

If equation 8.18, the relationship between ^2H and ^3H isotope effects, could be trusted, it would provide a way to overcome the difficulty in this analysis that the exact value of k_2/k_2^D is unknown, as comparison of the measured isotope effects for the two isotopes would allow it to be calculated. Unfortunately, however, it is doubtful whether this relationship applies accurately enough to enzyme reactions for such a calculation to offer a reliable advance on the more qualitative analysis suggested above.

8.8 Solvent isotope effects

A major class of isotope effects arises from isotopic substitution of the solvent (Schowen, 1972). The processes responsible for them are the same as those that determine pH behaviour, namely equilibria involving hydrons*, which will be discussed in Chapter 9. They also share several characteristics of temperature studies (Chapter 10), being experimentally easy to do, but difficult to interpret properly unless one can be sure which step of a mechanism is being studied.

Simply measuring the rate of a reaction in ^1H_2O and ^2H_2O reveals almost nothing about the mechanism, but measuring the rate as a function of the mole fraction of ^2H_2O in isotopic mixtures of various compositions can be more informative. Naively, one might expect that the value of any rate constant in such a mixture could be found by linear interpolation between the values in pure ^1H_2O and pure ^2H_2O, but nearly always the dependence is non-linear, and it is the shape of the curve that can reveal mechanistic information.

Consider a mixture of ^1H_2O and ^2H_2O in which the mole fraction of ^2H_2O is n, so that the $[^2H_2O]/[^1H_2O]$ ratio is $n/(1-n)$. For an exchangeable hydron in a reactant molecule AH, the deuteron/proton ratio will also be $[A^2H]/[A^1H] = n/(1-n)$ if the equilibrium constants for protons and

---

*In most of chemistry little confusion results from using the word "proton" both for the ^1H nucleus and for the nucleus of any hydrogen isotope, but this is not satisfactory for discussing hydrogen isotope effects. In contexts where greater precision is needed, the International Union of Pure and Applied Chemistry (1988) recommends the term *hydron* for any hydrogen nucleus without regard to isotope, *proton* for the ^1H nucleus and *deuteron* for the ^2H nucleus. This usage is followed in this section.

deuterons are the same at that position. There will normally be some selection, however, so that the actual ratio is $\phi n/(1-n)$, where ϕ is a *fractionation factor* for that exchangeable position. Such exchange of hydrons can occur not only in the reactant molecule itself but also in the transition state of a reaction in which it participates: in this transition state there is a similar deuteron/proton ratio $\phi^{\ddagger}n/(1-n)$ where $\phi^{\ddagger}$ is now the fractionation factor for the same exchangeable position. The total rate of reaction is now the sum of the rates for the protonated and deuterated species, and so the rate of reaction is proportional to $[1 + \phi^{\ddagger}n/(1-n)]/[1 + \phi n/(1-n)]$, more conveniently written as $(1-n+\phi^{\ddagger}n)/(1-n+\phi n)$. It follows that the observed rate constant k_n at mole fraction n of ^2H_2O may be expressed as a function of n and the ordinary rate constant k_{cat} in pure ^1H_2O as follows:

$$k_n = \frac{k_{cat}(1-n+n\phi^{\ddagger})}{1-n+n\phi} \tag{8.24}$$

To this point we have considered just one hydron, but a typical enzyme molecule contains many exchangeable hydrons, each of which can in principle contribute to the solvent isotope effect. If all hydrons (including the two hydrons in each solvent molecule) exchange independently, so that the final distribution of protons and deuterons may be calculated according to simple statistics, the effects of all the different hydrons are multiplicative and equation 8.24 can be generalized to the *Gross–Butler equation*:

$$k_n = \frac{k_0 \prod(1-n+n\phi_i^{\ddagger})}{\prod(1-n+n\phi_i)} \tag{8.25}$$

in which both products are taken over all exchangeable hydrons in the system.

In view of the large number of exchangeable hydrons and hence the large number of terms in each product in equation 8.25, this equation may seem too complicated to lead to any usable information. However, great simplification results from two considerations. First, equilibrium isotope effects are normally small (Section 8.6.3), because the effects on energy levels largely balance on the two sides of the equilibrium. This means that most or all of the fractionation factors ϕ_i in the denominator of equation 8.25 are close to unity. Second, for hydrons not directly affected in the reaction, these effects cancel in comparing the ground state with the transition state, very largely for secondary isotope effects (Section 8.6.2), and completely for more remote hydrons. This means that many of the fractionation factors in the numerator of equation 8.25 are also close to unity. In the end, therefore, most of the contribution to equation 8.25 comes from a small number of hydrons that are altered in the transition state, and to a good approximation one can often write the equation as follows:

$$k_n = k_0 \prod (1 - n + n\phi_i^{\ddagger}) \tag{8.26}$$

where the product is now taken over those hydrons with $\phi_i^{\ddagger}$ values significantly different from unity (the others can still of course be included, but they do not affect the result). For a one-hydron transition state, this gives a straight-line dependence of k_n on n, for a two-hydron transition state it gives a quadratic dependence, and so on. Thus in simple cases, the shape of the curve obtained provides a way of *counting* the hydrons involved in the rate-limiting step of a reaction. For this reason the type of experiment is often known as a *proton inventory*, or, more accurately, as a *hydron inventory*. A study of a phospholipase from the bacterium *Bacillus cereus* by Martin and Hergenrother (1999) provides an example. These authors first established that hydron transfer was involved in a slow step in the reaction catalysed by this enzyme, and then examined the rates in mixtures of ^1H_2O and ^2H_2O: they found a straightforward linear dependence of the mole fraction of ^2H, and concluded that the results were consistent with transfer of a single hydron in the rate-limiting step.

One should be careful about stretching this theory too far, as several assumptions were needed to arrive at equation 8.25, and further ones to get to equation 8.26. Moreover, the theory was developed in relation to an elementary rate constant of a reaction, but in enzyme applications it often has to be applied to less fundamental quantities. Many other complications can arise in enzyme reactions, so that it is by no means always true that equilibrium isotope effects are negligible or that reactions in ^2H_2O are slower than in ^1H_2O. Solvent isotope effects in hexokinase D (Pollard-Knight and Cornish-Bowden, 1984) offered examples of both exceptions: at low concentrations of glucose the reaction was about 3.5 times faster in ^2H_2O than in ^1H_2O, and as this inverse isotope effect persisted (and indeed increased) at low MgATP concentrations it must have been an equilibrium effect.

Problems

8.1 Sucrose glucosyltransferase catalyses the following reaction:

glucose 6-phosphate + fructose $\rightleftarrows$ sucrose + inorganic phosphate

In the absence of both sucrose and fructose, the enzyme also catalyses rapid ^{32}P exchange between glucose 1-phosphate and labelled inorganic phosphate. This exchange is strongly and competitively inhibited by glucose. The enzyme is a rather poor catalyst for the hydrolysis of glucose 1-phosphate. How may these results be explained?

8.2 The equation $k_{1H}/k_{3H} = (k_{1H}/k_{2H})^{1.442}$ shows how to calculate the expected value of a $^1H/^3H$ isotope effect when the corresponding $^1H/^2H$ isotope effect is known. How can the equation be modified to express the $^1H/^3H$ isotope effect in terms of a measured $^2H/^3H$ isotope effect?

8.3 The method described in Section 8.7 allows information about the relative magnitudes of the rate constants for the binding and chemical steps in a mechanism to be deduced. Chemical substitution in the reactive bond would also be expected to alter the rate constants for the chemical step, so why can it not be used to obtain the same sort of information as one can get from isotopically substituted substrates? Why is the method dependent on isotopic substitution?

Chapter 9

Effect of pH on Enzyme Activity

9.1 Enzymes and pH

Of the many problems that beset the first investigators of enzyme kinetics, none was more important than the lack of understanding of hydrogen-ion concentration, $[H^+]$. In aqueous chemistry, $[H^+]$ varies from about $1 M$ to about $10^{-14} M$, an enormous range that is commonly decreased to more manageable proportions by the use of a logarithmic scale, $pH = -\log[H^+]$. All enzymes are profoundly influenced by pH, and no substantial progress could be made in the understanding of enzymes until Michaelis and his collaborators made pH control a routine part of all serious enzyme studies. The concept of buffers for controlling the hydrogen-ion concentration, and the pH scale for expressing it, were first published by Sørensen (1909), in a classic paper on the importance of hydrogen-ion concentration in enzyme studies. Michaelis, however, was already working on similar lines (see Michaelis, 1958), and it was not long afterwards that the first of his many papers on effects of pH on enzymes appeared (Michaelis and Davidsohn, 1911). Although there are still some disagreements about the proper interpretation of pH effects in enzyme kinetics, the practical importance of pH continues undiminished: it is hopeless to attempt any kinetic studies without adequate control of pH.

It is perhaps surprising that it was left to enzymologists to draw attention to the importance of hydrogen-ion concentration and to introduce the use of buffers. We may reflect, therefore, on the special properties of enzymes that made pH control imperative before any need for it had been felt in the already highly developed science of chemical kinetics. With a few exceptions, such as pepsin and alkaline phosphatase, the enzymes that have been most studied are active only in aqueous solution at pH values in the range 5–9. Indeed, only pepsin has a physiologically important activity outside this middle range of pH. Now, in the pH range 5–9, the hydrogen-ion and hydroxide-ion concentrations are both in the range 10^{-5}–$10^{-9} M$, low enough to be very sensitive to impurities. Whole-cell extracts, and crude enzyme preparations in general, are well buffered by enzyme and other polyelectrolyte impurities, but this natural buffering is lost when an enzyme is purified, and must be replaced with artificial buffers. Until this effect was recognized, little progress in enzyme kinetics was possible. This situation can be contrasted with that in general chemistry: only a minority of reactions are studied in aqueous solution and of these most are studied at either very low or very high pH, with the concentration of either hydrogen or hydroxide ion high enough to be reasonably stable. Consequently, the early development of chemical kinetics was little hampered by the lack of understanding of pH.

The simplest type of pH effect on an enzyme, involving only a single acidic or basic group, is no different from the general case of hyperbolic inhibition and activation that was considered in Section 5.7.3. Conceptually, the protonation of a basic group on an enzyme is simply an example of the binding of a modifier at a specific site and there is therefore no need to repeat the algebra. However, there are several differences between protons and other modifiers that make it useful to examine protons separately. First, virtually all enzymes are affected by protons, so that the proton is far more important than any other modifier. It is far smaller than any other chemical species and has no steric effect; this means that certain phenomena, such as pure non-competitive inhibition (Section 5.2.2), are common with the proton as inhibitor but rare otherwise. The proton concentration can be measured and controlled over a much greater range than that available for any other modifier and therefore one can expect to be able to observe any effects that may exist. Finally, protons normally bind to many sites on an enzyme, so that it is often insufficient to consider binding at one site only.

9.2   Acid–base properties of proteins

Students encounter various definitions of acids and bases during courses of chemistry or biochemistry. Unfortunately it is not always made clear that they are not all equivalent or that the only one that matters for understanding the properties of enzymes, and indeed most other molecules found in living systems, is that of Brønsted (1923): "An acid is a species having a tendency to lose a proton, and a base is a species having a tendency to add on a proton". Apart from its emphasis on the proton, this definition is noteworthy in referring to *species*, which includes ions as well as molecules. Unfortunately, biochemists have conventionally classified the ionizable groups found in proteins according to the properties of the amino acids in the pure uncharged state. Accordingly, aspartate and glutamate, which are largely responsible for the

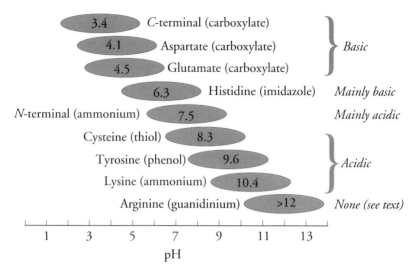

Figure 9.1. Ionizable groups in proteins. For each type of amino acid or terminal residue the type of chemical group responsible for the ionization is shown in parentheses, and the oval region extends for about two pH units around the labelled pK_a value at 25 °C for a group in a "typical" environment in a protein, based on data given by Steinhardt and Reynolds (1969). Individual residues in special environments may be "perturbed", and may have pK_a values substantially different from those shown here. The right-hand column shows the Brønsted character of the group at pH 7. Arginine is largely outside this classification as its typical pK_a value is too far from 7 for it to ionize significantly in the neutral range. The ammonio group is more commonly called an amino group, but this name refers to the uncharged (basic) form, which is not usually the predominant form in proteins.

basic properties of proteins under physiological conditions, are commonly referred to as "acidic". Of the so-called "basic" amino acids, histidine can act either as an acid or as a base under physiological conditions, lysine acts primarily as an acid, and arginine is largely irrelevant to the acid–base properties of proteins, because it does not lose its proton at pH values below 12. On the other hand, two of the so-called "neutral" amino acids, cysteine and tyrosine, do make appreciable contributions to the acid–base properties of proteins. If the object of the standard terminology had been to make discussion of these properties as difficult as possible to understand it could hardly have been more successful. An attempt at a more rational classification is given in Figure 9.1. It has almost nothing in common with that found in most general biochemistry textbooks, but is instead based on the Brønsted definition.

Some of the groups included in Figure 9.1, such as the C-terminal carboxylate and the ε-ammonio group of lysine, have pK_a values so far from 7 that it might seem unlikely that they would contribute to the catalytic properties of enzymes. The values given in the table are average values for "typical" environments, however, and may differ substantially from the pK_a

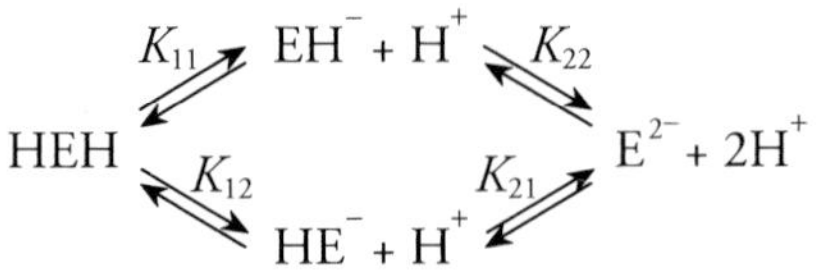

Figure 9.2. Dissociation of a dibasic acid.

values of individual groups in special environments, such as the vicinity of the active site. Such pK_a values are said to be *perturbed*. A clear-cut example is provided by pepsin, which has an isoelectric point of 1.4. As there are four groups that presumably bear positive charges at low pH (one lysine residue, two arginine residues and the N-terminal), there must be at least four groups with pK_a values well below the range expected for carboxylic acids. Although the enzyme contains a phosphorylated serine residue, this can only partly account for the low isoelectric point, and there must be at least three perturbed carboxylic acid groups. A possible explanation is that if two acidic groups are held in close proximity the singly protonated state should be stabilized with respect to the doubly protonated and doubly deprotonated states (see Knowles and co-workers, 1970).

9.3   Ionization of a dibasic acid

9.3.1   Expression in terms of group dissociation constants

The pH behaviour of many enzymes can be interpreted as a first approximation in terms of a simple model due to Michaelis (1926), in which only two ionizable groups are considered. The enzyme may be represented as a dibasic acid, HEH, with two non-identical acidic groups, as shown in Figure 9.2. With the dissociation constants defined as shown, the concentrations of all forms of the enzyme can be represented at equilibrium in terms of the hydrogen-ion concentration, $[H^+]$, which will be written for algebraic convenience simply as h in this chapter:

$$[EH^-] = [HEH]\frac{K_{11}}{h} \tag{9.1}$$

$$[HE^-] = [HEH]\frac{K_{12}}{h} \tag{9.2}$$

$$[E^{2-}] = [HEH]\frac{K_{11}K_{22}}{h^2} = [HEH]\frac{K_{12}K_{21}}{h^2} \tag{9.3}$$

Two points should be noted about these relationships. First, although K_{11} and K_{21} both define the dissociation of a proton from the same group, HE^- is more negative than HEH by one unit of charge, and so it ought to be less acidic, with $K_{11} > K_{21}$, not $K_{11} = K_{21}$; for the same reason we should expect that $K_{12} > K_{22}$. Second, the concentration of E^{2-} must be the same whether it is derived from HEH via EH^- or via HE^-; the two expressions for $[E^{2-}]$ in equation 9.3 must therefore be equivalent, with $K_{11}K_{22} = K_{12}K_{21}$.

If the total enzyme concentration is $e_0 = [HEH] + [EH^-] + [HE^-] + [E^{2-}]$, then

$$[HEH] = \frac{e_0}{1 + \dfrac{K_{11} + K_{12}}{h} + \dfrac{K_{11}K_{22}}{h^2}} \tag{9.4}$$

$$[EH^-] = \frac{e_0 K_{11}/h}{1 + \dfrac{K_{11} + K_{12}}{h} + \dfrac{K_{11}K_{22}}{h^2}} \tag{9.5}$$

$$[HE^-] = \frac{e_0 K_{12}/h}{1 + \dfrac{K_{11} + K_{12}}{h} + \dfrac{K_{11}K_{22}}{h^2}} \tag{9.6}$$

$$[E^{2-}] = \frac{e_0 K_{11}K_{22}/h^2}{1 + \dfrac{K_{11} + K_{12}}{h} + \dfrac{K_{11}K_{22}}{h^2}} \tag{9.7}$$

These expressions show how the concentrations of the four species vary with h, and, by extension with pH, and a typical set of curves is shown in Figure 9.3, with arbitrary values assumed for the dissociation constants.

9.3.2 Molecular dissociation constants

In a real experiment, the curves can never be defined as precisely as those in Figure 9.3, because the four dissociation constants cannot be evaluated. The reason for this can be seen by recognizing that $[EH^-]/[HE^-] = K_{11}/K_{12}$ is a constant independent of h. Thus no amount of variation in h will produce a change in $[EH^-]$ that is not accompanied by an exactly proportional change in $[HE^-]$. Consequently, it is impossible to know how much of any given property is contributed by EH^- and how much by HE^-, and for practical purposes we must therefore treat EH^- and HE^- as a single species, with

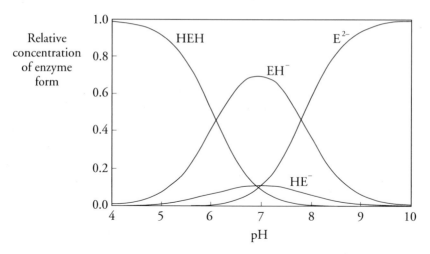

Figure 9.3. Relative concentrations of enzyme forms as a function of pH. The curves are calculated for an enzyme HEH with two ionizable groups, with the following group dissociation constants: $pK_{11} = 6.1$, $pK_{12} = 6.9$, $pK_{21} = 7.0$, $pK_{22} = 7.8$.

concentration given by

$$[EH^-] + [HE^-] = \frac{e_0(K_{11} + K_{12})/h}{1 + \dfrac{K_{11} + K_{12}}{h} + \dfrac{K_{11}K_{22}}{h^2}} = \frac{e_0}{\dfrac{h}{K_{11} + K_{12}} + 1 + \dfrac{K_{11}K_{22}}{(K_{11} + K_{12})h}}$$

(9.8)

which may be written as follows:

$$[EH^-] + [HE^-] = \frac{e_0}{\dfrac{h}{K_1} + 1 + \dfrac{K_2}{h}}$$

(9.9)

if new constants K_1 and K_2 are defined as follows:

$$K_1 = K_{11} + K_{12} = \frac{([EH^-] + [HE^-])h}{[HEH]}$$

(9.10)

$$K_2 = \frac{K_{11}K_{22}}{K_{11} + K_{12}} = \frac{[E^{2-}]h}{[EH^-] + [HE^-]}$$

(9.11)

These constants are called *molecular dissociation constants*, to distinguish them from K_{11}, K_{12}, K_{21} and K_{22}, which are *group dissociation constants*. They have the practical advantage that they can be measured, whereas the conceptually preferable group dissociation constants cannot, because it is impossible to evaluate K_{11}/K_{12}.

The expressions for [HEH] and $[E^{2-}]$ can also be written in terms of molecular dissociation constants, as follows:

$$[\text{HEH}] = \frac{e_0}{1 + \dfrac{K_1}{h} + \dfrac{K_1 K_2}{h^2}} \tag{9.12}$$

$$[E^{2-}] = \frac{e_0}{\dfrac{h^2}{K_1 K_2} + \dfrac{h}{K_2} + 1} \tag{9.13}$$

Functions of the sort expressed by equations 9.9 and 9.12–13 were thoroughly discussed by Michaelis (1926), and are often called *Michaelis functions*.

9.3.3 Bell-shaped curves

We shall now examine equation 9.9 in more detail, because many enzymes display the bell-shaped pH-activity profile characteristic of this equation. The curves for EH^- and HE^- in Figure 9.3 are of this form, and a representative set of bell-shaped curves for different values of $(pK_2 - pK_1)$ is given in Figure 9.4. Notice that the shapes of the curves are not all the same: the maximum becomes noticeably flat as $(pK_2 - pK_1)$ increases, whereas the profile approaches an

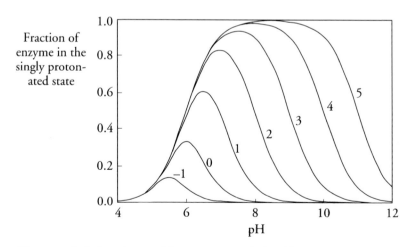

Figure 9.4. Bell-shaped curves. The curves were calculated from equation 9.9, with $pK_1 = 6.0$ and $pK_2 = 5.0$–11.0. Each curve is labelled with the value of $pK_2 - pK_1$, the quantity that determines its shape. (Notice that the plateau around the maximum becomes flatter as this difference increases.)

Table 9.1. Relationship between the width at half height and the pK difference for bell-shaped pH profiles. A convenient method for calculating the pK difference from the width at half height is to define the width at half height as $2 \log q$ and then calculate $pK_2 - pK_1$ as $2 \log(q - 4 + 1/q)$, as suggested by H. B. F. Dixon (1979).

Width at half height	$pK_2 - pK_1$	Width at half height	$pK_2 - pK_1$	Width at half height	$pK_2 - pK_1$
1.14[1]	$-\infty$	2.1	1.73	3.1	3.00
1.2	−1.27	2.2	1.88	3.2	3.11
1.3	−0.32	2.3	2.02	3.3	3.22
1.4	0.17	2.4	2.15	3.4	3.33
1.5	0.51	2.5	2.28	3.5	3.44
1.6	0.78	2.6	2.41	3.6	3.54
1.7	1.02	2.7	2.53	3.7	3.65
1.8	1.22	2.8	2.65	3.8	3.76
1.9	1.39	2.9	2.77	3.9	3.86
2.0	1.57	3.0	2.88	4.0[2]	3.96

[1]The width at half height of a curve defined by equation 9.9 cannot be less than 1.14.

[2]When the width at half height is greater than 4 the pK difference does not differ from the width by more than 1%.

inverted V as it becomes small or negative. The maximum is noticeably less than 1 unless $(pK_2 - pK_1)$ is greater than about 3. Consequently the values of the pH at which $[EH^-] + [HE^-]$ has half its maximum value are not equal to pK_1 and pK_2. However, the mean of these two pH values *is* equal to $(pK_1 + pK_2)/2$, and is also the pH at which the maximum occurs.

The relationship between the width at half height of the curve and $(pK_2 - pK_1)$ is shown in Table 9.1. This table allows measurement of the pH values where the ordinate is half-maximal to be converted into molecular pK values. Nonetheless, even if pK_1 and pK_2 are correctly estimated, the values of the group dissociation constants remain unknown, unless plausibility arguments are invoked, with untestable assumptions (H. B. F. Dixon, 1976).

Although Table 9.1 allows for negative values of $(pK_2 - pK_1)$, these are only possible if deprotonation is cooperative, in other words if loss of a proton from one group increases the tendency to lose it from the other. This is not impossible, but it requires strong compensatory interactions to overcome the electrostatic effects, and in general is unlikely. The smallest value that $(pK_2 - pK_1)$ can have without cooperativity is +0.6, corresponding to a width of 1.53 at the half height. In practice therefore we should expect bell-shaped curves any sharper than this to be unusual.

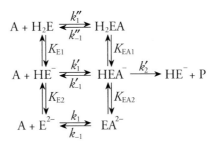

Figure 9.5. Enzyme with two dissociable groups. Only the singly protonated enzyme–substrate complex HEA^- is assumed to be catalytically active, but the substrate is able to bind to any ionization state of the enzyme.

9.4   Effect of pH on enzyme kinetic constants

9.4.1   Underlying assumptions

The bell-shaped activity curves that are often observed for the enzyme kinetic constants V and V/K_m can be accounted for by a simple extension of the theory of the ionization of a dibasic acid. (The treatment of K_m is more complicated, as we shall see.) The basic mechanism is as shown in Figure 9.5. The free enzyme is again treated as a dibasic acid, H_2E, with two molecular dissociation constants, K_{E1} and K_{E2}, and the enzyme–substrate complex H_2E dissociates similarly, with molecular dissociation constants K_{EA1} and K_{EA2}. Only the singly ionized complex HEA^- is able to react to give products, though substrate binding can occur in any ionic state. As we shall now be working solely with molecular dissociation constants, we no longer need to distinguish between the two possible forms of singly deprotonated enzyme: HE^- now refers to both of the species that were represented in Figure 9.2 and the previous equations as two species HE^- and EH^-, and the enzyme–substrate complex HEA^- is to be understood similarly.

Before proceeding further, I emphasize that Figure 9.5 incorporates some assumptions that may be oversimplifications. The protonation steps are represented by equilibrium constants, not by pairs of rate constants, with the implication that they are equilibria, equivalent to assuming that they are fast compared with the other steps. This may seem to be a reasonable assumption, in view of the simple nature of the reaction, and will often be true, but it is certainly not always true, especially if protonation causes a compulsory change of conformation. In simple treatments the question is often evaded by omitting the steps for binding of substrate to H_2E and E^{2-}: the protonation steps then become dead-end reactions, and hence necessarily equilibria according to the arguments of Section 4.7.3. However, there is usually no basis for supposing that substrate cannot bind directly to the different ionic states of the enzyme, and so the assumption is made more for convenience than because it is likely to be true. Omitting the first and third substrate-binding steps is thus just a way of evading a legitimate question, not a way of answering it.

In addition, Figure 9.5 implies that the catalytic reaction involves only two steps, as in the simplest Michaelis–Menten mechanism. If several steps are postulated, with each intermediate capable of protonation and deprotonation, the form of the final rate equation is not changed, but the interpretation of experimental results becomes more complicated as each experimental dissociation constant is the mean of the values for the different intermediates, weighted in favour of the predominant ones. (Compare the effect of introducing an extra step into the simple Michaelis–Menten mechanism, Section 2.7.1.) Lastly, it may not always be true that only HEA^- can undergo the chemical reaction to give products, but this is likely to be a reasonable assumption for many enzymes because most enzyme activities do approach zero at high and low pH values. If H_2EA could also react, with, for example, 10% of the activity of HEA^-, then we should expect the enzyme to have finite activity at low pH, more than 10% of the activity at the maximum, in fact, because, as should be evident from the discussion in Section 9.3.3, the activity at the maximum is less than the value for an enzyme completely in the most active ionic state. This sort of behaviour is not unknown, but it is sufficiently rare to make it reasonable to take it as a working hypothesis that only the singly protonated state is reactive.

9.4.2   pH dependence of V and V/Km

Recognizing that Figure 9.5 may be an optimistic representation of reality, let us consider the rate equation that it predicts. If there were no ionizations, and HE^- and HEA^- were the only forms of the enzyme, then the mechanism would be the ordinary Michaelis–Menten mechanism, with a rate given by the following equation:

$$v = \frac{k'_2 e_0 a}{\dfrac{k'_{-1} + k'_2}{k'_1} + a} = \frac{\tilde{V} a}{\tilde{K}_m + a} \tag{9.14}$$

in which $\tilde{V} = k'_2 e_0$ and $\tilde{K}_m = (k'_{-1} + k'_2)/k'_2$ are the *pH-independent parameters*. In reality, however, the free enzyme does not exist solely as HE^-, nor the enzyme–substrate complex solely as HEA^-. The full rate equation is of exactly the same form as equation 9.14:

$$v = \frac{V a}{K_m + a} \tag{9.15}$$

but the parameters V and K_m are not equal to $\widetilde{V}$ and $\widetilde{K}_m$; instead they are functions of h, and the expressions for V and V/K_m are of the same form as equation 9.9:

$$V = \frac{\widetilde{V}}{\dfrac{h}{K_{EA1}} + 1 + \dfrac{K_{EA2}}{h}} \tag{9.16}$$

$$\frac{V}{K_m} = \frac{\widetilde{V}/\widetilde{K}_m}{\dfrac{h}{K_{E1}} + 1 + \dfrac{K_{E2}}{h}} \tag{9.17}$$

Notice that the pH behaviour of V reflects the ionization of the enzyme–substrate complex, whereas that of V/K_m reflects that of the free enzyme (or the free substrate, see Section 9.5). With either parameter the pH dependence follows a symmetrical bell-shaped curve of the type discussed in the previous section.

9.4.3 pH-independent parameters and their relationship to "apparent" parameters

The relationship between true and pH-independent parameters just introduced exactly parallels that between true and expected values considered in relation to non-productive binding (Section 5.9.1); however, it also parallels the distinction made between apparent and true values (respectively!) in many other contexts. This may seem to be inconsistent: why do we say that the "true" values of the Michaelis–Menten parameters in an inhibition experiment are those that apply in the absence of inhibitor, whereas in a pH-dependence study they are the values that apply at the particular pH considered?

It is quite possible to insist on a greater degree of consistency, but this creates more problems than it solves, and as in other areas of chemistry it is more profitable to sacrifice strict consistency in favour of convenience: in thermodynamics, for example, we define the standard state of most species in solution as a concentration of $1\,M$, but we make an exception of the solvent, taking its standard state as the concentration that exists; in biochemistry, but not chemistry, we extend this to the proton, recognizing that equilibrium constants and standard Gibbs energy changes make much more biochemical sense if they refer to a standard state at pH 7 (for a fuller discussion of this point, see Alberty and Cornish-Bowden, 1993).

Applying this idea to the pH-dependence of Michaelis–Menten parameters, there is an important difference between the proton and other inhibitors. For most inhibitors, it is perfectly feasible to study the reaction in the absence of the inhibitor, and perfectly reasonable, therefore, to define the true parameters as those that apply in that case. It is not possible, however, to observe enzyme-catalysed reactions in the absence of protons; moreover, because protons usually activate enzymes as well as inhibit them (whereas other enzyme effectors more often inhibit than activate) one cannot even determine the proton-free behaviour by extrapolation. Ultimately all distinctions between constants and variables in biochemistry are matters of convenience rather than laws of nature, and apart from a few fundamental physical constants this is true of most "constants" in science: even in physical chemistry the temperature, for example, is not part of the definition of a standard state, and so equilibrium "constants" vary with the temperature.

9.4.4   pH dependence of K_m

The variation of K_m with pH is more complicated than that of the other parameters, as it depends on all four pK values:

$$K_m = \frac{\tilde{K}_m \left(\dfrac{h}{K_{E1}} + 1 + \dfrac{K_{E2}}{h} \right)}{\dfrac{h}{K_{EA1}} + 1 + \dfrac{K_{EA2}}{h}} \tag{9.18}$$

Nonetheless, it is possible in principle to obtain all four pK values by plotting $\log K_m$ against pH and applying a theory developed by M. Dixon (1953a). To understand this approach it is best to regard K_m as V divided by V/K_m, or to regard $\log K_m$ as $\log V - \log V/K_m$. Then an examination of how $\log V$ and $\log V/K_m$ vary with pH leads naturally to an understanding of the pH dependence of K_m.

It is obvious from inspection that at high h (low pH) equation 9.16 simplifies to $V = \tilde{V} K_{EA1}/h$ and that at low h (high pH) it simplifies to $V = \tilde{V}h/K_{EA2}$. If K_{EA1} and K_{EA2} are well separated there is also an intermediate region in which $V \approx \tilde{V}$. It follows that the plot of $\log V$ against pH should take the form shown at the top of Figure 9.6, approximating to three straight-line segments intersecting at $pH = pK_{EA1}$ and $pH = pK_{EA2}$. The behaviour of V/K_m, shown in the middle part of Figure 9.6, is similar, except that the intersections occur at $pH = pK_{E1}$ and $pH = pK_{E2}$. Despite the complicated appearance of equation 9.18, therefore, the form of the plot of $\log K_m$ against pH follows simply by subtraction, and must be as shown at the bottom of the Figure.

The plot approximates to a series of straight-line segments, each with slope +1, 0 or −1 (slopes of +2 or −2 are also possible, though they do not occur in Figure 9.6). As one reads across this segmented version of the plot from left to right, each increase in slope corresponds to a pK_a on the free enzyme,

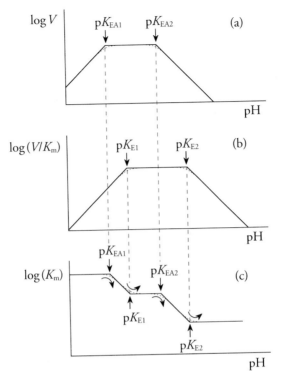

Figure 9.6. Interpretation of pH profiles according to the theory of M. Dixon (1953a). Although plots of the logarithm of any kinetic parameter always give smooth curves, as shown by the dotted lines, they can usefully be interpreted as if they consisted of straight-line segments. (a) Changes of slope in the plot of $\log V$ against pH reflect ionizations of the enzyme–substrate complex; (b) changes of slope in the plot of $\log(V/K_m)$ against pH reflect ionizations of the free enzyme; (c) in principle all ionizations affect K_m, in a way most easily rationalized by regarding $\log(K_m)$ as $\log V - \log(V/K_m)$. Notice that in the plot for $\log(K_m)$ each clockwise turn in the line (reading from left to right) corresponds to an ionization of the enzyme–substrate complex, whereas each anticlockwise turn corresponds to an ionization of the free enzyme.

either pK_{E1} or pK_{E2}, and each decrease in slope corresponds to a pK on the enzyme–substrate complex, either pK_{EA1} or pK_{EA2}.

The plots shown in Figure 9.6 are idealized, in the sense that the pK values are well separated (by at least 1 pH unit in each example) and the data span a wide enough range of pH for all four changes of slope to be easily seen. In a real experiment it would be unusual to have accurate data over a wide enough range to estimate all four pK values. However, the interpretation of changes in slope is the same even if only part of the plot is available.

9.4.5 Experimental design

To have any hope of yielding meaningful mechanistic information, pH-dependence curves should refer to parameters of the Michaelis–Menten equation or another equation that describes the behaviour at each pH. (Better still, they should refer to individual identified steps in the mechanism, but this is not often realizable.) In other words, a *series* of initial rates should be measured at *each* pH value, so that V and V/K_m can be determined at each pH value. The pH dependence of v is of little value by itself because competing effects on V and V/K_m can make any pK_a values supposedly measured highly misleading. In this respect measurements of the effects of pH should follow the same principles as measurements of the effects of changes in other environmental influences on enzymes, such as temperature, ionic strength, concentrations of inhibitors and activators, and so on. These last variables should of course be properly controlled, as in any kinetic experiment, but the ionic strength deserves special mention because the use of several different buffer systems necessary to span a broad range of pH may make it difficult to maintain a constant ionic strength: if this cannot be maintained constant then separate experiments need to be done to check that variations in ionic strength cannot explain any effects attributed to variations in pH.

As just noted, a single buffer system cannot be used to vary the pH over a wide range. In particular, a buffer based on a single ionizing group should not be used at pH values more than one unit from the pK_a of that group, which imposes a maximum range of two pH units for the range accessible with any such buffer. Thus two or more buffers are usually needed to characterize an enzyme, and the ranges used for the different ones should overlap sufficiently for effects due to the identity of the buffer (rather than to the pH itself) to be obvious. More detail on the use of buffers may be found in articles such as that of Price and Stevens (2002) or the book by Beynon and Easterby (1996).

Preliminary characterization of the pH behaviour of an enzyme is sometimes done in a way that is even less meaningful than measuring v as a function of pH, by measuring extent of reaction after a fixed time as a function of pH. It is becoming rare to find data published in this form, but in the older literature some potentially interesting information is rendered virtually useless by this type of experimental design (for example Inouye and co-workers, 1966). At best, an extent of reaction may give an indication of the initial rate, but it is likely to be complicated by variations in the degree of curvature (from whatever cause) in the different experiments.

9.5   Ionization of the substrate

Many substrates ionize in the pH range used in kinetic experiments. If substrate ionization is possible one ought to consider whether observed pK values

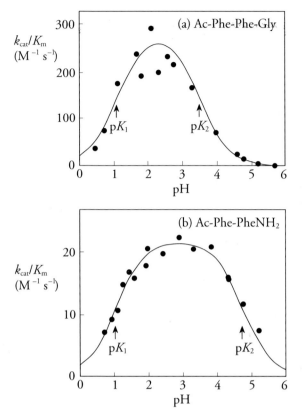

Figure 9.7. pH dependence of k_{cat}/K_m for the pepsin-catalysed hydrolysis of (a) acetyl-L-phenylalanyl-L-phenylalanylglycine, and (b) acetyl-L-phenylalanyl-L-phenylalaninamide (Cornish-Bowden and Knowles, 1969).

refer to the enzyme or to the substrate. The theory is similar to that for enzyme ionization and the results given above require only slight modification. The pH dependence of V, and decreases in slope in plots of $\log K_m$ against pH, still refer to the enzyme–substrate complex; but the pH dependence of V/K_m, and increases in slope in plots of $\log K_m$ against pH, may refer *either* to the free enzyme *or* to the free substrate. One may sometimes decide which interpretation is correct by studying another substrate that does not ionize. For example (Figure 9.7a), the pH dependence of k_{cat}/K_m for the pepsin-catalysed hydrolysis of acetyl-L-phenylalanyl-L-phenylalanylglycine shows pK values of 1.1 and 3.5, of which the latter may well be due to ionization of the substrate. That this interpretation is correct is confirmed by consideration of a substrate that does not ionize, acetyl-L-phenylalanyl-L-phenylalaninamide: this has essentially the same value for pK_1, 1.05, but a higher value of pK_2, 4.75 (Figure 9.7b), which presumably refers to an ionization of the enzyme.

9.6   More complicated pH effects

One of the main reasons for doing pH-dependence studies is to measure pK values and deduce from them the chemical nature of the groups on the enzyme that participate in catalysis. Although this is widely done, it demands more caution than it sometimes receives, because simple treatments of pH effects make several assumptions that may not always be valid.

It is not only the quantitative assumptions that are suspect; the qualitative interpretation of a pH profile can also be misleading. For example, although a bell-shaped pH curve may indicate a requirement for two groups to exist in particular ionic states, as discussed in Sections 9.4–5, this is not the only possibility: in some circumstances a single group that is required in different states for two steps of the reaction may give similar behaviour (H. B. F. Dixon, 1973; Cornish-Bowden, 1976a). This is an example of a change of rate-limiting step with pH (Jencks, 1969), and as it can lead to the mistake of assigning a pK value to a group that does not in fact exist in the enzyme or its substrate this type of pK is sometimes called a *mirage* pK (Brocklehurst, 1994).

Other complications are that a single protonation or deprotonation of the fully active state may lead to only partial loss of activity, and loss of activity may require more than one protonation or deprotonation.

For a fuller discussion of these and other more complicated pH effects, see Tipton and H. B. F. Dixon (1979) or Brocklehurst (1996). An article by Knowles (1976) is particularly valuable as a guide to some of the mistakes that can result from lack of taking sufficient care in the analysis of pH studies, and some of his points are brought up to date in a later review by Brocklehurst (1994).

Problems

9.1 If K_m depends on a single ionizing group, with pK_a values pK_E in the free enzyme and pK_{EA} in the enzyme–substrate complex, its dependence on the hydrogen-ion concentration h is $K_m = \tilde{K}_m (K_E + h)/(K_{EA} + h)$.
(a) At what pH does a plot of K_m against pH show a point of inflexion?
(b) At what pH does a plot of $1/K_m$ against pH show a point of inflexion?
[If you find your result difficult or impossible to believe, calculate K_m and $1/K_m$ at several pH values in the range 3–10, assuming $pK_E = 6.0$, $pK_{EA} = 7.0$, and plot both against pH. For a discussion of the principles underlying this problem, see Fersht (1999).]

9.2 Interpretation of a plot of $\log K_m$ against pH is most easily done in the light of the relationship $\log K_m = \log V - \log(V/K_m)$, in which K_m, V and V/K_m are not only dimensioned quantities, but they have three different dimensions. Does this relationship violate the rules discussed in Section 1.3, and, if so, to what extent is the analysis implied by Figure 9.6 invalid?

9.3 A bell-shaped pH profile has half-maximal ordinate values at pH values 5.7 and 7.5. Estimate the molecular pK_a values. If there is independent reason to believe that one of the group pK_a values is 6.1, what can be deduced about the other three? [This problem is less trivial than it may appear at first sight.]

9.4 Draw a more realistic scheme for pH dependence by modifying Figure 9.5 as follows: (a) allow both substrate and product to bind to all three forms of free enzyme, and assume that the rate constants for these binding reactions are independent of the state of protonation; (b) assume that the catalytic process is a three-step reaction in which all steps are reversible and the second step, the interconversion of HEA and HEP, occurs for the singly protonated complexes only. Assuming that all protonation reactions are at equilibrium in the steady state, use Cha's method (Section 4.6) to derive an expression for K_m as a function of the hydrogen-ion concentration. {The solution has a complicated appearance, which can be simplified by defining $f(h) = 1/[(h/K_1) + 1 + (K_2/h)]$.} Under what circumstances is K_m independent of pH? If it is independent of pH, what value must it have?

Chapter 10

Temperature Effects on Enzyme Activity

10.1 Temperature denaturation

In principle, the theoretical treatment discussed in Section 1.7 of the temperature dependence of simple chemical reactions applies equally to enzyme-catalysed reactions, but in practice there are several complications that must be properly understood if any useful information is to be obtained from temperature-dependence measurements.

First, almost all enzymes become denatured if they are heated much above physiological temperatures, and the conformation of the enzyme is altered, often irreversibly, with loss of catalytic activity. Denaturation is chemically a complicated and only partly understood process, and only a simplified account will be given here. I shall limit it to reversible denaturation, and will assume that an equilibrium exists at all times between the active and denatured enzyme and that only a single denatured species needs to be taken into account. However, I emphasize that limiting it to reversible denaturation is for the sake of simplicity, not because irreversible effects are unimportant in practice. The example discussed in Section 10.2 will deal with irreversible denaturation, but in a qualitative way.

Denaturation does not involve rupture of covalent bonds, but only of hydrogen bonds and other weak interactions that are involved in maintaining the active conformation of the enzyme. Although an individual hydrogen bond is far weaker than a covalent bond (about $20 \, kJ \, mol^{-1}$ for a hydrogen bond compared with about $400 \, kJ \, mol^{-1}$ for a covalent bond), denaturation generally involves the rupture of many of them. More exactly, it involves the replacement of many intramolecular hydrogen bonds with hydrogen bonds between the enzyme molecule and solvent molecules. The standard enthalpy of reaction, $\Delta H^{0'}$, is often very high for denaturation, typically $200-500 \, kJ \, mol^{-1}$, but the

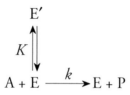

Figure 10.1. Simple mechanism for enzyme denaturation.

rupture of many weak bonds greatly increases the number of conformational states available to an enzyme molecule, and so denaturation is also characterized by a large standard entropy of reaction, $\Delta S^{0\prime}$.

The effect of denaturation on observed enzymic rate constants can be seen by considering the simple example of an active enzyme E in equilibrium with an inactive form E', as shown in Figure 10.1. For simplicity, the catalytic reaction is represented as a simple second-order reaction with rate constant k, as is usually observed at low substrate concentrations. The equilibrium constant for denaturation, K, varies with temperature according to the van't Hoff equation (Section 1.7.1):

$$-RT \ln K = \Delta G^{0\prime} = \Delta H^{0\prime} - T\Delta S^{0\prime} \tag{10.1}$$

where R is the gas constant, T is the absolute temperature and $\Delta G^{0\prime}$, $\Delta H^{0\prime}$ and $\Delta S^{0\prime}$ are the standard Gibbs energy, enthalpy and entropy of reaction, respectively. This relationship can be rearranged to provide an expression for K:

$$K = \exp\left(\frac{\Delta S^{0\prime}}{R} - \frac{\Delta H^{0\prime}}{RT}\right) \tag{10.2}$$

The rate equation k for the catalytic reaction may be governed by the integrated Arrhenius equation:

$$k = A \exp(-E_a/RT) \tag{10.3}$$

where A is a constant and E_a is the Arrhenius activation energy. The rate of the catalytic reaction is given by $v = k[E][A]$, but to use this equation the concentration $[E]$ of active enzyme has to be expressed in terms of the total concentration $e_0 = [E] + [E']$, and so

$$v = \frac{ke_0a}{1 + K} \tag{10.4}$$

The observed rate constant, k_{obs}, may be defined as $k/(1 + K)$, and varies with the temperature according to the following equation:

$$k_{obs} = \frac{A \exp(-E_a/RT)}{1 + \exp\left(\dfrac{\Delta S^{0\prime}}{R} - \dfrac{\Delta H^{0\prime}}{RT}\right)} \qquad (10.5)$$

At low temperatures, when $\Delta S^{0\prime}/R$ is small compared with $\Delta H^{0\prime}/RT$, the exponential term in the denominator is insignificant, and so k_{obs} varies with temperature in the ordinary way according to the Arrhenius equation. At temperatures above $\Delta H^{0\prime}/\Delta S^{0\prime}$, however, the denominator increases steeply with temperature and the rate of reaction decreases rapidly to zero.

10.2   Temperature optimum

Although this model is oversimplified, it does show why the Arrhenius equation appears to fail for enzyme-catalysed reactions at high temperatures. In the older literature, it was common for *optimum temperatures* for enzymes to be reported, and this is still occasionally seen today. However, the temperature at which k_{obs} is a maximum has no particular significance, as the temperature dependence of enzyme-catalysed reactions is often found in practice to vary with the experimental procedure. In particular, the longer a reaction mixture is incubated before analysis the lower the "optimum temperature" is likely to be (Figure 10.2). The explanation of this effect is that denaturation often occurs fairly slowly, so that the denaturation reaction cannot properly be treated as an equilibrium. The extent of denaturation therefore increases with the time of incubation. This ought not to be a problem with modern experimental techniques, because in continuously assayed reaction mixtures time-dependent processes are usually obvious.

Even if most reports of temperature optima are thus likely to be artefacts, there are some that can survive sceptical examination. For example, Thomas and Scopes (1998) studied 3-phosphoglycerate kinases from various bacteria and found reversible decreases in activity with increased temperature before reaching temperatures at which irreversible denaturation became appreciable. Subsequently Daniel, Danson and Eisenthal (2001) studied this and various other examples and proposed a plausible model to account for them, in which the active form of the enzyme exists in equilibrium with an inactive form, with a temperature dependence of the equilibrium that increases the proportion of inactive form with temperature. The maximum in the activity occurs because at low temperatures the normal Arrhenius increase in activity is more important

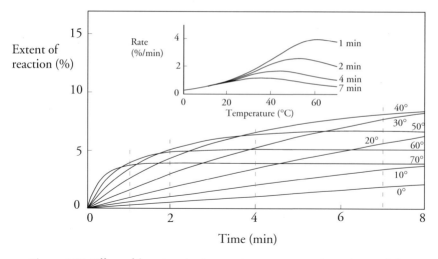

Figure 10.2. Effect of heat inactivation on the temperature dependence of the rates of enzyme-catalysed reactions. If the "initial rate" is taken to be the mean rate during some fixed period, such as 1, 2, 4 or 7 min (as indicated by the dotted lines), the temperature dependence will show a bell-shaped curve with a maximum at a temperature that varies with the period used, as illustrated in the inset.

than the effect of temperature on the equilibrium, whereas at high temperatures the reverse is true. In practice, as the authors point out, irreversible temperature denaturation of both forms of the enzyme is likely to occur as well, but this does not invalidate the major point that a genuine maximum in the activity curve may occur even at zero time.

10.3   Application of the Arrhenius equation to enzymes

Because of denaturation, straightforward results can usually be obtained from studies of the temperature dependence of enzymes only within a fairly narrow range of temperature, say 0–50 °C at best, but even within this range there are important hazards to be avoided. First, the temperature dependence of the initial rate commonly gives curved Arrhenius plots, from which little useful information can be obtained. Such plots often show artefacts (see, for example, Silvius, Read and McElhaney, 1978), and a *minimum* requirement for a satisfactory temperature study is to measure a series of rates at each temperature, so that Arrhenius plots can be drawn for the separate Michaelis–Menten parameters, V, K_m and V/K_m. These plots are also often curved, and there are so many possible explanations of this — for example, a change in conformation of the enzyme, the existence of the enzyme as a mixture of isoenzymes or an effect of temperature on a substrate — that it is dangerous to conclude much from the shape of an Arrhenius plot unless it can be correlated with other temperature effects that can be observed independently. Massey, Curti and Ganther (1966), for example, found sharp changes in slope at about 14 °C in Arrhenius plots for D-amino acid oxidase; at the same temperature, other techniques, such as sedimentation velocity and ultraviolet spectroscopy, indicated a change in conformational state of the enzyme. It is then clearly reasonable to interpret the kinetic behaviour as a consequence of the same change in conformation.

In general one can attach little significance to studies of the temperature dependence of any Michaelis–Menten parameter unless the mechanistic meaning of this parameter is known. If K_m is a function of several rate constants, its temperature dependence is likely to be a complicated combination of competing effects, and of little significance or interest; but if K_m is known with reasonable certainty to be a true dissociation constant, its temperature dependence can provide useful thermodynamic information about the enzyme.

Most of the "activation energies" for enzyme-catalysed reactions that have appeared in the literature have little value, but it would be wrong to suggest that no useful information can be obtained from studies of temperature dependence; if proper care is taken they can lead to valuable information about enzyme reaction mechanisms. For example, Bender, Kézdy and Gunter (1964) made a classic study of α-chymotrypsin that differed subtantially from typical temperature-dependence studies: they obtained convincing evidence of the particular steps in the mechanism that were affected, they compared results for numerous different substrates, for an enzyme about which much was known already, and they interpreted them with a proper understanding of chemistry.

10.4 Entropy–enthalpy compensation

It should be evident from the preceding discussion that application of the Arrhenius or van't Hoff equations to enzymes over a restricted range of temperature is error-prone and liable to lead to results of dubious value. Rather surprisingly, however, estimates of enthalpies and entropies of activation or of reaction frequently lead to remarkably good correlations between the two parameters (as for example in Figure 10.3), whether these are measured for a series of mutants of different proteins, or for proteins extracted from a series of different organisms, or from other series for which there seems no obvious reason to expect a correlation. This has led to the popular idea of *entropy–enthalpy compensation*, whereby variations in enthalpy are supposedly compensated for by variations in entropy to produce approximately the same rate of reaction or equilibrium constant at a particular temperature.

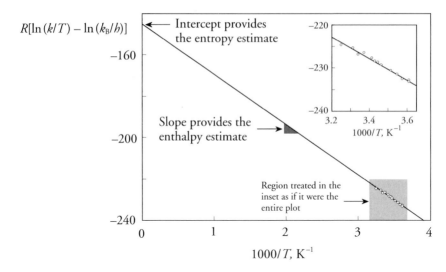

Figure 10.3. Modified Arrhenius plot for a mutant form of morphinone reductase. The main plot shows the data plotted so that the vertical axis appears explicitly in the plot; notice the extremely long extrapolation needed to estimate the entropy of activation. The inset shows the same data with the severely restricted abscissa scale conventionally (and misleadingly) used for such plots.

When the data are obtained from calorimetric measurements, and the heat is measured directly, there may be some value in this concept, but I shall not discuss that aspect (though see Sharp, 2001), being more concerned with entropy and enthalpy values obtained from temperature dependence data using the Arrhenius or the van't Hoff equation. The fallacy in this approach was exposed many years ago (Exner, 1964), and there would be no reason to mention it at all if it were not that it continues to be used experimentally (for example Richieri and co-workers, 1997; Craig and co-workers, 2001) and discussed in textbooks (for example Gutfreund, 1995).

The essential problem with concluding compensation from measurements of temperature dependence is that the data span too narrow a range of temperatures to allow derivation of two independent thermodynamic parameters. In the three examples cited these ranges were 10–45 °C (283–318 K), 3–35 °C (278–308 K) and 0–18 °C (273–291 K) respectively, and even the longest of these ranges corresponds to a $1000/T$ range of only 0.39 K^{-1}, even though $1000/T$ had a value of 3.14 K^{-1} at the highest temperature used. What this means is that extrapolating the experimental points to $1000/T = 0$ to obtain the intercept, and hence the entropy value, is an extrapolation of eight times the range of the data. Let us look at this in more detail for an example from a study by Craig and co-workers (2001) of morphinone reductase. Figure 10.3 shows a modified Arrhenius plot for a mutant in which arginine-238 was replaced by glutamate: the inset shows the data in the way such experiments are nearly always presented in the literature; the main plot shows the same data over a wide enough scale for the ordinate axis to appear explicitly in the plot. Note that the extrapolation needed to obtain the enthalpy of activation from the ordinate intercept is about nine times the range of the data. Although the cluster of experimental points around an abscissa value of 3.4 define the ordinate value of the line precisely at the temperature corresponding to this value, the slope is much less well defined, and any small error in it will necessarily be magnified into a large error in the ordinate intercept. In other words the data define *one* piece of information, the activity at 21 °C, quite well, but one cannot calculate two supposedly independent numbers from this single one without generating a spurious correlation.

Even completely random data can generate an apparently impressive compensation plot if the range of temperatures is similar to that in real biological experiments. For example, assigning Arrhenius energies of activation falling randomly in the range 30–160 kJ/mol to 100 "samples" with activities at 18 °C that varied arbitrarily over a tenfold range resulted in an excellent compensation plot (Cornish-Bowden, 2002), and even when the activities at 18 °C spanned a millionfold range the correlation remained striking.

The slope of a plot of entropy against enthalpy has the dimensions of temperature, and the temperature that results from measuring it is sometimes called the *compensation temperature* or the *isokinetic temperature*, as it is the temperature at which the supposed compensation is perfect. Conclusions drawn from measurements of such temperatures are as open to argument as the concept of compensation itself, and further information may be found in a recent article of McBane (1998).

Problems

10.1 The following measurements of V at temperature t were made for an enzyme-catalysed reaction over a temperature range in which no thermal inactivation could be detected. Are they consistent with interpretation of V as $k_2 e_0$, where e_0 is constant and k_2 is the rate constant for a single step in the mechanism?

t (°C)	V (mM/min)	t (°C)	V (mM/min)
5	0.32	30	11.9
10	0.75	35	19.7
15	1.67	40	30.9
20	3.46	45	46.5
25	6.68	50	68.3

10.2 Open a telephone directory at a random page and take the first ten telephone numbers listed on the page. Then ignore all but the last four digits of each number, and write them down as a pair of two-digit numbers. For example, if the first number is 049191 6619, then write down 66 and 19. For each of the ten pairs of numbers, add the smaller of the two numbers to 100 and the larger to 200, giving 119 and 266 for the example considered. Now treat the ten pairs of numbers as measurements of the catalytic constant at 5 °C and 30 °C of samples of an enzyme isolated from ten different species. For each enzyme estimate the Arrhenius activation energy E_a from the two-point straight line, and hence calculate the enthalpy and entropy of activation from equations 1.49–50, assuming $R = 8.31\,\mathrm{J\,mol^{-1}\,K^{-1}}$, $T = 300\,\mathrm{K}$ and $RT/N_A h = 6.25 \times 10^{12}\,\mathrm{s^{-1}}$. Plot the resulting entropies of activation against the enthalpies, and comment on any relationship that is apparent.

Chapter 11

Control of Enzyme Activity

11.1 Function of cooperative and allosteric interactions

11.1.1 Futile cycles

All living organisms require a high degree of control over metabolism, to permit orderly change without precipitating catastrophic progress towards thermodynamic equilibrium. Less obviously, it is unlikely that enzymes with the properties described in earlier chapters can provide the degree of control necessary.

The interconversion of fructose 6-phosphate and fructose 1,6-bisphosphate illustrates the essential problem, and the enzyme properties needed for effective regulation to be possible. The conversion of fructose 6-phosphate into fructose 1,6-bisphosphate requires ATP:

$$\text{fructose 6-phosphate} + \text{ATP} \rightarrow \text{fructose 1,6-bisphosphate} + \text{ADP} \quad (11.1)$$

It is catalysed by phosphofructokinase and is the first unique step in glycolysis, the first step that does not also form part of other metabolic processes. It is thus a suitable step for regulating glycolysis, and, although the modern view (Section 12.5) is that it is an oversimplification to seek a unique site of regulation of any pathway, regulation of phosphofructokinase is certainly important for glycolysis in most cells.

Under cellular conditions the reaction catalysed by phosphofructokinase is essentially irreversible, and a different reaction, irreversible in the opposite direction, converts fructose 1,6-bisphosphate back into fructose 6-phosphate in gluconeogenesis. This is a hydrolytic reaction, catalysed by fructose bisphosphatase:

$$\text{fructose 1,6-bisphosphate} + \text{water} \rightarrow \text{fructose 6-phosphate} + \text{phosphate}$$

$$(11.2)$$

The occurrence of antiparallel pairs of irreversible reactions is of the greatest importance in metabolic regulation: it means that the *direction* of net flux between two metabolites can be determined by differential regulation of the activities of the two enzymes. A single reversible reaction could not be controlled in this way, because a catalyst cannot affect the direction of flux through a reaction, which is determined solely by thermodynamic considerations. The catalyst affects only the rate at which equilibrium can be attained.

If both phosphofructokinase and fructose bisphosphatase reactions were to proceed in an uncontrolled way at similar rates, there would be no net interconversion of fructose 6-phosphate and fructose 1,6-bisphosphate, but continuous hydrolysis of ATP, potentially resulting in death. This is often called a *futile cycle*, and to prevent it either the two processes must be segregated into different cells (or different compartments of the same cell), or both enzymes must be regulated in such a way that each is active only when the other is inhibited. Many potential cycles are indeed controlled by compartmentation, but this is not possible for all in all circumstances, and so there remains a need for the second option. For example, tissues such as liver carry out both glycolysis and gluconeogenesis, and have both enzymes present.

The term "futile cycle" has fallen into disfavour, because it is now realized that cycling is quite widespread and that it is by no means necessarily harmful. As discussed in Section 12.9.2, cycling between active and inactive forms of an enzyme can be an extremely sensitive mechanism for regulating catalytic activity that easily repays the small cost in hydrolysed ATP. Even when the main result of cycling is to generate heat it is hardly "futile" if it enables a warm-blooded animal to maintain the temperature needed for life, or if it enables a bumblebee to fly (and gather nectar) on a cold day.

11.1.2   Inadequacy of Michaelis–Menten kinetics for regulation

We must now ask whether an enzyme that obeys the ordinary laws of enzyme kinetics can be regulated precisely enough to prevent futile cycling. For an enzyme that obeys the Michaelis–Menten equation, $v = Va/(K_m + a)$, a simple calculation shows that the rate is $0.1V$ when $a = K_m/9$, and that it is $0.9V$ when $a = 9K_m$. In other words, an enormous increase in substrate concentration, 81-fold, is required to bring about a comparatively modest increase in rate from 10% to 90% of the limit.

Essentially the same result applies to linear inhibition. Suppose that a reaction is inhibited competitively according to the equation $v = Va/[K_m(1 + i/K_{ic}) + a]$. The ratio of inhibited to uninhibited rate may be written as $v/v_0 = (K_m + a)/[K_m(1 + i/K_{ic}) + a]$, which has values of 0.1 when $i = 9K_{ic}(1 + a/K_m)$ and 0.9 when $i = K_{ic}(1 + a/K_m)/9$: again an 81-fold change in concentration is needed to span the middle 80% of the range of possible rates.

Even with two or more inhibitors acting in concert, the qualitative conclusion is the same: an inordinately large change in the environment is necessary to bring about even a modest change in rate. The requirements for effective regulation of metabolism are exactly the opposite, however: on the one hand, the concentrations of major metabolites must be maintained within small tolerances, and on the other hand, reaction rates must be capable of changing greatly — probably more than the range of 10–90% of the limit just considered — in response to fluctuations within these small tolerances.

11.1.3 Cooperativity

Clearly, the ordinary laws of enzyme kinetics are inadequate for supplying the degree of control needed for metabolism. Instead, many enzymes with important roles in metabolic regulation respond with exceptional sensitivity to changes in metabolite concentrations. This property is known as *cooperativity*, because it arises from "cooperation" between the active sites of polymeric enzymes. As illustrated in Figure 11.1, the plot of rate against substrate concentration shows a characteristic *sigmoid* (S-shaped) curve quite different

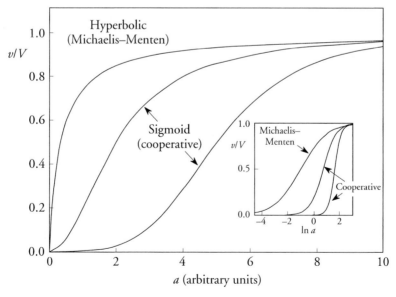

Figure 11.1. Comparison of a hyperbolic curve of rate as a function of substrate concentration with two sigmoid (cooperative) curves, calculated with Hill coefficients (see below, Section 11.2.1) of 2 and 4. Note that all of the curves are sigmoid when replotted as a function of $\ln a$ (inset) but the cooperative curves are steeper.

from the rectangular hyperbola given by the Michaelis–Menten equation. The steepest part of the curve is shifted from the origin to a positive concentration, typically one within the physiological range for the metabolite concerned. A major theme of this chapter is to examine the main theories that have been proposed to explain behaviour of the sort seen in Figure 11.1.

11.1.4 Allosteric interactions

The interconversion of fructose 6-phosphate and fructose 1,6-bisphosphate illustrates another important aspect of metabolic regulation, namely that the immediate and ultimate products of a reaction are usually different. Although ATP is a substrate of the phosphofructokinase reaction, the effect of glycolysis as a whole is to generate ATP, with a large overall stoicheiometry if glycolysis is considered as the route into the tricarboxylate cycle and electron transport. Thus ATP must be regarded as a product of glycolysis, even though it is a substrate of one of the main enzymes at which glycolysis is regulated. Hence ordinary product inhibition of phosphofructokinase works in the opposite direction from what is required for efficient regulation, and to permit a steady supply of metabolic energy phosphofructokinase ought also to be inhibited by the ultimate product of the pathway, ATP, as, in fact, it is.

   This type of inhibition cannot be provided by the usual mechanisms that involve binding the inhibitor as a structural analogue of the substrate: in some cases these would bring about an unwanted effect; in others the ultimate product of a pathway might bear little structural resemblance to any of the reactants of the regulated enzyme; for example, histidine is quite different in structure from phosphoribosyl pyrophosphate, its metabolic precursor. To permit inhibition or activation by metabolically appropriate effectors, many regulated enzymes have evolved sites for effector binding that are separate from the catalytic sites. Monod, Changeux and Jacob (1963) proposed that they should be called *allosteric* sites, from the Greek for *different shape**, to emphasize the structural dissimilarity between substrate and effector, and enzymes that possess them are called *allosteric enzymes*.

   Many allosteric enzymes are also cooperative, and vice versa. This agrees with the fact that both properties are important in metabolic regulation, but it does not mean that the two terms are interchangeable: they describe two different properties and should be clearly distinguished. In many cases, they were recognized separately: haemoglobin was known to be cooperative for

---

*Strictly "another solid", but "different shape" better conveys the essential idea that the inhibition does not result from any structural similarity between effector and substrate.

Jacques Lucien Monod (1910–1972)

Jacques Monod had a distinguished career in bacteriology, where he introduced the chemostat and developed the Monod equation for bacterial growth, which has the same form as the Michaelis–Menten equation. He later became one of the founding fathers of molecular biology, a field that in its early days was not as remote from classical enzymology as it later became. In proposing the first plausible explanation of the regulatory properties of certain enzymes he stimulated the explosion of interest in these properties after 1965. From his mother, an American from Milwaukee, he acquired a perfect command of English, both spoken and written, and his writing was outstandingly clear and idiomatic. Politically active throughout his life, he participated actively in the Resistance to the Nazi occupation of France, and was briefly a member of the French Communist Party, though not for long, as he quickly found that his spirit of independent thought was incompatible with the discipline expected of its members.

more than 60 years before the allosteric effect of 2,3-bisphosphoglycerate was described; the first enzyme in the biosynthetic pathway to histidine was one of the first allosteric enzymes to be known, but it has not been reported to be cooperative.

11.2 The development of models to explain cooperativity

11.2.1 The Hill equation

It is often convenient to express the degree of cooperativity of an enzyme in terms of the following equation:

$$v = \frac{Va^h}{K_{0.5}^h + a^h} \tag{11.3}$$

This is known as the *Hill equation*, as Hill (1910) proposed a similar equation as an empirical description of the cooperative binding of oxygen to haemoglobin. The parameter V fulfils the same role as the *limiting rate* in the Michaelis–Menten equation, and may be known by the same name. However, although $K_{0.5}$, like K_m in the Michaelis–Menten equation, defines the value of the substrate concentration a at which $v = 0.5V$, it should *not* be called the Michaelis constant or given the symbol K_m, because these refer specifically to the Michaelis–Menten equation, and equation 11.3 is not equivalent to the Michaelis–Menten equation (except, trivially, when $h = 1$). The Hill equation is often written in a form resembling the following equation:

$$v = \frac{Va^h}{K + a^h} \tag{11.4}$$

with a constant K in the denominator that is not raised to the power h: although this does not prevent the equation from being used as a calculating device, it has the disadvantage that K has dimensions of a concentration to the power h, so for example if h has the value 3.4 then K has units of $mM^{3.4}$, and it is difficult to give such a quantity a physical meaning.

Although attempts are sometimes made to derive the equation from a model, it is best to follow the example of Hill: he regarded his equation as purely empirical, and explicitly disavowed any physical meaning for the exponent h, which is now usually called the *Hill coefficient*. When h is an integer the equation can be a limiting case of physical models of substrate binding, but h is usually not found experimentally to be an integer, and except in the limit realistic models do not predict that it should be. It is incorrect, therefore, to treat it as an estimate of the number of substrate-binding sites on the enzyme, though for some models it does provide a lower limit for this number. For haemoglobin the number of oxygen-binding sites is 4 (though this was not known in Hill's time), but typical experimental values of h for haemoglobin are about 2.7.

If equation 11.3 is rearranged as follows:

$$\frac{v}{V - v} = \frac{a^h}{K^h_{0.5}} \tag{11.5}$$

then $v/(V - v)$ may be regarded in the absence of direct binding information as a measure of $[EA]/[E]$, and when logarithms are taken of both sides:

$$\ln[v/(V - v)] = h \ln a - h \ln K_{0.5} \tag{11.6}$$

it may be seen that a plot of $\ln[v/(V - v)]$ against $\ln a$ should be a straight line with slope h^*. This plot, illustrated in Figure 11.2, is called a *Hill plot*, and provides a simple means of evaluating h and $K_{0.5}$. It has been found to fit a

---

*A plot of $\log[v/(V - v)]$ against $\log a$ would have the same slope, and hence the same value of the Hill coefficient h, because this does not depend on the type of logarithms used (as long as this is the same for both coordinates).

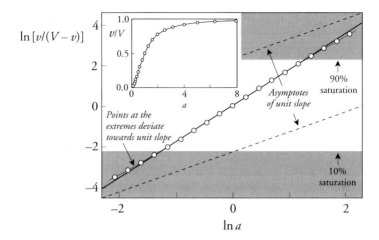

Figure 11.2. Hill plot, with the corresponding plot in linear coordinates shown as an inset. The lines are calculated from the Hill equation (equation 11.3) but the points are calculated from a realistic binding function. As the Hill equation is at best an approximation one should always expect the observations to tend towards unit slope at the extremes, whatever the slope in the central region. However, even in the absence of experimental error the deviations from linearity are often small, especially in the part of the curve between 10 and 90% saturation, shown as the unshaded region in this plot.

wide variety of cooperative kinetic data remarkably well for v/V values in the range 0.1–0.9, though deviations always occur at the extremes (as indicated in the Figure), because equation 11.3 is at best only an approximation to a more complicated relationship.

The Hill coefficient is widely used as an index of cooperativity, the degree of cooperativity increasing with h. A non-cooperative (Michaelis–Menten) enzyme has $h = 1$, and so positive cooperativity means h greater than 1. *Negative cooperativity*, with h less than 1, also occurs for some enzymes, though it is less common (at least for pure enzymes), and it is less clear what physiological role it may fulfil. The Hill equation was originally developed as an irreversible equation (equation 11.4) and for mechanistic discussions of cooperative enzymes it is still nearly always considered only in irreversible form. However, it is coming to be an indispensable tool also for computer modelling of metabolic systems, and for this purpose it is also needed in a reversible form, as will be discussed in Section 12.10.

11.2.2 An alternative index of cooperativity

The *cooperativity index*, R_a, of Taketa and Pogell (1965) is less widely used than the Hill coefficient, but it has a more obvious experimental meaning and as it is always treated as purely empirical, it is not confused by attempts to relate it to models of dubious validity. It is defined as the ratio of a values that give $v/V = 0.9$ and $v/V = 0.1$. The relationship between R_a and h can be found by substituting these two values of v/V successively into equation 11.3:

$$0.9 = a_{0.9}^h/(K_{0.5}^h + a_{0.9}^h) \tag{11.7}$$

$$0.1 = a_{0.1}^h/(K_{0.5}^h + a_{0.1}^h) \tag{11.8}$$

and solving for the two values of a:

$$a_{0.9} = 9^{1/h} K_{0.5} \tag{11.9}$$

$$a_{0.1} = K_{0.5}/9^{1/h} \tag{11.10}$$

Then R_a is easily obtained as $a_{0.9}/a_{0.1}$:

$$R_a = 81^{1/h} \tag{11.11}$$

It follows that $R_a = 81$ characterizes a non-cooperative enzyme; cooperative enzymes have values less than 81; negatively cooperative enzymes have values greater than 81. This relationship is only as accurate as equation 11.3, of course, but that is adequate for most purposes. Some representative values are listed in Table 11.1.

11.2.3   Assumption of equilibrium binding in cooperative kinetics

In discussing non-cooperative kinetics I have emphasized (Section 2.7.1) that one cannot assume that substrate binding is at equilibrium, so one cannot assume that the Michaelis constant K_m is the same as the thermodynamic substrate dissociation constant. In principle the arguments apply with almost equal force to enzymes that deviate from Michaelis–Menten kinetics. The only mitigating feature is that for cooperative enzymes one can postulate that arriving at high catalytic activity has been less important during evolution than arriving at useful regulatory behaviour, so that it is less certain that catalytic rate constants have evolved to be as high as chemically possible within the normal structural constraints. With a few enzymes, such as aspartate transcarbamoylase (Newell, Markby and Schachman, 1989), binding has been studied in conditions of equilibrium and found to follow the same cooperative behaviour as observed kinetically. It is anyway virtually impossible to obtain usable rate equations for cooperative systems unless some simplifying assumptions are made, and the one that is made almost

Table 11.1. Relationship between the two indexes of cooperativity. The Table shows the relationship between the Hill coefficient h and the cooperativity index R_a. The values are calculated on the assumption that the Hill equation (equation 11.3) holds exactly. Very few individual enzymes show cooperativity greater than $h = 4$. However, there is almost no limit to the sensitivity of response that can result from the operation of interconvertible enzyme cascades, and the lowest lines of the Table are included to facilitate discussion of such systems.

h	R_a	Description
0.5	6560	
0.6	1520	
0.7	533	Negatively cooperative
0.8	243	
0.9	132	
1.0	81.0	Non-cooperative
1.5	18.7	
2.0	9.00	
2.5	5.80	Positively cooperative
3.0	4.33	(normal range for individual enzymes)
3.5	3.51	
4.0	3.00	
5.0	2.41	
6.0	2.08	
8.0	1.73	
10	1.55	Extreme cooperativity
15	1.34	(beyond range for individual enzymes)
20	1.25	
50	1.092	
100	1.045	
1000	1.0044	

always is that binding is at equilibrium. Normally the only time such an assumption is not made is in deriving models in which the cooperativity is wholly kinetic in origin (Section 11.8). In the rest of this chapter I shall assume equilibrium binding when discussing the kinetics of oligomeric enzymes.

11.2.4 The Adair equation

Suppose that an enzyme has two active sites that bind substrate independently and at equilibrium with dissociation constants K_{s1} and K_{s2}, as shown in

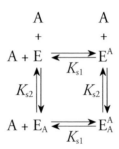

Figure 11.3. Binding of substrate at two independent sites.

Figure 11.3. If the same chemical reaction takes place independently at the two sites with rate constants k_1 and k_2, then each site will independently obey Michaelis–Menten kinetics with a Michaelis constant equal to the appropriate dissociation constant, and the total rate will be the sum of the rates for the two sites:

$$v = \frac{k_1 e_0 a}{K_{s1} + a} + \frac{k_2 e_0 a}{K_{s2} + a} \tag{11.12}$$

The limiting rate is the sum of the limiting rates of the two sites: $V = k_1 e_0 + k_2 e_0$. If we assume for simplicity that the catalytic rate constants are equal, so that we can write $k_1 e_0 = k_2 e_0 = V/2$, then equation 11.12 may be written as follows:

$$\frac{v}{V/2} = \frac{a}{K_{s1} + a} + \frac{a}{K_{s2} + a} \tag{11.13}$$

This corresponds to the analysis of proton dissociation in terms of group dissociation constants (Section 9.3.1), but exactly the same model can be expressed in terms of molecular dissociation constants (compare Section 9.3.2):

$$\frac{v}{V} = \frac{\dfrac{a}{K_1} + \dfrac{a^2}{K_1 K_2}}{1 + \dfrac{2a}{K_1} + \dfrac{a^2}{K_1 K_2}} \tag{11.14}$$

in which K_1 is the dissociation constant for the first molecule of substrate (regardless of site), and K_2 is that for the second, so $K_1 = 2[E][A]/[EA]$ and $K_2 = [EA][A]/2[EA_2]$. The relationship of these constants to the site dissociation constants is taken up in equation 11.18, below. Equation 11.14 is known as the *Adair equation*, as Adair (1925) expressed the binding of oxygen to haemoglobin in terms of an equation of the same general form, written for four binding sites rather than two, as haemoglobin can bind up to four molecules of oxygen:

$$y = \frac{\dfrac{a}{K_1} + \dfrac{3a^2}{K_1K_2} + \dfrac{3a^3}{K_1K_2K_3} + \dfrac{a^4}{K_1K_2K_3K_4}}{1 + \dfrac{4a}{K_1} + \dfrac{6a^2}{K_1K_2} + \dfrac{4a^3}{K_1K_2K_3} + \dfrac{a^4}{K_1K_2K_3K_4}} \tag{11.15}$$

As well as being written for four sites, this equation has v/V replaced by y, a quantity known as the *fractional saturation*, the fraction of the total number of binding sites that are occupied by ligand. This is because in a true binding experiment there is no rate and hence no limiting rate. However, as it is difficult to measure binding directly with sufficient accuracy, it is not uncommon to measure other quantities, either rates or spectroscopic signals, and to interpret them as measures of binding. So far as rates are concerned, I have already discussed in the previous section whether binding can be assumed to be at equilibrium in the steady state, and will not labour the point here. There is, however, a second complication about regarding v/V as a measure of y: it depends on assuming that each active site has the same catalytic constant. We made this assumption in going from equation 11.12 to equation 11.13, but there is no reason to expect it to be true in general, and it becomes progressively more implausible as one moves to models with greater numbers of sites and greater differences between their binding constants. One has the same complication if one treats a spectroscopic signal as a measure of binding; it is not quite so severe, however, because although the variations between catalytic constants between different sites may be large one can reasonably assume that the spectroscopic effects of binding may be broadly similar for different sites even if they are not identical.

It is useful to examine equation 11.15, the equation for four sites, as it illustrates rather better than equation 11.14 the general form of the Adair equation for an arbitrary number n of sites, showing more clearly that the numerical coefficients in both numerator (1, 3, 3, 1) and denominator (1, 4, 6, 4, 1) are the binomial coefficients for $n - 1$ and n respectively: the numerator coefficients are $(n - 1)!/i!(n - 1 - i)!$ for $i = 0$ to $n - 1$, and the denominator coefficients are $n!/i!(n - i)!$ for $i = 0$ to n. So the Adair equation can be written in general as follows

$$y = \frac{\displaystyle\sum_{i=1}^{n} (n - i)! a^i / i!(n - i - 1)! \prod_{j=1}^{i} K_j}{1 + \displaystyle\sum_{i=1}^{n} n! a^i / i!(n - i)! \prod_{j=1}^{i} K_j} \tag{11.16}$$

However, although this type of expression may be found in highly theoretical discussions of cooperativity it is usually regarded as excessively abstract for more experimental contexts. For educational purposes it is often more effective to sacrifice generality for the sake of simplicity, and in this chapter I shall normally write equations that assume particular numbers of binding sites rather than aim for complete generality.

The coefficients in the Adair equation are often regarded as "statistical factors". Thus, there are four ways of binding one substrate molecule to an enzyme molecule with four vacant sites, but only one way of removing the single substrate molecule from a complex with one molecule bound: this produces the factor 4/1 (or 4) in the denominator. There are three ways of binding a second molecule of ligand, but two ways of removing one, so the 4 is multiplied by 3/2 to give 6, and so on.

The particular way of writing the Adair equation used here, in both equations 11.14 and 11.15, is chosen to give dissociation constants that are equal if all of the sites are identical and do not interact, a useful property for quantifying the degree of departure from this simple assumption. Constants defined in this way are sometimes called *intrinsic constants*, but as the same name is also sometimes applied to the quantities I have here called group dissociation constants, I shall not use it here. Various other ways of writing the Adair equation may be found in the literature, with different numerical coefficients (so that the dissociation constants are no longer equal to one another in the simplest case), or with association instead of dissociation constants, or with products of association constants written with special symbols, for example $\psi_3 \equiv 1/K_1K_2K_3$. The use of association constants is particularly common in the haemoglobin literature and in other papers with a binding rather than a kinetic emphasis.

Returning now for simplicity to the two-site enzyme, the relationship between the molecular dissociation constants of equation 11.14 and the two group dissociation constants of equation 11.13 follows from a comparison of equation 11.14 with a multiplied-out version of equation 11.13:

$$\frac{v}{V} = \frac{\dfrac{(K_{s2} + K_{s1})a}{2K_{s1}K_{s2}} + \dfrac{a^2}{K_{s1}K_{s2}}}{1 + \dfrac{(K_{s2} + K_{s1})a}{K_{s1}K_{s2}} + \dfrac{a^2}{K_{s1}K_{s2}}} \tag{11.17}$$

Thus:

$$\frac{1}{K_1} = \frac{1}{2}\left(\frac{1}{K_{s1}} + \frac{1}{K_{s2}}\right), \quad K_2 = \tfrac{1}{2}(K_{s1} + K_{s2}) \tag{11.18}$$

so K_1 is the harmonic mean of the group dissociation constants and K_2 is the arithmetic mean.

It is obvious from the ordinary idea of a mean that K_1 and K_2 are equal to each other and to the group dissociation constants if these are equal to each other. More interesting is the relationship between them when the group dissociation constants are different. This can be examined by considering the ratio K_2/K_1 defined by equation 11.18:

$$\frac{K_2}{K_1} = \frac{1}{4}\left(2 + \frac{K_{s2}}{K_{s1}} + \frac{K_{s1}}{K_{s2}}\right) \tag{11.19}$$

As this contains three terms, of which the second and third move in opposite directions when the ratio K_{s2}/K_{s1} varies, one might think at first sight that the value could be either greater than or less than 1 for different values of this ratio. In fact it is simpler than that, because the larger of any positive number and its reciprocal is further from 1 than the smaller (for example, 3 is further from 1 than 1/3 is), and consequently the sum defined by equation 11.19 cannot be less than 1, or:

$$K_1 \leq K_2 \tag{11.20}$$

Observationally, this means that the second molecule will always bind more weakly than the first, exactly, of course, as we expect from everyday experience with large objects: it is easier to detach something that is weakly attached than something that is tightly attached.

The implications of equation 11.20 for cooperativity are not easy to derive algebraically, even if one uses the cooperativity index of Taketa and Pogell (Section 11.2.2), because solving equation 11.14 for a after setting v/V to 0.1 or 0.9 leads to expressions whose meanings are not transparent. However, it is easy to show numerically, by calculating curves with various different values of K_2/K_1, that as long as equation 11.20 is obeyed the result is always *negative* cooperativity, that is to say $R_a > 81$ or $h < 1$ (compare equation 11.11). In conclusion, therefore, the hypothesis embodied in Figure 11.3 is incapable of explaining positive cooperativity. The problem is not with the assumption of binding at two sites, which is reasonable enough, but with the assumption that this binding is *independent*, neither binding process having any influence on the other. Putting this the other way around, we can say that whenever we observe positive cooperativity we can be sure that the binding at different sites is *not* independent. (Negative cooperativity can also imply interactions between the sites, but, unlike positive cooperativity, it does not have to imply this.)

11.2.5 Mechanistic and operational definitions of cooperativity

The Adair equation, equation 11.14, is more general than the model from which we derived it, as it can still define the behaviour even if equation 11.20 is not obeyed. It allows a *mechanistic* definition of cooperativity that may be compared with the purely empirical definition in terms of R_a (Section 11.2.2) or the pseudo-mechanistic definition in terms of h (Section 11.2.1). The three definitions are qualitatively equivalent, at least for two-site enzymes: if we define positive cooperativity as meaning that $K_2 < K_1$, then this also means that $R_a < 81$ and that $h > 1$. It becomes more complicated when there are more than two sites, as it is quite possible with equations such as equation 11.15 to have relationships such as $K_1 > K_2 \approx K_3 < K_4$, as observed for the binding of oxidized NAD to glyceraldehyde 3-phosphate dehydrogenase from yeast (Cook and Koshland, 1970). Mechanistically this is clearly a mixture of positive and negative cooperativity, but such a mixture is not possible for R_a, as it is a single number that has to be either greater than or less than 81, and cannot be both.

This is clearly rather unsatisfactory, and one may ask if any operational definition of cooperativity is possible that recognizes that nature is complicated. Whitehead (1978) pointed out that the Hill coefficient provides just such a definition. Consider the quantity $Q = a(1 - y)/y$: this is a constant equal to the dissociation constant if there is only one binding site, or if there are n identical independent sites (when the Adair constants satisfy the relationship $K_1 = K_2 = \ldots = K_n$). If, however, Q decreases as a increases, so that dQ/da is negative, then it is clear that the binding is getting progressively stronger as more ligand binds: it is then reasonable to say that the system is positively cooperative at the particular value of a at which a negative value of dQ/da has been observed. Non-cooperative and negatively cooperative systems can be defined similarly. For any binding function, the sign of dQ/da is opposite to that of $(h - 1)$: dQ/da is negative, zero or positive according to whether h is greater than, equal to, or less than 1; consequently a definition of cooperativity in terms of the Hill coefficient is exactly equivalent to the more rational definition proposed by Whitehead. This conclusion is entirely independent of any consideration of whether the Hill equation has any physical or descriptive validity.

If there are more than two sites, the definition of cooperativity in terms of the Hill coefficient is not necessarily equivalent to a definition in terms of Adair constants. Cornish-Bowden and Koshland (1975) explored the relationship between the two definitions when applied to four-site proteins, and found a fair but not exact correspondence. For the data of Figure 11.4, for example, the curve has a slope greater than 1 at low ligand concentrations, is equal to 1 close

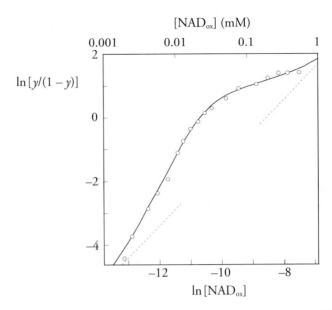

Figure 11.4. Hill plot for the binding of oxidized NAD to yeast glyceraldehyde 3-phosphate dehydrogenase. The plot shows data of Cook and Koshland (1970) recalculated as described by Cornish-Bowden and Koshland (1975). The shape of the curve suggests that the Adair constants (in equation 11.15) satisfy the relationship $K_1 > K_2 \approx K_3 < K_4$, in agreement with the following values found by curve fitting (Cornish-Bowden and Koshland, 1970): $K_1 = 0.217\,\text{mM}$, $K_2 = 0.0067\,\text{mM}$, $K_3 = 0.013\,\text{mM}$, $K_4 = 0.286\,\text{mM}$.

to half-saturation, and is less than 1 at high ligand concentrations, in agreement with the relationship $K_1 > K_2 \approx K_3 < K_4$ mentioned at the beginning of this section, which was found by fitting the data to the Adair equation. Examination of many calculated Hill plots indicates that this sort of correspondence usually applies. To summarize, although definitions of cooperativity based on the Hill plot and the Adair equation are not exactly equivalent, they are qualitatively similar and no great harm comes from continuing to use both as appropriate: the Hill-plot definition applies more generally, but the Adair-equation definition has greater physical meaning in the circumstances where it can be used.

Unfortunately there is no easy way of converting a set of Adair constants to the parameters of the Hill equation that best approximate the same curve. Not only is there no exact correspondence between the measures of cooperativity, as just discussed, but there is no analytical expression even for the half-saturation concentration ($K_{0.5}$ in equation 11.3) except in certain special cases; the best one can usually say is that it will be within the range of the Adair constants (smaller than the largest and larger than the smallest). This is especially unfortunate when one considers that the half-saturation concentration, like the Hill coefficient itself, is an important experimental parameter that facilitates comparison of one curve with another.

11.3   Analysis of binding experiments

11.3.1   Equilibrium dialysis

As noted already, the equation for binding a ligand to a protein molecule has the same form as the Michaelis–Menten equation, and the equation for binding to several non-interacting sites is of the same form as the kinetic equation for a mixture of enzymes that catalyse the same reaction. In principle, the kinetic case differs in that the limiting rate is unknown and must be treated as an experimental quantity to be measured, whereas in binding experiments it is known at the outset that saturation means one molecule of ligand bound per site. In reality this difference is more theoretical than real, however, because the protein molarity is often not known accurately enough for the exact limit to be predicted, and the number of binding sites per molecule may not be known either (and is sometimes, indeed, the principal piece of information that binding experiments are intended to provide).

In practice, therefore, binding is often measured by *equilibrium dialysis*, by setting up an equilibrium across a membrane that is permeable to ligand but not to protein, and measuring the concentrations of ligand on the two sides. The free concentration a_{free} is assumed to be the same on both sides, and is measured directly on the side without the protein, so the concentration a_{bound} of ligand bound to the protein can be obtained by subtracting the known value of a_{free} from the total on that side. Before proceeding further, we must note an important experimental characteristic of equilibrium dialysis. In an ordinary steady-state kinetic experiment it is the kinetic *activity* of the enzyme that is measured, and as long as the specific activity is high the molarity of the enzyme can be small without preventing us from making accurate measurements. By contrast, in equilibrium dialysis, the measured quantity, a_{bound}, can never exceed the total molarity of binding sites, and is obtained, moreover, by subtraction; it is important, therefore, to ensure that the protein and ligand concentrations are similar in magnitude. As this means that equilibrium dialysis and steady-state kinetic experiments are done in very different ranges of protein concentration, they may sometimes give apparently inconsistent results, for example if the enzyme associates at high concentrations.

If there is just one site, a_{bound} is the fractional saturation y multiplied by the protein concentration. However, we shall not assume that the number of sites is known, but will treat a_{bound} as the observed quantity whose dependence on a_{free} is to be determined:

$$a_{bound} = \frac{n e_0 a_{free}}{K + a_{free}} \qquad (11.21)$$

In this equation n is the number of binding sites, all of which are assumed to bind with the same dissociation constant K and without interaction.

11.3.2 The Scatchard plot

As equation 11.21 has exactly the form of the Michaelis–Menten equation it can in principle be analysed by the same methods. The one used almost universally is the plot of a_{bound}/a_{free} against a_{bound}, commonly known as a *Scatchard plot* (Scatchard, 1949), which is related to the following rearrangement of the equation:

$$\frac{a_{bound}}{a_{free}} = \frac{ne_0}{K} - \frac{a_{bound}}{K} \tag{11.22}$$

This shows that if equation 11.21 is obeyed the plot should show a straight line with slope $1/K$ and intercepts ne_0/K on the ordinate and ne_0 on the abscissa. Comparison with the plots in Section 2.6 shows that it corresponds to the plot of v against v/a (Section 2.6.4) with the axes reversed. Curiously, the same biochemists who never think of using anything but a double-reciprocal plot to analyse kinetic data never use anything but a Scatchard plot for analysing binding data; the binding equivalent of the double-reciprocal plot, sometimes known as a *Klotz plot* (Klotz, Walker and Pivan, 1946), is now quite rare [though see Su and Robyt (1994), for examples], and I am not aware of any examples of use of the binding equivalent of the plot of a/v against a. This practice has nothing to do with the respective merits of the different plots, or different needs of kinetic and binding data, but is just a matter of fashion.

In the Scatchard plot neither of the plotted variables is a true independent variable; both are calculated by transforming the actual observations. This has an important consequence for the appearance of the plot, one that makes it easy to recognize a particular class of mistake that can be made in constructing a Scatchard plot (for example Scheibe and Wagner, 1992; Nyce and Metzger, 1997). Consider the simplest case where duplicate observations lead to two different values $a_{free,1}$ and $a_{free,2}$ at the same value of $a_{total} = a_{free} + a_{bound}$. In a direct plot of a_{free} against a_{total} these would of course produce two points with the same abscissa value but different ordinate values: the displacement of the two points with respect to one another would be parallel with the vertical axis. In the Scatchard plot the result is different, because both coordinates are affected by a change in the value of a_{free}: not only are the two ordinate values different, $(a_{total} - a_{free,1})/a_{free,1}$ and $(a_{total} - a_{free,2})/a_{free,2}$, but the two abscissa values are also different, $a_{total} - a_{free,1}$ and $a_{total} - a_{free,2}$. In fact the displacement is along a line through the origin, not parallel with either axis. It is correspondingly difficult, and for practical purposes impossible, to choose any particular series of abscissa values to be plotted; one must take the values as they appear. It follows, therefore, that if one sees a plot in which replicate points are displaced vertically, or where two or more experiments in different conditions lead to the same series of abscissa values, or where the abscissa values are regularly spaced, it is virtually certain that the plot has been drawn incorrectly and that conclusions based on it cannot be trusted.

As the Scatchard plot is frequently used for analysing data more complicated than can be expressed by equation 11.21 it also is important to understand its properties when there are multiple non-equivalent binding sites. The simplest case to consider is a protein with two classes of sites, n_1 sites with dissociation constant K_1 and n_2 sites with dissociation constant K_2. The binding equivalent of equation 11.12 is then as follows:

$$a_{bound} = \frac{n_1 e_0 a_{free}}{K_1 + a_{free}} + \frac{n_2 e_0 a_{free}}{K_2 + a_{free}} \tag{11.23}$$

The question is now whether any simple graphical method exists that permits the ready estimation of the parameters. The correct answer to this question is no, but this is not the answer that most people who carry out binding experiments would give. For any supposed values of the parameters it is simple to calculate the expected shape of the line in a Scatchard plot. Unfortunately this is hardly ever done, and in consequence the parameters estimated from Scatchard plots rarely agree, even approximately, with the data from which they are derived. An example is shown in Figure 11.5. For non-interacting non-equivalent sites the curvature is always in the direction shown,

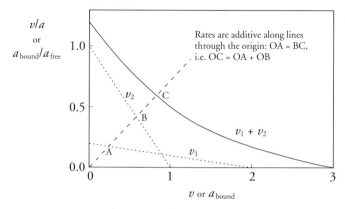

Figure 11.5. Interpretation of a curved Scatchard plot by the method of Rosenthal (1967). The resultant curve can be derived from two or more straight-line components by addition along lines through the origin. The analysis applies equally well to kinetic (v/a against v) and binding (a_{bound}/a_{free} against a_{bound}) data, and the axes are labelled for both kinds of experiment.

though the quantitative details vary. The most important point to notice is that the two straight lines, which represent the plots that would be obtained if only one of the two classes of sites were present, lie far from the curve at all points (compare the Michaelis–Menten hyperbola shown in Figure 2.1 and the discussion in Section 2.3.4). This means that one cannot hope to estimate *either* of these straight lines (and hence one cannot hope to estimate either pair of parameters) by drawing a straight line through some of the points, and the error that results from doing this may be large: for example, in Figure 11.5 naive extrapolation of the part of the curve at low a_{bound} would suggest two sites of high affinity rather than one. In general, there is only one binding parameter of interest that can be deduced in a simple way from a Scatchard plot, namely the *total* number of binding sites, the extrapolated intersection of the curve with the a_{bound} axis.

It is, in fact, quite easy to judge by eye whether the points on a Scatchard plot agree with the straight lines drawn to represent the separate classes of site, because the curve can be obtained by adding the values from them along lines through the origin. The relevant addition is illustrated in Figure 11.5 for an arbitrary point along the curve. The calculation applies even if there are more than two classes of site, so one can draw straight lines for any number of classes, and the final curve is the sum, along lines through the origin, of all the contributions.

I have considered non-interacting non-equivalent sites in some detail because the Scatchard plot is often used for them. Negative cooperativity produces similar curves (and hence cannot be distinguished from non-interacting non-equivalent sites by inspection of the plot), but positive cooperativity produces curvature in the opposite direction with, in extreme cases, a maximum in the value of a_{bound}/a_{free}.

11.4   Induced fit

Early theories of haemoglobin cooperativity assumed that the oxygen-binding sites on each molecule would have to be close enough together to interact electronically. This assumption was made explicit by Pauling (1935), but it was already implied in Hill's and Adair's ideas. Indeed, as long as the binding sites are close together there is no special mechanistic problem to be overcome for explaining cooperativity: no one, for example, feels any need for conformational changes or other exotic mechanisms to explain why the quinone molecule readily binds either zero or two hydrogen atoms but not one, or in other words to explain why the binding of hydrogen atoms to quinone has a Hill coefficient of 2.

However, when the three-dimensional structure of haemoglobin was determined (Perutz and co-workers, 1960), the haem groups proved to be 2.5–4.0 nm apart, too far to interact in any of the ways that had been envisaged. Nonetheless, long-range interactions occur in all positively cooperative proteins, and probably in most others as well, and all modern theories account for these in terms of protein flexibility. In that limited sense they derive from the theory of *induced fit* of Koshland (1958, 1959ab), and the purpose of this section is to examine the experimental and theoretical basis of this theory.

The high degree of specificity that enzymes display towards their substrates has impressed biochemists since the earliest studies of enzymes, long before anything was known about their physical and chemical structures. Fischer (1894) was particularly impressed by the ability of living organisms to discriminate totally between sugars that differed only slightly and at atoms remote from the sites of reaction. To explain this ability, he proposed that the active site of an enzyme was a negative imprint of its substrate(s), and that it would catalyse the reactions only of compounds that fitted precisely. This is similar to the mode of action of an ordinary key in a lock, and the theory is known as Fischer's *lock-and-key model* of enzyme action. For many years, it seemed to explain all of the known facts of enzyme specificity, but as more detailed research was done there were more and more observations that were difficult to account for in terms of a rigid active site of the type that Fischer had envisaged. For example, the occurrence of enzymes for two-substrate reactions that require the substrates to bind in the correct order provides one kind of evidence, as mentioned in Section 7.2.1. A more striking example, noted by Koshland, was the failure of water to react in several enzyme-catalysed reactions where the lock-and-key model would predict reaction. Consider, for example, the reaction catalysed by yeast hexokinase:

$$\text{glucose} + \text{MgATP} \rightleftharpoons \text{glucose 6-phosphate} + \text{MgADP} \qquad (11.24)$$

The enzyme from yeast is not particularly specific for its sugar substrates: it will accept not only glucose but other sugars, such as fructose and mannose. Water does not react, however, even though it can scarcely fail to saturate the active site of the enzyme, at a concentration of 56 M, about 7×10^6 times the Michaelis constant for glucose, and chemically it is at least as reactive as the sugars that do react.

Koshland argued that these and other observations provided strong evidence for a *flexible* active site; he proposed that the active site of an enzyme has the potential to fit the substrate precisely, but that it does not adopt the conformation that matches the substrate until the substrate binds. This conformational adjustment accompanying substrate binding brings about

Hermann Emil Fischer (1852–1919)

Considered by his father as too stupid to be a businessman and better suited to being a student, Emil Fischer became the greatest chemist of his age. In addition to his lock-and-key model of enzyme specificity, he made major contributions to organic chemistry, including work on carbohydrates (especially their stereochemistry), purines, amino acids (he was the discoverer of valine and proline), proteins and triacylglycerols. In 1902 he became the second winner of the Nobel Prize for Chemistry, "in recognition of his synthetic work in the sugar and purine groups". He was in poor health at the end of his life, caused in part by toxic effects of the heavy use of phenylhydrazine in his synthetic work, and after losing two of his sons in the First World War he committed suicide in 1919.

the proper alignment of the catalytic groups of the enzyme with the site of reaction in the substrate. With this hypothesis the properties of yeast hexokinase can easily be explained: water can certainly bind to the active site of the enzyme, but it lacks the bulk to force the conformational change needed for catalysis.

Koshland's theory is known as the *induced-fit* hypothesis, to emphasize its differences from Fischer's theory, which assumes that the fit between enzyme and substrate preexists and does not need to be induced. The lock-and-key analogy can be pursued a little further by likening Koshland's conception to a Yale lock, in which the key can fit only by realigning the tumblers, and in doing so it allows the lock to open. However, a better analogy is perhaps provided by an ordinary glove, with the *potential* of fitting a hand exactly, but fitting it in reality only when the hand is inserted. This analogy has the additional merit of illustrating the essential stereochemical character of enzyme structure: even though a left glove and a right glove may look similar, a left glove does not fit a right hand.

The induced-fit theory has had important consequences in several branches of enzymology (see Section 7.2.1, for example), but it was especially important for understanding the allosteric and cooperative properties of proteins, because it provided a simple and plausible explanation of long-range interactions. Provided that a protein combines rigidity with flexibility in a controlled and purposive way, like a pair of scissors, a substrate-induced conformational change at one point in the molecule may be communicated over several nanometres to any other point.

11.5 The symmetry model of Monod, Wyman and Changeux

11.5.1 Basic postulates of the symmetry model

Cooperative interactions in haemoglobin are not unique in requiring inter-actions between sites that are widely separated in space; the same is true of other cooperative proteins, and of allosteric effects in many enzymes. A striking example is provided by the allosteric inhibition of phosphoribosyl-ATP pyrophosphorylase by histidine: Martin (1963) found that mild treat-ment of this enzyme by Hg^{2+} ions destroyed the sensitivity of the catalytic activity to histidine, but affected neither the uninhibited activity nor the binding of histidine. In other words, the metal ion interfered with neither the catalytic site nor the allosteric site, but with the connection between them. Monod, Changeux and Jacob (1963) studied many examples of cooperative and allosteric phenomena, and concluded that they were closely related and that conformational flexibility probably accounted for both. Subsequently, Monod, Wyman and Changeux (1965) proposed a general model to explain both phenomena within a simple set of postulates. It has sometimes been called the *allosteric model*, but the term *symmetry model* emphasizes the prin-cipal difference between it and alternative models, and avoids the contentious association between allosteric and cooperative effects.

The symmetry model starts from the observation that each molecule of a typical cooperative protein contains several subunits. Indeed, this must be so for binding cooperativity at equilibrium, though it is not required in kinetic cooperativity, as I shall discuss in Section 11.8. For simplicity I shall describe the symmetry model in terms of the binding of a substrate A to a protein in which the number n of subunits is two, mentioning results for an unspecified number of subunits whenever those for $n = 2$ fail to express the general case adequately. Any number of subunits greater than one is possible, and any other kind of ligand (inhibitor or activator) can be considered instead of a substrate.

The symmetry model is based on the following postulates:

1. Each subunit can exist in two different conformations, R and T. These designations, nowadays regarded just as labels, originally stood for *relaxed* and *tense*, because the protein needs to relax to bind substrate, breaking some of the interactions that maintain its native structure in order to make new ones with the substrate.
2. All subunits of a molecule must be in the same conformation at any time; hence, for a dimeric protein, the conformational states R_2 and T_2 are the

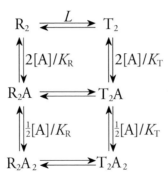

Figure 11.6. Symmetry model of Monod, Wyman and Changeux (1965), illustrated here for a two-site protein.

only ones permitted, the mixed conformation RT being forbidden. (This condition becomes much more restrictive for more than two subunits. For example, for $n = 4$ the allowed states are R_4 and T_4, and R_3T, R_2T_2 and RT_3 are all forbidden.)

3. The two states of the protein are in equilibrium, with an equilibrium constant $L = [T_2]/[R_2]$.

4. A ligand can bind to a subunit in either conformation, but the dissociation constants are different: $K_R = [R][A]/[RA]$ for each R subunit, $K_T = [T][A]/[TA]$ for each T subunit. The ratio K_R/K_T is sometimes written as c, but here we shall use the more explicit form.

11.5.2 Algebraic analysis

These postulates imply the set of equilibria between the various states shown in Figure 11.6, and the concentrations of the six forms of the protein are related by the following expressions:

$$[R_2A] = 2[R_2][A]/K_R \tag{11.25}$$

$$[R_2A_2] = \tfrac{1}{2}[R_2A][A]/K_R = [R_2][A]^2/K_R^2 \tag{11.26}$$

$$[T_2] = L[R_2] \tag{11.27}$$

$$[T_2A] = 2[T_2][A]/K_T = 2L[R_2][A]/K_T \tag{11.28}$$

$$[T_2A_2] = \tfrac{1}{2}[T_2A][A]/K_T = L[R_2][A]^2/K_T^2 \tag{11.29}$$

In each equation, the "statistical" factor 2, $\frac{1}{2}$ or 1 results from the definition of the dissociation constants in terms of individual sites although the expressions are written for complete molecules (compare Section 11.2.4). For example, $K_R = [R][A]/[RA] = 2[R_2][A]/[R_2A]$, because there are two vacant sites in each R_2 molecule and one occupied site in each R_2A molecule.

The fractional saturation y is defined as before as the fraction of sites occupied by ligand, and takes the following form:

$$y = \frac{[R_2A] + 2[R_2A_2] + [T_2A] + 2[T_2A_2]}{2([R_2] + [R_2A] + [R_2A_2] + [T_2] + [T_2A] + [T_2A_2])} \qquad (11.30)$$

In the numerator the concentration of each molecule is counted according to the number of occupied sites it contains (and so empty molecules are not counted at all), but in the denominator each molecule is counted according to how many sites it contains, whether occupied or not, and so each concentration is multiplied by the same factor 2. Substituting the concentrations from equations 11.25–29, this becomes

$$y = \frac{[A]/K_R + [A]^2/K_R^2 + L[A]/K_T + L[A]^2/K_T^2}{1 + 2[A]/K_R + [A]^2/K_R^2 + L + 2L[A]/K_T + L[A]^2/K_T^2}$$

$$= \frac{(1 + [A]/K_R)[A]/K_R + L(1 + [A]/K_T)[A]/K_T}{(1 + [A]/K_R)^2 + L(1 + [A]/K_T)^2} \qquad (11.31)$$

Generalizing this for more than two subunits, the corresponding equation for n unspecified is as follows:

$$y = \frac{(1 + [A]/K_R)^{n-1}[A]/K_R + L(1 + [A]/K_T)^{n-1}[A]/K_T}{(1 + [A]/K_R)^n + L(1 + [A]/K_T)^n} \qquad (11.32)$$

11.5.3   Properties implied by the binding equation

The shape of the saturation curve defined by equation 11.32 depends on the values of n, L and K_R/K_T, as may be illustrated by assigning some extreme values to these constants. If $n = 1$, with only one binding site per molecule, the equation simplifies to $y = [A]/(K_{RT} + [A])$, where $K_{RT} = (1 + L)/(1/K_R + L/K_T)$ is a composite dissociation constant that recognizes that both R and T forms participate in the binding. Despite the complicated expression for this dissociation constant, however, it is still a constant, and so no cooperativity is possible if $n = 1$.

If $L = 0$, the T form of the protein does not exist under any conditions, and the factor $(1 + [A]/K_R)^{n-1}$ cancels between the numerator and denominator, leaving $y = [A]/(K_R + [A])$, which predicts hyperbolic (non-cooperative) binding. A similar simplification occurs if L approaches infinity, when the R form does not exist: in this case, $y = [A]/(K_T + [A])$. It follows that both R and T forms are needed if cooperativity is to be possible, and the two forms must be functionally different from one another, so that $K_R \neq K_T$. If $K_R = K_T$ it is again possible to cancel the common factor $(1 + [A]/K_R)^{n-1}$, leaving a hyperbolic expression. This illustrates the reasonable expectation that if the ligand binds equally well to the two states of the protein, the relative proportions in which they exist are irrelevant to the binding behaviour.

Apart from these special cases, equation 11.32 predicts positive cooperativity, as may be seen by multiplying out the factors $(1 + [A]/K_R)^{n-1}$ and $(1 + [A]/K_T)^{n-1}$, and rearranging the result into the form of the Adair equation. For the dimer, equation 11.31 becomes

$$y = \frac{\left(\dfrac{1/K_R + L/K_T}{1 + L}\right)[A] + \left(\dfrac{1/K_R^2 + L/K_T^2}{1 + L}\right)[A]^2}{1 + 2\left(\dfrac{1/K_R + L/K_T}{1 + L}\right)[A] + \left(\dfrac{1/K_R^2 + L/K_T^2}{1 + L}\right)[A]^2} \tag{11.33}$$

Comparison of this with equation 11.14 shows the two Adair constants to be as follows:

$$K_1 = \frac{1 + L}{1/K_R + L/K_T}, \quad K_2 = \frac{1/K_R + L/K_T}{1/K_R^2 + L/K_T^2} \tag{11.34}$$

and their ratio is

$$\frac{K_2}{K_1} = \frac{(1/K_R + L/K_T)^2}{(1 + L)(1/K_R^2 + L/K_T^2)} = \frac{\dfrac{1}{K_R^2} + \dfrac{2L}{K_R K_T} + \dfrac{L^2}{K_T^2}}{\dfrac{1}{K_R^2} + \dfrac{L}{K_T^2} + \dfrac{L}{K_R^2} + \dfrac{L^2}{K_T^2}} \tag{11.35}$$

As the outer terms in the multiplied-out numerator and denominator are the same, it is only necessary to examine the middle terms, and as $2xy$ is less than $x^2 + y^2$ for any values of x and y it follows that $K_1 \geq K_2$, so the model predicts positive cooperativity in terms of the Adair equation. Similar relationships apply between all pairs of Adair constants in the general case of unspecified n, and so the model predicts positive cooperativity at all stages in the binding process.

As this conclusion is algebraic rather than intuitive, it is helpful to examine one last special case, in which K_T is infinite and A binds *only* to the R state. This is a natural application of the idea of induced fit, though it is not an essential characteristic of the symmetry model as proposed by Monod, Wyman and Changeux. When K_T is infinite equation 11.31 simplifies to

$$y = \frac{(1 + [A]/K_R)[A]/K_R}{L + (1 + [A]/K_R)^2} \tag{11.36}$$

Without the constant L in the denominator this would be an equation for hyperbolic binding, because the common factor $(1 + [A]/K_R)$ would cancel. When $[A]$ is sufficiently large L becomes negligible compared with the rest of the denominator, and the curve approaches a hyperbola. But when $[A]$ is small L dominates the denominator and causes y to rise very slowly from the origin as $[A]$ increases from zero. In other words, as long as L is significantly different from zero the curve of y against $[A]$ must be sigmoid.

When $K_R \neq K_T$, the degree of cooperativity, and hence the steepness of the curve, does not increase indefinitely as L increases, but passes through a maximum when $L^2 = K_T^n/K_R^n$. When this relationship is obeyed the half-saturation concentration ($K_{0.5}$ in equation 11.3) takes the simple form $K_{0.5} = \sqrt{K_R K_T}$. However, as one may see from the representative binding curves calculated from equation 11.31 shown in Figure 11.7, this is in general an unreliable estimate; the best one can say in general is that the half-saturation concentration is between K_R and K_T.

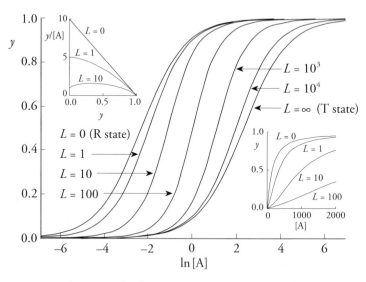

Figure 11.7. Binding curves for the symmetry model. The curves are calculated from equation 11.33, with $K_T/K_R = 100$ and values of L as indicated. In the corresponding Scatchard plots, shown in the upper-left inset for small L values, the extreme cases ($L = 0$ for the pure R state and also, but not shown, $L = \infty$ for the pure T state) yield straight lines and the intermediate cases yeild lines with downward curvature. The extreme curves are hyperbolic when y is plotted against $[A]$, whereas the intermediate curves are sigmoid, as illustrated for some smaller L values in the lower-right inset.

11.5.4 Heterotropic effects

Monod, Wyman and Changeux distinguished between *homotropic effects*, or interactions between identical ligands, and *heterotropic effects*, or interactions between different ligands, such as a substrate and an allosteric effector. Although the symmetry model requires homotropic effects to be positively cooperative, it places no corresponding restriction on heterotropic effects, and it can accommodate these with no extra complications; this is, indeed, one of its most satisfying features. If a second ligand B binds preferentially to the R state of the protein, the state preferred by A, at a different site from A (so that there is no competition between them), it facilitates binding of A by increasing the availability of molecules in the R state; it thus acts as a positive heterotropic effector, or allosteric activator. On the other hand, a ligand C that binds preferentially to the T state, which binds A weakly or not at all, has the opposite effect: it hinders the binding of A by decreasing the availability of molecules in the R state, and will thus act as a negative heterotropic effector, or allosteric inhibitor. If all binding is *exclusive*, which means that each ligand binds either to the R state or to the T state, but not to both, the resulting binding equation for A, as modified by the presence of B and C, is particularly simple:

$$y = \frac{(1 + [A]/K_R)[A]/K_R}{L^{app} + (1 + [A]/K_R)^2} = \frac{(1 + [A]/K_R)[A]/K_R}{L\left(\dfrac{1 + [C]K_{CT}}{1 + [B]/K_{BR}}\right)^2 + (1 + [A]/K_R)^2} \qquad (11.37)$$

The allosteric constant L is now replaced by an apparent value L^{app} that increases with the inhibitor concentration and decreases with the activator concentration, reflecting the capacity of inhibitors to displace the equilibrium away from the state that favours substrate binding, and of activators to displace it towards the same state. When ligands do not bind exclusively to one or other state, the behaviour is naturally more complicated, but one can still get a reasonable idea of the possibilities by examining Figure 11.7 in the light of equation 11.37.

High concentrations of allosteric effectors of either sort clearly tend to decrease the cooperativity, as they make the protein resemble either pure R or pure T, but there may be effects in the opposite direction at low concentrations if the value of L (the value of L^{app} in the absence of effectors) is not optimal. Consider, for example, the constants used for constructing Figure 11.7 with $L = 10$. Any concentration of activator tends to decrease L^{app}, taking it further from $L^{app} = 100$ (the square root of 10^4, and so the value for maximum cooperativity), and hence making it less cooperative.

However, adding an allosteric inhibitor initially increases the cooperativity, to a maximum at $L^{app} = 100$, but further increases in inhibitor concentration tend to decrease it. If the value of L were greater than 100 rather than less, it would be the activator that would increase the cooperativity at low concentrations, whereas the inhibitor would decrease the cooperativity at all concentrations. These tendencies are not entirely obvious from examination of the curves in Figure 11.7, because in the middle of the range the differences in steepness are not immediately apparent to the eye. However, one can get a correct impression of the directions in which the steepness changes from the fact that the curves at the extremes are noticeably less steep than the one in the middle.

A complication arises if we consider an ordinary (non-allosteric) competitive inhibitor that binds to the R state at exactly the same sites as the substrate A. This is considered in Problem 11.5 at the end of this chapter.

The binding properties of phosphofructokinase from *Escherichia coli* were thoroughly studied by Blangy, Buc and Monod (1968): over a wide range of concentrations of ADP and phospho*enol*pyruvate, an allosteric activator and inhibitor respectively, the binding of the substrate fructose 6-phosphate proved to agree well with the predictions of the symmetry model. Nonetheless, it cannot be regarded as a universal explanation of binding cooperativity, because it cannot explain the negative cooperativity observed for some enzymes, and some of its postulates are not altogether convincing. The central assumption of conformational symmetry is not readily explainable in structural terms, for example, and for many enzymes it is necessary to postulate the occurrence of a *perfect K system*, which means that the R and T states of the enzyme have identical catalytic properties despite having grossly different binding properties. These and other questionable aspects of the symmetry model have stimulated the search for alternatives.

11.6   The sequential model of Koshland, Némethy and Filmer

11.6.1   Postulates

Although the symmetry model incorporates the idea of purposive conformational flexibility, it departs from the theory of induced fit in permitting ligands to bind to both R and T conformations, albeit with different binding constants. Koshland, Némethy and Filmer (1966) showed that a more orthodox application of induced fit, known as the *sequential model*, could account for cooperativity equally well. Like Monod, Wyman and Changeux, they postulated the existence of two conformations, which they termed the A and B conformations, corresponding to the T and R conformations respectively. This inversion of the order in which they are usually spoken has sometimes been a source of confusion, and for that reason, and also to allow continued use of A as a symbol for substrate, as elsewhere in this book, the symbols T and R will be used here[*]. Unlike Monod, Wyman and Changeux, Koshland, Némethy and Filmer assumed that the R conformation was *induced* by ligand binding, so that substrate binds only to the R conformation, the R conformation exists only with substrate bound to it, and the T conformation exists only with substrate not bound to it.

Koshland, Némethy and Filmer postulated that cooperativity arose because the properties of each subunit were modified by the conformational states of the neighbouring subunits. The same assumption is implicit in the symmetry model, but it is emphasized in the sequential model, which is more concerned with the details of interaction, and avoids the arbitrary assumption that all subunits must exist simultaneously in the same conformation. Hence conformational hybrids, such as TR in a dimer, or T_3R, T_2R_2 and TR_3 in a tetramer, are not merely allowed, but follow directly from the assumption of strict induced fit.

Because the symmetry model was not concerned with the details of subunit interactions, there was no need in Section 11.5 to consider the geometry of subunit association, the quaternary structure of the protein. By contrast, the sequential model does require consideration of geometry, for any protein with more than two subunits, because different arrangements of subunits result in different binding equations. Here we shall consider a dimer for simplicity, and the geometry can then be ignored, but it cannot be ignored when extending the treatment to trimers, tetramers and so on.

The emphasis on geometry and the need to treat each geometry separately have given rise to the widespread but erroneous idea that the sequential model is more general and complicated than the symmetry model, but for any given geometry the two models are about equally complicated and neither is a special case of the other. Both can be generalized into the same general model (Haber and Koshland, 1967), by relaxing the symmetry requirement of the symmetry model and the strict induced-fit requirement of the sequential model, but it is questionable whether this is helpful, because the resulting equation is too complicated to use.

---

[*]Koshland, Némethy and Filmer also followed the "haemoglobin convention" of expressing their model in terms of association constants, but dissociation constants are used here to facilitate comparison with the model of Monod, Wyman and Changeux, and to maintain consistency with the rest of this book. As a result, most of the equilibrium constants are the reciprocals of the corresponding constants given in the original paper.

To see how a binding equation is built up in the sequential model, consider the changes that occur when a molecule T_2 binds one molecule of A to become RTA:

1. One subunit must undergo the conformational change $T \rightarrow R$, a change represented by the notional equilibrium constant $K_t = [T]/[R]$ for an isolated subunit. In the simplest version of the sequential model K_t is tacitly assumed to be large, so that the change occurs to a negligible extent if it is not induced by ligand binding.
2. One molecule of A binds to a subunit in the R conformation, represented by $[A]/K_A$, where K_A is the dissociation constant $[R][A]/[RA]$ for binding of A to an isolated subunit in the R conformation.
3. In a dimer there is one interface across which the two subunits can interact. In the initial T_2 molecule these are evidently two T subunits, so it is a $T:T$ interface, but in RTA it becomes a $T:R$ interface, a change represented by a notional equilibrium constant $K_{R:T} = [R:T]/[T:T]$. In the original discussion by Koshland, Némethy and Filmer there was some confusion about whether subunit interaction terms should be regarded as absolute measures of interface stability (so that there would be implied dimensions and logically a constant $K_{T:T}$ should be introduced to represent the stability of the $T:T$ interface), or whether they should be regarded as measures of the stability relative to a standard state (the $T:T$ interface). The latter interpretation is just as rigorous, simpler to apply (because it leads to constants that are inherently dimensionless, so there is no question of ignoring dimensions), and leads to simpler equations with fewer constants; it will be used here.
4. Finally, a statistical factor of 2 is required, because there are two equivalent ways of choosing one out of two subunits to bind A. (The word "equivalent" is essential here, because non-equivalent choices would lead to distinguishable molecules that would have to be treated separately.)

11.6.2 Algebraic analysis

Putting all this together we may write down the following expression for the concentration of RTA in terms of those of T_2 and A:

$$[RTA] = \frac{2[T_2][A]K_{R:T}}{K_t K_A} \tag{11.38}$$

Although using a dimer as example allows the sequential model to be explained with minimal complications, it leaves one or two essential aspects of the model unexplained, so we must pause briefly to consider what expression would result from applying the same rules to the formation of a molecule $R_2T_2A_2$ from a tetramer T_4. We cannot now ignore geometry, because there are at least three different possible arrangements. Here we shall suppose that they interact as if arranged at the corners of a square, and that of the two different ways in which two ligand molecules can be bound to such a molecule we are dealing with the one in which the two ligand molecules are on adjacent (rather than diagonal) subunits. This gives a concentration of

$$[R_2T_2A_2]_{\text{adjacent}} = \frac{4[T_4][A]^2 K_{R:T}^2 K_{R:R}}{K_t^2 K_A^2} \tag{11.39}$$

As there is now a new kind of interface between the two adjacent subunits in the R conformation that have ligand bound, we need a new kind of subunit interaction constant, $K_{R:R}$, to represent its stability relative to that of the $T:T$ interface, but otherwise equation 11.39 is constructed in just the same way as equation 11.38, from the same components.

Returning now to the dimer, we can write down an expression for the concentration of R_2A_2 according to the same principles:

$$[R_2A_2] = \frac{[T_2][A]^2 K_{R:R}}{K_t^2 K_A^2} \tag{11.40}$$

Substituting equations 11.38 and 11.40 into the expression for the fractional saturation, we have:

$$y = \frac{[RTA] + 2[R_2A_2]}{2([T_2] + [RTA] + [R_2A_2])} = \frac{\dfrac{[A]K_{R:T}}{K_t K_A} + \dfrac{[A]^2 K_{R:R}}{K_t^2 K_A^2}}{1 + \dfrac{2[A]K_{R:T}}{K_t K_A} + \dfrac{[A]^2 K_{R:R}}{K_t^2 K_A^2}} \tag{11.41}$$

The sequential model is closely concerned with subunit interactions, and the essential question that an equation such as equation 11.41 answers is how binding of a ligand is affected by the stability of the mixed interface $R:T$ relative to the mean stability of the interfaces $R:R$ and $T:T$ between subunits in like conformations. Inspection of equation 11.41 shows that making $K_{R:T}$ smaller increases the importance of the outer terms with respect to the inner, but this can be made clearer by defining a constant $c^2 = K_{R:T}^2/K_{R:R}$ to express this relative stability. (It may seem surprising at first sight that there is no mention of the $T:T$ interface in this definition, but remember that both $K_{R:T}$ and $K_{R:R}$ already define the stabilities of the $R:T$ and $R:R$ interfaces relative to the $T:T$ interface.)

If $K_{R:T}$ is replaced by $cK_{R:R}^{1/2}$ using this definition, it then becomes clear that $\overline{K} = K_t K_A / K_{R:R}^{1/2}$ always occurs as a unit; although its three components are conceptually distinct, they cannot be separated experimentally by means of binding measurements. The equation can thus be simplified in appearance without loss of generality by writing it in terms of c and $\overline{K}$:

$$y = \frac{c[A]/\overline{K} + [A]^2/\overline{K}^2}{1 + 2c[A]/\overline{K} + [A]^2/\overline{K}^2} \tag{11.42}$$

The definition of c is the same for all quaternary structures: it applies not only to dimers, but also to trimers, tetramers, and so on, regardless of how the subunits are arranged. The definition of $\overline{K}$ is a little more complicated: it always contains $K_t K_A$ as an inseparable unit (as follows from steps 1 and 2 in the description above of how any binding process is decomposed into different notional components); on the other hand, the power to which $K_{R:R}$ is raised in the denominator varies with the number of subunits and with the number of R:R interfaces that the fully liganded molecule contains. However, this has little importance: the important point is that regardless of quaternary structure and geometry the range of binding behaviour possible for a single ligand in the sequential model is determined by two parameters, one to represent the stability of the R:T interface with respect to the R:R and T:T interfaces, the other an average dissociation constant for the complete binding process from fully unliganded to fully liganded protein. This is the geometric mean of the Adair dissociation constants, as may be seen for the dimer by writing these explicitly, after comparing equation 11.42 with equation 11.14:

$$K_1 = \overline{K}/c; \quad K_2 = c\overline{K} \tag{11.43}$$

with the ratio as follows:

$$K_2/K_1 = c^2 \tag{11.44}$$

11.6.3 Properties implied by the binding equation

It is now clear that the degree of cooperativity, and hence the shape of the binding curve, depends only on the value of c. As illustrated in Figure 11.8, values of $c < 1$ generate positive cooperativity and values of $c > 1$ generate negative cooperativity. The effect of varying $\overline{K}$ is not shown (to avoiding making the Figure too complicated), but can be stated simply: it has no effect on the shapes of the curves when $\ln[A]$ is the abscissa (and affects only the scaling with other variables as abscissa), but simply causes them to be shifted to the right (if $\overline{K}$ increases) or left (if $\overline{K}$ decreases); in other words, $\overline{K}$ has no effect on the degree of cooperativity.

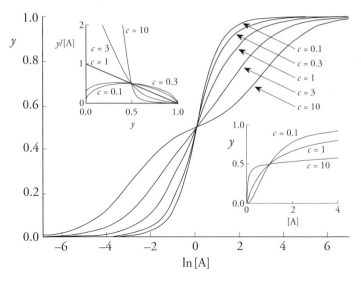

Figure 11.8. Binding curves for the sequential model. The curves are calculated from equation 11.42, with $\overline{K} = 1000$ and the values of c used as labels. The value of $\overline{K}$ does not affect the shape of a curve, but only its location along the abscissa. The corresponding Scatchard plots are shown in the inset at upper left, and a selection of plots of y against [A] in the inset at lower right.

It would be convenient if there were a simple correspondence between the parameters of the sequential model and those of the Hill equation (equation 11.3), but although, as we have seen, the degree of cooperativity depends only on c, there is no one-to-one relationship between h and c, as h varies with the degree of saturation and c does not. By contrast, the relationship between $\overline{K}$ and $K_{0.5}$, the half-saturation concentration, is as simple as one could ask: they are identical.

As incorrect statements are sometimes found in the literature one should notice that in the sequential model the shape of the curve is defined by fewer parameters than in the symmetry model (one instead of two). Thus the capacity of the sequential model to explain negative cooperativity whereas the symmetry model cannot is not a consequence of the large number of constants considered in deriving the sequential model (K_t, K_A, $K_{R:T}$ and $K_{R:R}$).

More subtle misconceptions about the sequential model are implied by some authors' use of names such as "Adair–Koshland model" or "Pauling–Koshland model" for it. Although crediting it to Adair correctly implies that the sequential model is a special case of the Adair model it also incorrectly implies that the symmetry model is not. In reality, *both* models are expressible in terms of Adair constants (equations 11.34 and 11.43), as, indeed, any valid equation to describe binding of a ligand to a macromolecule at equilibrium must be. The simplest test of meaningfulness that one can apply to a proposed equation written for this purpose is adherence to the Adair equation;

equations that are not special cases of the Adair equation, such as some that were proposed for lactate dehydrogenase (Weber and Anderson, 1965; Anderson and Weber, 1965), generally violate the principle of microscopic reversibility (Section 4.6).

The reference to Pauling is misleading for a different reason. Although some of the mathematics in the sequential model is the same as that applied by Pauling (1935) to haemoglobin, the underlying concepts are different: Pauling was working at a time when it was reasonable to suppose that the oxygen-binding sites of haemoglobin were close enough together in space to interact in an ordinary chemical way, and there was no implication of conformational interactions.

For a positively cooperative dimer there is no difference at all between the binding curves that the two models can predict, as any value less than 1 of the ratio K_2/K_1 of Adair constants that one can give is equally consistent with the other. It is thus impossible to distinguish between them on the basis of binding experiments with a dimer. In principle they become different for trimers and higher oligomers, because the symmetry model then allows the binding curves plotted as in Figures 11.7–8 to become unsymmetrical about the half-saturation point, whereas the curves generated by the sequential model are always symmetrical with respect to rotation through 180° about this point. However, the departures from symmetry are quite small, and highly accurate data are needed to detect them. Moreover, the greatest degree of cooperativity occurs in the symmetry model when $Lc^n = 1$, and as this is also the condition for a symmetrical binding curve in the symmetry model one may expect that for at least some enzymes evolution will have eliminated any asymmetry that might have existed.

11.7   Association–dissociation models of cooperativity

Various groups (Frieden, 1967; Nichol, Jackson and Winzor, 1967) independently suggested that cooperativity might in some circumstances result from the existence of an equilibrium between protein forms in different states of aggregation, such as a monomer and a tetramer. If a ligand has different dissociation constants for the two forms, then this model predicts cooperativity even if there is no interaction between the binding sites in the tetramer. Conceptually the model is rather similar to the symmetry model, and the cooperativity arises in a similar way, but the equations are more complicated, because they need to take account of the dependence of the degree of association on the protein concentration. Consequently, in contrast to the equations for the symmetry and sequential models, this concentration does not cancel from the expressions for the saturation curves.

This type of model is much more amenable to experimental verification than the other models we have considered, because the effects of protein concentration ought to be easily observable. They are, indeed, observed for a number of enzymes, such as glutamate dehydrogenase (Frieden and Colman, 1967) and glyceraldehyde 3-phosphate dehydrogenase (Ovádi and co-workers, 1979), and other examples are noted by Kurganov (1982), who discusses association–dissociation models in detail.

11.8   Kinetic cooperativity

All the models discussed in the earlier part of this chapter have essentially been equilibrium models that can be applied to kinetic experiments only by assuming that v/V is a true measure of y. However, cooperativity can also arise for purely kinetic reasons, in mechanisms that would show no cooperativity if binding could be measured at equilibrium. This was known from the studies of Ferdinand (1966) and Rabin (1967) and others when the classic models of cooperativity were being developed, but there did not at that time seem to be experimental examples of cooperativity in monomeric enzymes. As a result it was widely assumed that even if multiple binding sites were not strictly necessary for generating cooperativity they provided the only mechanisms actually found in nature, and the purely kinetic models were given little attention. However, rat-liver hexokinase D provided an example of positive cooperativity in a monomeric enzyme, making it clear that models for such properties would need to be considered seriously.

Hexokinase D is an enzyme found in the liver and pancreatic islets of vertebrates. Because of a mistaken perception that it is more specific for glucose than the other vertebrate hexokinases (see Cárdenas, Rabajille and Niemeyer, 1984a), it is frequently known in the literature as "glucokinase", but this name will not be used here. It is monomeric over a wide range of conditions, including those used in its assay (Holroyde and co-workers, 1976; Cárdenas, Rabajille and Niemeyer, 1978), but it shows marked deviations from Michaelis–Menten kinetics when the glucose concentration is varied at constant concentrations of the other substrate, $MgATP^{2-}$ (Niemeyer and co-workers, 1975; Storer and Cornish-Bowden, 1976b). When replotted as a series of Hill plots, the data show h values ranging from 1.5 at saturating $MgATP^{2-}$ to a low value, possibly 1.0, at vanishingly small $MgATP^{2-}$ concentrations. On the other hand, there are no deviations from Michaelis–Menten kinetics with respect to $MgATP^{2-}$ itself.

Other examples of cooperativity in monomeric enzymes are not abundant, but they exist (see Cornish-Bowden and Cárdenas, 1987), and indicate that mechanisms that generate kinetic cooperativity can no longer be ignored. I shall consider two such mechanisms in this section. The older is due to Ferdinand (1966), who pointed out that the steady-state rate equation for the random-order ternary-complex mechanism (Section 7.3.2) is much more complicated than equation 7.6 if it is derived without assuming substrate-binding steps to be at equilibrium; he suggested that a model of this kind, which he called a *preferred-order* mechanism, might provide an explanation for the cooperativity of phosphofructokinase. Although it is clear enough from consideration of the method of King and Altman that deviations from Michaelis–Menten kinetics ought to occur with this mechanism, this explanation is rather abstract and algebraic, and cannot easily be expressed in conceptually simple terms. The point is that both pathways for substrate binding may make significant contributions to the total flux through the reaction, but the relative magnitudes of these contributions change as the substrate concentrations change. Thus the observed behaviour corresponds approximately to one pathway at low concentrations, but to the other at high concentrations.

Ricard, Meunier and Buc (1974) developed an alternative model of kinetic cooperativity from earlier ideas of Rabin (1967) and Whitehead (1970). Their model is known as a *mnemonical* model (from the Greek for memory), because it depends on the idea that the enzyme changes conformation relatively slowly, and is thus able to "remember" the conformation that it had during a recent catalytic cycle. It is shown (in a simplified form) in Figure 11.9.

Its essential characteristics are as follows: it postulates that there are two forms E and E′ of the free enzyme that differ in their affinities for A, the first substrate to bind; in addition equilibration between E, E′, A and EA must be slow relative to the maximum flux through the reaction. With these postulates,

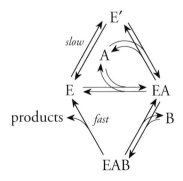

Figure 11.9. The mnemonical model for kinetic cooperativity.

the behaviour of hexokinase D is readily explained. As the concentration of B is lowered, the rate at which EA is converted into EAB and thence into products must eventually become slow enough for E, E′, A and EA to equilibrate. Consequently at vanishingly small concentrations of B, the binding of A should behave like an ordinary equilibrium, with no cooperativity, because there is only a single binding site. At high concentrations of B, on the other hand, it becomes possible for EA to be removed so fast that it cannot equilibrate and the laws of equilibria cease to apply (Storer and Cornish-Bowden, 1977). Deviations from Michaelis–Menten kinetics are then possible because at low concentrations of A the two forms of free enzyme can equilibrate but at high concentrations they cannot.

In the mnemonical mechanism as proposed by Ricard, Meunier and Buc (1974) the same form of EA complex is produced from both forms of free enzyme when substrate binds to them. However, this is not a necessary feature of the model, and the *slow-transition* model developed a little earlier by Ainslie, Shill and Neet (1972) supposes that two different conformational states exist during the whole catalytic cycle, with transitions between them possible at any point that occur at rates that are slow compared with the catalytic rate. A model of this kind applied to hexokinase D proved capable of explaining its properties at least as well as the mnemonical model outlined above (Cárdenas, Rabajille and Niemeyer, 1984b). In general, any model that allows substrate to bind in two or more parallel steps will generate a rate equation with terms in the square or higher power of the concentration of the substrate concerned, so there is no limit to the models of kinetic cooperativity that can be devised. Unfortunately it is quite difficult in practice to distinguish between them or to assert with much confidence that one fits the facts better than another. Certainly, both the mnemonical and slow-transition models are able to explain the behaviour of hexokinase D adequately. A thorough discussion of the kinetic cooperativity of hexokinase D is provided by Cárdenas (1995).

Problems

11.1 Watari and Isogai (1976) proposed a plot of $\log[v/a(V - v)]$ against $\log a$ as an alternative to the Hill plot. What is the slope of this plot (expressed in terms of the Hill coefficient h)? What advantage does the plot have over the Hill plot?

11.2 Write down an equation for the rate of a reaction catalysed by a mixture of two enzymes, each of which obeys Michaelis–Menten kinetics, one with limiting rate V_1 and Michaelis constant K_{m1} and the other with V_2

and K_{m2}. Differentiate this equation twice with respect to a and show that the resulting second derivative is negative at all values of a. What does this imply about the shape of a plot of v against a?

11.3 Derive an expression for the Hill coefficient in terms of K_1 and K_2 for an enzyme that obeys equation 11.14, and hence show that the definition of cooperativity in terms of h is identical to one in terms of K_1 and K_2 for this system. At what value of a is h a maximum or a minimum, and what is its extreme value?

11.4 The near coincidence of the two asymptotes in Figure 11.4 is presumably just that, a coincidence. What relationship between the dissociation constants (as defined by equation 11.15) does it imply?

11.5 Consider an inhibitor that binds to a protein that obeys the simplest form of the symmetry model (equation 11.36) as a simple (non-allosteric) analogue of the substrate, binding only to the R form, to the same sites as the substrate, in such a way that substrate and inhibitor cannot be bound simultaneously to the same site. What effect would you expect such an inhibitor to have on substrate binding at (a) low, and (b) high concentrations of both?

11.6 In the sequential model there is no one-to-one relationship between the Hill coefficient h and the parameter c that defines the degree of cooperativity. There is, however, a one-to-one relationship between c and the extreme value of h: in the simplest form of the model this extreme occurs when the ligand concentration $[A]$ is equal to the mean dissociation constant $\overline{K}$. Taking h as the slope (at any value of $[A]$) of a plot of $\ln[y/(1-y)]$ against $\ln[A]$, show that h can be expressed as $[A]/[y(1-y)]$ multiplied by the slope of a plot of y against $[A]$ (regardless of the model assumed for the cooperativity). Then, for the specific model expressed by equation 11.42, derive an expression for h at half saturation ($[A]=\overline{K}$), and use it to express the extreme value of h in terms of c.

Chapter 12

Kinetics of Multi-Enzyme Systems

12.1 Enzymes in their physiological context

12.1.1 Enzymes as components of systems

Most of this book has been concerned with the properties of enzymes considered one at a time, even though in living organisms virtually all enzymes act as components of systems; their substrates are products of other enzymes, and their products are substrates of other enzymes. For most of the history of enzymology there has been little to connect the sorts of kinetic measurements people make with the physiological roles of the enzymes they study: after an enzyme has been identified from some physiological observation, the first thing that an enzymologist does is to purify it, or at least separate it from its physiological neighbours. Nearly all kinetic studies of enzymes are thus made on enzymes that have been deliberately taken out of physiological context. This may well be necessary for understanding the chemical mechanisms of enzyme catalysis, but one cannot obtain a full understanding of how enzymes fulfil their roles in metabolic pathways if they are only examined under conditions where all other aspects of the pathway are suppressed.

One might have expected the discovery of feedback inhibition and the associated properties of cooperative and allosteric interactions to have re-established the importance of enzymes as physiological elements; in reality it increased the separation between the practice of enzymology and the physiology of enzymes, because it started to seem natural to think that a few enzymes, such as phosphofructokinase, could be classified as "regulatory enzymes", and the rest could be largely ignored in discussions of physiological regulation. The most extreme form of this idea is to think that all that one needs to do to understand the regulation of a pathway is to identify the regulatory step, usually assumed to be unique, and study all the interactions of the enzyme catalysing it.

The mechanisms discussed in the previous chapter constitute an essential component in the total understanding of enzyme physiology, but they leave an important question unanswered: how do we know that an effect on the activity of any enzyme will be translated into an effect on the flux of metabolites through a pathway? This can only be answered by moving away from the study of enzymes one at a time and towards a *systemic* treatment, that is to say one that considers how the components of a system affect one another.

This is far from trivial, even in the steady state, even though the analysis of the steady-state kinetic behaviour of an individual isolated enzyme may now be regarded as a solved problem. Analytical expressions for the steady-state rates do not in general exist, even for two-enzyme systems, and the difficulties rapidly increase as more enzymes are added. There is nothing in ordinary enzyme kinetics to justify an assumption about how even complete knowledge of the rate equation for a particular "regulatory enzyme" would allow any quantitative prediction of the effect that a change in its activity would have on the flux through the pathway in which it is embedded. In this chapter, therefore, we examine the relationships between the kinetics of pathways and the kinetic properties of their component enzymes.

12.1.2   Moiety conservation

In Section 7.1 I emphasized that the distinction between substrates and coenzymes has no meaning when discussing enzyme mechanisms, but I noted that a different view was possible in discussing metabolism. The point is that if all one's attention is focussed on a single two-substrate enzyme, hexokinase for example, it is irrelevant that one of its substrates, glucose, is involved in relatively few *other* metabolic reactions, whereas the other, ATP, is involved in many. However, as soon as one comes to place hexokinase in its physiological context this difference between the two substrates becomes highly relevant. The common coenzymes, ATP and ADP, the oxidized and reduced forms of NAD, and so on, all exist as members of pairs of metabolites such that the sum of the concentrations of the two members is constant, or at least it changes on a slower time scale than the one being considered, whereas their ratio is a variable. Thus in many contexts the appropriate variable to consider is the ratio of ATP and ADP concentrations, not the two independent concentrations. Notice that each pair of concentrations treated in this way results in a decrease of one in the total number of variables in the system: although this is not a point that will be discussed any further in this chapter, it is important to take account of it in computer models of metabolism (Section 12.10), which are typically underdetermined if there are no constraints on the values of variables (in other words they do not contain enough information to define a unique steady state).

When ATP is converted into ADP and back again, the ADP fragment common to both molecules remains unchanged. In this context it is called a *moiety* and the organization of metabolism to maintain its concentration constant is called *moiety conservation*. Much of the discussion in this chapter can be adapted to take account of moiety conservation, treating the ratio of the two concentrations concerned as a third kind of variable (Hofmeyr, Kacser and van der Merwe, 1986), to add to the metabolic fluxes and metabolite concentrations that will be the primary focus of the chapter.

12.1.3   Characterization of enzymes in permeabilized cells

For many years enzymes have not only usually been purified before characterizing their kinetic behaviour, but they have also usually been studied in unnatural buffers chosen for reasons of reproducibility, stability or just convenience. In mechanistic work this is not a major issue, as it does not require a great leap of faith to suppose that a mechanism established in artificial conditions applies qualitatively to catalysis in the living cell, even if some quantitative details may be incorrect. However, for understanding in quantitative terms how an enzyme fulfils its physiological role such details become too important to set aside. There is a need, therefore, for experiments to be done in conditions closer to natural ones than those that exist in a typical spectrophotometric assay.

One possibility might seem to be to use cell extracts obtained as the water-soluble fraction after rupturing the cells and centrifuging to separate the solid debris. However, an extract of this kind is not much more natural than a solution of the purified enzyme in a phosphate buffer, and any properties associated with cell membranes or other elements of cell organization are irretrievably lost when the cell is destroyed, even though they may be essential for the proper function of the enzyme *in vivo* (Clegg and Jackson, 1988). A better solution is to use permeabilized cells (Reeves and Sols, 1973; Serrano, Gancedo and Gancedo, 1973), obtained by using drugs such as digitonin to create pores that allow small molecules to diffuse freely while leaving the basic structure in place and keeping proteins and other large molecules confined to their natural locations. Experiments in these conditions are said to be done *in situ*, an intermediate stage that combines much of the naturalness of experiments *in vivo* with much of the experimental control possible in experiments *in vitro*.

Methylglyoxal metabolism in yeast illustrates questions studied *in situ* that would have been difficult to study in purified extracts. Methylglyoxal

Henrik Kacser (1918–1995)

Henrik Kacser was born in Câmpina, Roumania, of an Austrian mother and a Hungarian father, who worked there as an oil engineer. He was educated in Berlin and Belfast, where he trained as a chemist. He moved to Edinburgh in 1952, and spent his entire career there. By 1973 his research work, with just one publication in the preceding eight years, must have seemed to his colleagues to be over, but this would have been an illusion: although his paper with James Burns did not immediately have very much impact, it came to be seen as the start of a new field, encompassing the most successful approach to the analysis of biochemical systems that has yet appeared. From 1973 until his sudden death, Kacser was not merely active but increasingly active, as the undisputed leader of the field that he had created.

is a toxic intermediate formed spontaneously in the decomposition of triose phosphates, and is converted into the harmless product lactate by the action of two enzymes, lactoylglutathione lyase and hydroxyacylglutathione hydrolase. Use of permeabilized cells allowed both to be characterized under conditions approaching those in the living cell, including substrate concentrations close to those *in vivo* (Martins and co-workers, 2001a). The next step was to determine metabolic fluxes from methylglyoxal to lactate in a complete system, again in permeabilized cells, in order to construct a computer model of the pathway (Martins, Cordeiro and Ponces Freire, 2001b). As the lyase requires glutathione for activity (regenerated in the reaction catalysed by the hydrolase, so it acts catalytically in the pathway as a whole, though not in either reaction considered by itself) one could use glutathione to control whether the pathway was active or inactive, and thus measure other fluxes, such as the glycolytic flux, with and without the methylglyoxal pathway.

12.2   Metabolic control analysis

Several overlapping systems have developed during the past three decades for analysing the behaviour of metabolic systems, but here I shall refer to only one of them, *metabolic control analysis*, which originated from work of Kacser and Burns (1973) and Heinrich and Rapoport (1974), and is now by far the most widely known and used. In its simplest form, it is concerned with the steady states of systems of enzymes that connect a series of metabolites, with two or

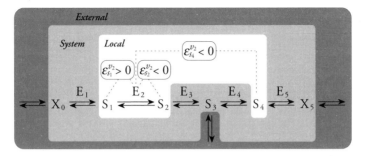

Figure 12.1. A metabolic pathway of five enzymes. Although ultimately a metabolic system consists of an entire cell, or even an entire organism, some degree of isolation is needed in order to analyse it. In the example shown, the heavily shaded part of the diagram, showing connections to X_0, S_3 and X_5, is considered to be "outside the system", so that X_0 is a source metabolite and X_5 is a sink metabolite even though in a larger system they would be intermediates. A higher degree of isolation, represented by the unshaded part of the diagram, is needed for considering how the activity of a single enzyme, E_2 in this example, depends on interactions with various metabolites. The elasticities represented by the symbol ε are discussed in Section 12.3.

more reservoirs of metabolites whose concentrations are fixed independently of the enzymes in the system, and can thus be considered as "external" to it. The reservoirs include at least one source, from which metabolites flow, and at least one sink, into which they flow. Neither of these flows need be irreversible, and the classification into sources and sinks is not absolute: both might well be considered as internal metabolites of a larger system under different circumstances. Nonetheless, in any one analysis it is essential to be precise about which metabolites are considered internal and which external, and it is accordingly helpful to use symbols that indicate the difference: in common with many current papers I shall use S (Substrate) and X (eXternal).

In the example shown in Figure 12.1, the system is considered to be the pathway from a source X_0 to a sink X_5, with internal metabolites S_1, S_2, S_3 and S_4: the heavily shaded part of the scheme, including the external connections to these metabolites and to S_3, is considered to be outside the system.

In addition to the metabolites connected by the enzymes, there can be any number of external effectors with fixed concentrations. Few reactants in a living organism are external, but there are so many reactions to be considered that the entire system is difficult to comprehend. To make metabolism manageable for analysis, therefore, the system must be defined as just a part of the whole organism, and the metabolites at the interfaces with the rest of the organism must be defined as external.

In the simplest version of metabolic control analysis considered here, each rate must be proportional to the concentration of exactly one enzyme, and no enzyme can act on more than one reaction in the system. However, these are not absolute restrictions as they can easily be relaxed at the cost of some additional complications in the analysis. Various reviews (for example, Fell, 1992; Cornish-Bowden, 1995a) may be consulted for fuller information, as well as textbooks (Fell, 1997; Heinrich and Schuster, 1996) and other books (Cornish-Bowden and Cárdenas, 1990, 2000) with contributions from many of the groups active in the field.

12.3   Elasticities

12.3.1   Definition of elasticity

Enzyme kinetic behaviour is usually expressed in terms of rate equations such as the reversible Michaelis–Menten equation (equation 2.46):

$$v = \frac{k_A e_0 a - k_P e_0 p}{1 + \dfrac{a}{K_{mA}} + \dfrac{p}{K_{mP}} + \dfrac{i}{K_i}} \tag{12.1}$$

shown here for the interconversion of metabolites A and P in the presence of an inhibitor I with concentration i and (competitive) inhibition constant K_i.

The reversible form should be preferred for considering physiological states, because products are normally always present. In metabolic simulations one should be cautious before writing irreversible equations, because doing so can generate entirely false results about the behaviour of a pathway; for example, irreversible equations can suggest that no steady state is possible in conditions where more realistic, albeit more complicated, equations yield a stable steady state. This contrasts with the usual conditions in the cuvette, where it is easy to ensure irreversibility. In a metabolic pathway the essential characteristic is not reversibility as such, but sensitivity of the early steps to conditions at the end: if feedback loops or other mechanisms supply this information to the early enzymes then it may make little difference whether or not the small degree of reversibility of reactions with large equilibrium constants is taken into account (Cornish-Bowden and Cárdenas, 2001a). It is also important to distinguish between the possibility of a reverse reaction, a property of the negative term in the numerator of the rate expression, and product sensitivity, a property of positive terms in the denominator. It is perfectly possible, and indeed common in practice, for an enzyme-catalysed reaction to be virtually irreversible and yet significantly inhibited by its product: the negative numerator term can be negligible even though the positive denominator product terms are not. It is then safe to omit the small negative numerator term, but dangerous to omit denominator product terms.

To investigate enzyme mechanisms by the sort of kinetic analysis that has occupied most of this book, especially Chapter 7, it is clearly necessary to express the kinetic behaviour in terms of an equation that resembles equation 12.1. However, in metabolic control analysis one is not much interested in enzyme mechanisms, not because they are not important but because a different aspect of the system is at issue. The sort of question asked is not "how can the variation of v with a be explained?" but "how much will v change if there is a small change in a?", or even "how much will a change if there is a small change in v?". This last form of the question reminds us that in living systems the distinction between independent and dependent variables is much less clear cut than we are used to in the laboratory. In a typical steady-state kinetic experiment, we normally decide what the concentrations are going to be and then measure the rates that result; in the cell, both rates and concentrations are properties of the whole system, and although it may sometimes be possible to regard the rates as determined by the metabolite concentrations, or the metabolite concentrations as decided by the rates, the reality is that both are dependent variables. This point will be taken up in more detail in Section 12.3.4.

Ordinary kinetic equations like equation 12.1 can certainly show how a rate responds to a small concentration change, but they do so in a way that is inconveniently indirect. However, partial differentiation with respect to a gives the following expression:

$$\frac{\partial v}{\partial a} = \frac{k_A e_0 \left[1 + p \left(\frac{1}{K_{mP}} + \frac{k_P}{k_A K_{mA}} \right) + \frac{i}{K_i} \right]}{\left(1 + \frac{a}{K_{mA}} + \frac{p}{K_{mP}} + \frac{i}{K_i} \right)^2} \tag{12.2}$$

As it stands, this derivative has the dimensions of reciprocal time, and as one is usually more interested in relative derivatives than absolute ones, it is usual to convert it into a relative form by multiplying by a/v:

$$\frac{\partial \ln v}{\partial \ln a} = \frac{a \partial v}{v \partial a}$$

$$= \frac{1 + p \left(\frac{1}{K_{mP}} + \frac{k_P}{k_A K_{mA}} \right) + \frac{i}{K_i}}{\left(1 - \frac{k_P p}{k_A a} \right) \left(1 + \frac{a}{K_{mA}} + \frac{p}{K_{mP}} + \frac{i}{K_i} \right)}$$

$$= \frac{1}{1 - \frac{k_P p}{k_A a}} - \frac{\frac{a}{K_{mA}}}{1 + \frac{a}{K_{mA}} + \frac{p}{K_{mP}} + \frac{i}{K_i}}$$

$$= \frac{1}{1 - \Gamma/K_{eq}} - \frac{\alpha}{1 + \alpha + \pi + \iota} \tag{12.3}$$

in which $\Gamma = p/a$ is the "mass action ratio", K_{eq} is the equilibrium constant by virtue of the Haldane relationship (Section 2.7.2), and $\alpha = a/K_{mA}$, $\pi = p/K_{mP}$ and $\iota = i/K_i$ are the concentrations scaled by the appropriate Michaelis or inhibition constants. This equation may appear complicated, but when it is rearranged into the difference between two fractions, as in the last forms shown, it is seen that both fractions have simple interpretations: the first measures the "disequilibrium", the departure of the system from equilibrium; the second measures the degree of saturation of the enzyme with the reactant considered. However, complicated or not, metabolic control analysis is not usually concerned with the algebraic form of this derivative, but with its numerical value: if it is zero, v does not vary with a; if it is positive, v increases as a increases; if it is negative, v decreases as a increases. Accordingly it is given a name, the *elasticity*, and symbol, ε, to express its central importance in metabolic control analysis:

$$\varepsilon_a^v \equiv \frac{\partial \ln v}{\partial \ln a} \equiv \frac{a \partial v}{v \partial a} \tag{12.4}$$

The superscript v in the symbol may seem sufficiently obvious to be superfluous, but metabolic control analysis is always concerned with systems of more than one enzyme, and so one needs a superscript to specify which rate is being considered.

If one examines equation 12.4 in the light of Section 1.2 of this book, one may reasonably feel that this has been a long-winded way of introducing a new name for an old concept, as the elasticity is familiar to all biochemists as the *order of reaction*. The only difference is that whereas the recommendations (International Union of Biochemistry, 1982) discourage the use of this term when its value is not a constant, suggesting *apparent order* instead, there is no suggestion either in equation 12.4 or in the way that it was derived that the quantity defined should be a constant. However, this recommendation has not been widely followed; it was designed to avoid conflict with the then newly revised recommendations of the International Union of Pure and Applied Chemistry (1981), but few biochemists find variable orders objectionable. The description of the curve defined by the Michaelis–Menten equation in Section 2.3.4 can be interpreted as meaning that there is a gradual transition from first order with respect to substrate at very low concentrations, passing through non-integral orders to approach zero order at saturation, with an order of 0.5 at half-saturation. It is easy to confirm, by putting $p = 0$ in equation 12.3, that the elasticity behaves in exactly the same way; it is, in fact, identical to the quantity normally understood by biochemists as the order of reaction.

The term elasticity comes from econometrics, where it designates a quantity similar to that in control analysis, but opposite to it in sign: the elasticity for a commodity is a percentage *decrease* in demand divided by the percentage *increase* in its price that provoked the decreased demand. This is an obscure precedent for a biochemical term, and the term *kinetic order* used in *biochemical systems theory* (Savageau, 1976), an alternative approach that covers much of the same territory as metabolic control analysis, is much better. Nonetheless, I shall retain the term elasticity in this chapter, as it, together with its synonym *elasticity coefficient*, is in universal use in metabolic control analysis.

Returning to equation 12.1, we can differentiate it with respect to each concentration in turn to obtain the following complete set of elasticities, which will now be expressed as differences between disequilibrium and saturation terms (as in the last form of equation 12.3), as they are easiest to understand in this form:

$$\varepsilon_a^v = \frac{1}{1 - \Gamma/K_{eq}} - \frac{\alpha}{1 + \alpha + \pi + \iota} \tag{12.5}$$

$$\varepsilon_p^v = \frac{-\Gamma/K_{eq}}{1 - \Gamma/K_{eq}} - \frac{\pi}{1 + \alpha + \pi + \iota} \tag{12.6}$$

$$\varepsilon_{e_0}^v = \begin{cases} 1 & \text{if E catalyses the reaction} \\ 0 & \text{if E does not catalyse the reaction} \end{cases} \tag{12.7}$$

$$\varepsilon_i^v = -\frac{\iota}{1 + \alpha + \pi + \iota} \tag{12.8}$$

The second form of equation 12.7 does not follow from equation 12.1, which did not consider the possibility of more than one enzyme in the system. It simply states that an enzyme has an elasticity of zero with respect to a reaction that it does not catalyse: an obvious point, perhaps, but worth making explicitly as it is important in the theory of metabolic control analysis.

12.3.2 Common properties of elasticities

Although equations 12.5–8 were derived from a specific model, the reversible Michaelis–Menten equation, and their exact forms are dependent on this model, they illustrate a number of points that apply fairly generally, some of them universally:

1. Reactant elasticities are normally positive when the direction of disequilibrium is such that the reactant is a substrate, negative when it is a product. ("Normally" here means that substrate inhibition and product activation generate exceptions to this generalization.) Note, however,

that the passage from positive to negative as the reaction passes from one side of equilibrium to the other is not via zero, as one might naively guess, but via infinity: reactant elasticities are infinite at equilibrium! This characteristic underlines the danger of writing irreversible rate equations in computer simulation. With irreversible reactions, in the absence of cooperativity and substrate inhibition, substrate elasticities are normally in the range 0 to 1: values close to zero are characteristic of high substrate concentrations, and infinite elasticities are impossible. It follows that the numerical values of elasticities for such reactions are entirely different from those likely to be found in a living cell.

2. Enzymes have unit elasticities for their own reactions, and zero elasticities for other reactions. These generalizations are not universally true, as they depend on the assumption that each rate is proportional to the total concentration of one enzyme only. They fail if an enzyme associates (with itself or with other enzymes in the system) to produce species with altered kinetic properties. Much of metabolic control analysis assumes the truth of these generalizations, and the equations become considerably more complicated when they fail.

3. Elasticities for non-reactant inhibitors are always negative. Conversely, elasticities for non-reactant activators are always positive. The qualification "non-reactant" can be ignored as long as one remembers that a product inhibitor is transformed into a substrate when the direction of flux changes. In addition, elasticities for non-reactant inhibitors and activators are independent of the degree of disequilibrium (the qualification is now indispensable).

12.3.3   Enzyme kinetics viewed from control analysis

From the point of view of metabolic control analysis, measuring elasticities is what enzymologists have been doing since the time of Michaelis and Menten, even if the term itself is unfamiliar. Nonetheless, there are important differences in emphasis, and the measurements made in traditional experiments may not be useful for metabolic control analysis. In ordinary studies of enzymes, experiments are usually designed to reveal information about the mechanism of action. (Even experimenters whose interests are primarily physiological usually follow procedures that were originally designed to shed light on mechanisms.) Because different mechanisms of action often predict patterns of behaviour that differ only slightly, if at all, one is often forced to design experiments carefully to illuminate any small deviations from expected behaviour that may exist, and the experiments themselves must be done with great attention to accuracy. Kinetic analysis frequently involves extrapolation of observations to infinite or zero concentrations (compare Section 7.4.1). Moreover, experiments are rarely done with anything approaching a complete system, as it is rare for an enzyme in a cuvette to encounter even half of the metabolites that might influence its activity in the cell; if any additional enzymes are present they are either trace contaminants with negligible effect on the enzyme of interest, or they are coupling enzymes deliberately added in quantities designed to be optimal for the assay, without any relation to the concentrations that may exist in the cell.

All of these characteristics are quite inappropriate for metabolic control analysis. Although one is still interested in describing the kinetic behaviour of an enzyme, the objective is not to understand the mechanism but to integrate the kinetic description into a description of the kinetic behaviour of a system — at the simplest level a system of a few enzymes constituting a pathway, but ultimately a complete organ or organism. To a good approximation, properties that are at the limits of accuracy of one's equipment, and consequently are difficult to measure, are not important in the behaviour of the system: if mechanistic differences don't produce major differences in kinetic behaviour they don't matter.

On the other hand, one can no longer afford to simplify the experiment by omitting metabolites that affect the kinetics: all reactants and effectors should be present at concentrations as close as possible to those that occur in the cell. This includes products, of course, and implies that reactions need to be studied under reversible conditions. Even if the equilibrium constant strongly favours reaction in one direction, the conditions should be at least in principle reversible; apart from anything else product inhibition may be significant even if the complete reverse reaction is not.

Despite this emphasis on a complete realistic reaction mixture, elasticity measurements remain unnatural in one respect: they refer to an enzyme isolated from its pathway, treating all concentrations of metabolites that influence its activity as constants, and ignoring the effects that other enzymes in the pathway may have on these concentrations.

12.3.4   Rates and concentrations as effects, not causes

As I have discussed, rates and intermediate concentrations are both properties of a whole metabolic system, and one cannot regard either of them as independent variables unless they are explicitly defined as such when specifying the system under study. One can better understand the importance of this point by considering a situation opposite to the one usually considered to apply in the spectrophotometer. Suppose that an enzyme behaves in the ordinary way, following Michaelis–Menten kinetics, but an experiment is set up so that the

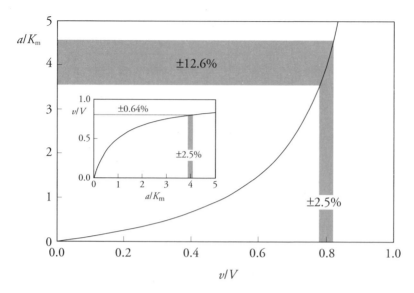

Figure 12.2. A different view of the Michaelis–Menten relationship. The inset (a relabelled form of Figure 2.1) shows the usual view, with v shown as a function of a, and in this view the near-saturation domain is one in which small variations in a lead to even smaller variations in v. However, in a metabolic system it is no less correct to regard a as a function of v, in which case one may describe the same domain as one on the verge of catastrophe, as small changes in v may produce huge changes in a.

rate is set by the experimenter and the substrate concentration that results is measured. It is then appropriate to write the Michaelis–Menten equation as an expression for a in terms of v rather than as equation 2.15:

$$\frac{a}{K_m} = \frac{v/V}{1 - v/V} \tag{12.9}$$

For displaying the results of such an experiment one should reverse the usual axes, plotting a against v rather than *vice versa*. Thus instead of Figure 2.1 we should draw the Michaelis–Menten hyperbola as shown in Figure 12.2. This is exactly the same curve as that in Figure 2.1, but its psychological impact is quite different, as one can see by trying to describe the behaviour of the enzyme at substrate concentrations around $5K_m$, or rates around $0.8V$. Looking at the curve in Figure 2.1 one would say that this was rather an uninteresting region where nothing much would happen if the conditions changed, which would usually be taken to mean that a changed. But exactly the same starting point in Figure 12.2 is close to a catastrophe, because only a 20% increase in v will bring the same enzyme, with the same kinetic properties, to a state where no steady state is possible, as v will reach V.

Neither Figure 2.1 nor Figure 12.2 truly represent conditions in the cell, because neither a nor v can really be manipulated independently of the other.

However, Figure 12.2 may often be closer to the reality, because many enzymes in the middles of pathways may do little other than process substrates as fast or as slowly as they receive them: such enzymes may in practice operate at whatever rate is demanded, adjusting the concentrations around them accordingly. Near-saturation is thus not a boring state where nothing much happens, but a state on the verge of catastrophe, a point of view that Atkinson (1977) in particular has emphasized.

Consideration of the linear inhibition types (Chapter 5) provides an even more striking example (Cornish-Bowden, 1986). As long as one regards a kinetic equation as an expression for the dependence of a rate on one or more concentrations, there is little difference between even the extreme types of inhibition, competitive and uncompetitive; the differences between the various degrees of mixed inhibition are even smaller. As a result, most inhibitors in the literature are described as competitive, regardless of the magnitude of any uncompetitive component, which passes unnoticed by many experimenters. However, as soon as one considers an enzyme that must adjust the concentrations around it to suit the rate that is set externally, the situation changes dramatically, and even the sleepiest experimenter would notice the difference between competitive and uncompetitive inhibition. The equations corresponding to equation 12.9 are

$$a = \frac{K_m \left(1 + \dfrac{i}{K_{ic}}\right)}{\dfrac{V}{v} - 1} \tag{12.10}$$

for competitive inhibition and

$$a = \frac{K_m}{\dfrac{V}{v} - 1 - \dfrac{i}{K_{iu}}} \tag{12.11}$$

for uncompetitive inhibition. These two equations are not just minor variations on one another; they are utterly and irreconcilably different from one another. As i has a linear effect on a in equation 12.10, the change in a is always proportionately smaller than the change in i that provokes it, however large the change in i may be. In equation 12.11 the presence of i in a negative denominator term means that the denominator can become zero, making it impossible to achieve a steady state. This can happen, moreover, at quite moderate inhibitor concentrations. If the enzyme is half-saturated in the absence of inhibitor, for example, then $V/v = 2$ and it is sufficient for $i = K_{iu}$ for no steady state to exist. This means that if i exceeds K_{iu} the substrate concentration will increase indefinitely and no steady state will be reached.

It follows, therefore, that competitive inhibition is almost an irrelevance in the living cell, as quite small changes in substrate concentration can overcome changes in inhibitor concentration. For this reason efforts to produce pharmacologically useful compounds by searching for analogues of natural substrates (substances likely to be competitive inhibitors) often result in disappointment. Uncompetitive inhibitors, by contrast, may be expected to have potentially devastating effects on living cells, and this may be one reason why it is difficult to find clear examples of naturally occurring uncompetitive inhibitors. It probably also explains the effectiveness of the commercial herbicide Glyphosate (or "Roundup"), an uncompetitive inhibitor of 3-phosphoshikimate 1-carboxyvinyltransferase (Boocock and Coggins, 1983).

Before leaving this example it is worth noting that the qualitative difference between equations 12.10 and 12.11 derives from the negative coefficient of i in equation 12.11; the difference in numerators has much less importance. The crucial question, therefore, is whether a significant uncompetitive component is present at all: if it is, even if it is quantitatively less than the competitive component in ordinary constant-substrate conditions, it will generate obvious uncompetitive effects.

12.4   Control coefficients

12.4.1   Definitions

To this point we have only been discussing the ordinary kinetic behaviour of isolated enzymes, albeit in terminology rather different from that used in mechanistic studies. The objective of metabolic control analysis is now to determine how the kinetic behaviour of a sequence of enzymes composing a pathway can be explained in terms of the properties of the individual isolated enzymes. If a system such as the one defined in Figure 12.1 is set up, the concentrations of the reservoirs X_0 and X_5 are constant, as are the kinetic properties of the enzymes, but the individual enzyme rates v_i and the concentrations of the internal metabolites S_j are free to vary. Even if these concentrations are initially arbitrary, they will tend to vary so that each approaches a steady state. (A steady state does not necessarily exist, and if one does exist it is not necessarily unique: for simplicity, however, I shall assume that there is a unique steady state.) Taking S_1 as an example, a steady state implies that the rate v_1 at which it is supplied must be equal to the rate v_2 at which it is consumed. A steady state in s_2 likewise implies $v_2 = v_3$ and so on; when all the metabolites are in steady state all the enzyme rates in a pathway as simple as that in Figure 12.1 must be equal to one another, with a value J that is called the *flux*. If there are branches or other complications there can be several different fluxes in the same pathway, and the relationships are more complicated. However, the principles remain straightforward and obvious: the total flux into each branch-point metabolite is equal to the total flux out of it.

Enzyme rates are *local* properties, because they refer to enzymes isolated from the system. Steady-state fluxes and metabolite concentrations, by contrast, are *systemic* properties*. Elasticities (defined in Section 12.3.1) are also local properties, but there are systemic properties analogous to them that are called *control coefficients*. Suppose that some change in an external parameter u (undefined for the moment, but see Section 12.4.2) brings about a change in a local rate v_i when the enzyme E_i is isolated: what is the corresponding effect on the system flux J when E_i is embedded in the system? This is not known *a priori*, and the ith flux control coefficient is defined by the following ratio of derivatives:

$$C_i^J = \frac{\partial \ln J}{\partial \ln u} \bigg/ \frac{\partial \ln v_i}{\partial \ln u} = \frac{\partial \ln J}{\partial \ln v_i} \qquad (12.12)$$

The simpler form shown at the right is not strictly correct, because v_i is not a true independent variable of the system, but it is acceptable as long as it is remembered that there is always an implied external parameter u even if it is not shown explicitly. This definition corresponds to the way Heinrich and Rapoport (1974) defined their *control strength*; in apparent contrast, the *sensitivity coefficient* of Kacser and Burns (1973) was defined in terms of the effect of changes in enzyme concentration on flux (both of these terms have been superseded in current work by *control coefficient*):

$$C_i^J = \frac{\partial \ln J}{\partial \ln e_i} \qquad (12.13)$$

These definitions may appear to be different, but provided that equation 12.7 is true, so that each enzyme rate is proportional to the total enzyme concentration, equations 12.12 and 12.13 are equivalent. Equation 12.12 has the advantage of avoiding the widespread misunderstanding that metabolic control analysis is limited to effects brought about by changes in enzyme concentration. Initially it was usual to follow Kacser and Burns in using definitions similar to equation 12.13, but there is now a widespread view that control coefficients ought not to be defined in terms of any specific parameter (though see Section 12.4.2), and that equation 12.12 should be

---

*The distinction between *rate* and *flux* made in metabolic control analysis has no obvious correspondence with the distinction between *rate* and *chemiflux* (also often shortened to *flux*) made in radioactive tracer experiments (Section 8.5.1).

regarded as the fundamental definition of a control coefficient. The quantity defined by equation 12.13 is then better regarded as an example of a *response coefficient*, which happens to be numerically equal to the corresponding control coefficient only because the connecting elasticity is assumed to be unity (see Section 12.7 below).

These definitions of a flux control coefficient now allow a precise statement of the circumstances in which an enzyme could be said to catalyse the rate-limiting step of a pathway. Such a description would be reasonable if any variation in the activity of the enzyme produced a proportional variation in the flux through the pathway, and in terms of equations 12.12–13 this would mean that such an enzyme had $C_i^J = 1$. For example, if increasing the activity of phosphofructokinase twofold in a living cell caused a twofold increase in the glycolytic flux then phosphofructokinase would have $C_i^J = 1$ and one could call it the rate-limiting enzyme for glycolysis. In fact, however, Heinisch (1986) found experimentally that increasing the activity of phospho-fructokinase 3.5-fold in fermenting yeast had *no detectable effect* on the flux to ethanol. Similar experiments have subsequently been carried out with other supposedly rate-limiting enzymes, with similar results, and there are theoretical reasons for expecting it to be very rare for any enzyme to have complete flux control (Section 12.5).

A concentration control coefficient is the corresponding quantity that defines effects on metabolite concentrations, for example, for a metabolite S_j with concentration s_j:

$$C_i^{s_j} = \frac{\partial \ln s_j}{\partial \ln u} \bigg/ \frac{\partial \ln v_i}{\partial \ln u} = \frac{\partial \ln s_j}{\partial \ln v_i} \tag{12.14}$$

In this equation the simpler form at the right is subject to the same reservations as the corresponding form in equation 12.12, implying the existence of a parameter u even if this is not explicit.

12.4.2   The perturbing parameter

As noted already, the easiest way to perturb the activity of an enzyme in a system without perturbing anything else is to vary its concentration e_0, and so the most obvious interpretation of the perturbing parameter u that appeared in equation 12.12 is that it is identical to e_0. At first sight this may seem to be just one of many possibilities, but in practice it is virtually the only one, because most inhibitors and activators change not only the activity of an enzyme but also its sensitivity to its substrates and products: in other words, they alter not only the apparent limiting rate of an enzyme but also its Michaelis constants, as extensively discussed in Chapters 5 and 6. Pure non-competitive inhibitors would be the exception, but, as discussed in those chapters, these are not available for almost any enzyme, apart from highly unspecific effectors such as protons that would not fulfil the required role of perturbing just one enzyme activity. Finding a pure non-competitive inhibitor for just one enzyme in a system would be difficult enough; finding a whole series to allow perturbation of any enzyme at will is likely to be a fantasy for the foreseeable future. In practice, therefore, identifying the parameter u with the enzyme concentration corresponds closely with experimental reality, and varying enzyme concentrations by genetic or other means remains the only practical way of perturbing one enzyme activity at a time.

None of this means that ordinary inhibitors and activators cannot be used to probe the control structure of a pathway. On the contrary, Groen and co-workers (1982) used them very effectively for this purpose in a pioneering study of mitochondrial respiration, but the analysis was more complicated than simply treating each inhibitor titration as a perturbation of the activity of one enzyme.

12.5   Summation relationships

The fundamental properties of the control coefficients are expressed by two *summation relationships*, of which the first, due to Kacser and Burns (1973), defines the sum of flux control coefficients:

$$\sum_{i=1}^{n} C_i^J = 1 \tag{12.15}$$

and the second, due to Heinrich and Rapoport (1975), defines the sum of concentration control coefficients:

$$\sum_{i=1}^{n} C_i^{s_j} = 0 \tag{12.16}$$

in which n is the number of enzymes in the system, and s_j is the concentration of any internal metabolite. If the pathway is branched there will be more than one flux: equation 12.15 then holds with J defined as any of these.

Various proofs of these relationships exist, of which that of Reder (1988) is probably the most rigorous and general. However, as it assumes a knowledge of matrix algebra, I shall not give it here, preferring the original "thought experiments" of Kacser and Burns (1973), which are easier to understand. Suppose we make small changes de_i in all enzyme concentrations in any system in which the reaction rates are all proportional to the concentrations of the enzymes that catalyse them. The total effect on any flux J may be written as the sum of the individual effects:

$$dJ = \frac{\partial J}{\partial e_1}de_1 + \frac{\partial J}{\partial e_2}de_2 + \frac{\partial J}{\partial e_3}de_3 + \ldots \tag{12.17}$$

Dividing all terms by J, multiplying each term on the right-hand side by unity (expressed as a ratio of equal enzyme concentrations), and introducing the definitions of the flux control coefficients, this becomes

$$\begin{aligned}\frac{dJ}{J} &= \frac{e_1}{J}\frac{\partial J}{\partial e_1}\frac{de_1}{e_1} + \frac{e_2}{J}\frac{\partial J}{\partial e_2}\frac{de_2}{e_2} + \frac{e_3}{J}\frac{\partial J}{\partial e_3}\frac{de_3}{e_3} + \ldots \\[2mm] &= \frac{\partial \ln J}{\partial \ln e_1}\frac{de_1}{e_1} + \frac{\partial \ln J}{\partial \ln e_2}\frac{de_2}{e_2} + \frac{\partial \ln J}{\partial \ln e_3}\frac{de_3}{e_3} + \ldots \\[2mm] &= C_1^J\frac{de_1}{e_1} + C_2^J\frac{de_2}{e_2} + C_3^J\frac{de_3}{e_3} + \ldots \tag{12.18}\end{aligned}$$

As we have assumed nothing about the magnitudes of the changes de_i apart from saying that they are small, we can give them any small values we like, so let us assume that each enzyme concentration changes in the same proportion, so that each de_i/e_i has the same value α. A moment's reflection should show that such a change is equivalent to changing the time scale of the measurements: thus it should change all steady-state fluxes through the system by exactly a factor of α. It follows therefore that equation 12.18 can be written as follows:

$$\alpha = C_1^J\alpha + C_2^J\alpha + C_3^J\alpha + \ldots \tag{12.19}$$

and dividing by α gives a result equivalent to equation 12.15:

$$1 = C_1^J + C_2^J + C_3^J + \ldots \tag{12.20}$$

Applying the same logic to the concentration control coefficients, the only difference is that changing the time scale should leave all concentrations unchanged, and putting a zero on the left-hand side gives the equivalent to equation 12.16:

$$0 = C_1^{s_j} + C_2^{s_j} + C_3^{s_j} + \ldots \tag{12.21}$$

The essence of equation 12.15 is that control of flux through a pathway is shared by all the enzymes in the system, which need not contain any step catalysed by an enzyme whose properties determine the kinetic behaviour of the whole system, that is to say a rate-limiting enzyme as defined in Section 12.4.1. If all flux control coefficients are positive, the idea of sharing control is completely straightforward: no enzyme can have a control coefficient greater than 1, and if any enzyme has one approaching 1 those of the others must be correspondingly small. Although flux control coefficients are normally positive in unbranched pathways, exceptions can occur if substrate inhibition or product activation dominate the behaviour of some enzymes. With branched pathways the idea of sharing is less clear, because flux control coefficients are then often negative and they may also be greater than 1. However, one often (though not universally) finds that the following generalizations apply: any enzyme has a positive flux control coefficient for the flux through its own reaction; numerically significant negative flux control coefficients are not common, occurring mainly for enzymes and fluxes that occur in different branches immediately after a branch point.

12.6 Relationships between elasticities and control coefficients

12.6.1 Connectivity properties

For an unbranched pathway of n enzymes there is one summation relationship for flux control coefficients, and $(n - 1)$ summation relationships between concentration control coefficients, but there are n flux control coefficients and $n(n - 1)$ concentration control coefficients, or n^2 control coefficients altogether. In effect, therefore, the summation relationships provide n equations relating n^2 unknowns. To calculate all of the unknowns, a further $n(n - 1)$ equations are needed. These come from the *connectivity properties*, which will now be described.

If an enzyme concentration e_i and a metabolite concentration s_j change simultaneously by de_i and ds_j respectively in such a way as to produce no effect on the rate v_i through the enzyme in question, the changes must be related as follows:

$$\frac{dv_i}{v_i} = \frac{de_i}{e_i} + \varepsilon_{s_j}^{v_i} \frac{ds_j}{s_j} = 0 \qquad (12.22)$$

and so

$$\frac{de_i}{e_i} = -\varepsilon_{s_j}^{v_i} \frac{ds_j}{s_j} \qquad (12.23)$$

A corresponding equation can be written for each value of i, so we can readily calculate the small changes that have to be made to all the enzyme concentrations to produce a particular change in one metabolite concentration, leaving all other metabolite concentrations and all rates (and hence all fluxes) unchanged. If there is no change in flux for a particular series of enzyme perturbations, equation 12.18 may be written as follows:

$$0 = C_1^J \frac{de_1}{e_1} + C_2^J \frac{de_2}{e_2} + C_3^J \frac{de_3}{e_3} + \ldots \tag{12.24}$$

and by substituting equation 12.23 for each value of i into this and cancelling the common factor $-ds_j/s_j$ from all the terms it becomes

$$C_1^J \varepsilon_{s_j}^{v_1} + C_2^J \varepsilon_{s_j}^{v_2} + C_3^J \varepsilon_{s_j}^{v_3} + \ldots = 0 \tag{12.25}$$

This now expresses the *connectivity property* between flux control coefficients and elasticities, which was discovered by Kacser and Burns (1973). In a real pathway some of the elasticities will normally be zero, as it is unlikely that every metabolite has a significant effect on every enzyme. Consequently some of the terms in sums such as that in equation 12.25 will usually be missing, but in this section I shall include all the terms that could in principle occur.

As no metabolite concentration except s_j has been changed, an equation similar to equation 12.24 applies to any metabolite S_k for which $k \neq j$:

$$0 = C_1^{s_k} \frac{de_1}{e_1} + C_2^{s_k} \frac{de_2}{e_2} + C_3^{s_k} \frac{de_3}{e_3} + \ldots \tag{12.26}$$

but if $k = j$ there is a change ds_j/s_j and so

$$\frac{ds_j}{s_j} = C_1^{s_j} \frac{de_1}{e_1} + C_2^{s_j} \frac{de_2}{e_2} + C_3^{s_j} \frac{de_3}{e_3} + \ldots \tag{12.27}$$

and, substituting equation 12.23 as before, these lead to the connectivity properties between concentration control coefficients and elasticities (Westerhoff and Chen, 1984):

$$C_1^{s_k} \varepsilon_{s_j}^{v_1} + C_2^{s_k} \varepsilon_{s_j}^{v_2} + C_3^{s_k} \varepsilon_{s_j}^{v_3} + \ldots = \begin{cases} -1 & k = j \\ 0 & k \neq j \end{cases} \tag{12.28}$$

For an unbranched pathway of n enzymes there are $(n-1)$ equations similar to equation 12.25, one for each metabolite, and $(n-1)^2$ equations similar to equation 12.28, one for each combination of two metabolites, and together these provide the $n(n-1)$ additional equations that must be combined with the n summation relationships to provide the n^2 equations needed to calculate all the control coefficients from the elasticities.

12.6.2 Control coefficients in a three-step pathway

Although complications arise with branched pathways, these do not alter the essential point that a sufficient number of independent relationships between control coefficients and elasticities exist for it to be possible in principle to calculate all of the control coefficients. This not only establishes that the steady-state properties (control coefficients) of a complete system follow from the properties of its components (elasticities), but it also shows how the calculation can be done. The actual solution of n^2 simultaneous equations is complicated if n is not trivially small, and in current practice the problem is treated as one of matrix algebra (Fell and Sauro, 1985; Fell, 1992). I shall not enter into details here, but instead will examine the results of such a calculation for a simple example.

$$X_0 \xrightleftharpoons[]{E_1} S_1 \xrightleftharpoons[]{E_2} S_2 \xrightleftharpoons[]{E_3} X_3 \tag{12.29}$$

For the three-step pathway shown in equation 12.29 the three flux control coefficients may be expressed in terms of elasticities as follows:

$$C_1^J = \frac{\varepsilon_{S_1}^{v_2} \varepsilon_{S_2}^{v_3}}{\varepsilon_{S_1}^{v_2} \varepsilon_{S_2}^{v_3} - \varepsilon_{S_1}^{v_1} \varepsilon_{S_2}^{v_3} + \varepsilon_{S_1}^{v_1} \varepsilon_{S_2}^{v_2} - \varepsilon_{S_1}^{v_2} \varepsilon_{S_2}^{v_1}} \tag{12.30}$$

$$C_2^J = \frac{-\varepsilon_{S_1}^{v_1} \varepsilon_{S_2}^{v_3}}{\varepsilon_{S_1}^{v_2} \varepsilon_{S_2}^{v_3} - \varepsilon_{S_1}^{v_1} \varepsilon_{S_2}^{v_3} + \varepsilon_{S_1}^{v_1} \varepsilon_{S_2}^{v_2} - \varepsilon_{S_1}^{v_2} \varepsilon_{S_2}^{v_1}} \tag{12.31}$$

$$C_3^J = \frac{\varepsilon_{S_1}^{v_1} \varepsilon_{S_2}^{v_2} - \varepsilon_{S_1}^{v_2} \varepsilon_{S_2}^{v_1}}{\varepsilon_{S_1}^{v_2} \varepsilon_{S_2}^{v_3} - \varepsilon_{S_1}^{v_1} \varepsilon_{S_2}^{v_3} + \varepsilon_{S_1}^{v_1} \varepsilon_{S_2}^{v_2} - \varepsilon_{S_1}^{v_2} \varepsilon_{S_2}^{v_1}} \tag{12.32}$$

The numerators are placed over the corresponding terms in the denominators to emphasize that not only is the denominator identical in the three expressions, but also that it consists of the sum of the three numerators, in accordance with the summation relationship, equation 12.15. Each term in the denominator consists of a product of elasticities, one elasticity for each internal metabolite in the system; each numerator consists of those denominator terms that do not contain the activity of the enzyme whose control coefficient is being expressed; for example, equation 12.32 refers to E_3, so the numerator does not contain any elasticities with superscript v_3.

The minus signs in equations 12.30–32 arise naturally from the algebra. We should not allow them to mislead us into thinking that any of the individual terms in the equations are negative. Under "normal" conditions (defined as in Section 12.3.2 as conditions where there is no substrate inhibition or product activation) the combination of positive substrate elasticities with negative product elasticities makes all of the terms in the three equations positive.

There are corresponding relationships for each metabolite concentration, for example, for s_1:

$$C_1^{s_1} = \frac{\varepsilon_{s_2}^{v_3} - \varepsilon_{s_2}^{v_2}}{\varepsilon_{s_1}^{v_2}\varepsilon_{s_2}^{v_3} - \varepsilon_{s_1}^{v_1}\varepsilon_{s_2}^{v_3} + \varepsilon_{s_1}^{v_1}\varepsilon_{s_2}^{v_2} - \varepsilon_{s_1}^{v_2}\varepsilon_{s_2}^{v_1}} \tag{12.33}$$

$$C_2^{s_1} = \frac{\varepsilon_{s_2}^{v_1} - \varepsilon_{s_2}^{v_3}}{\varepsilon_{s_1}^{v_2}\varepsilon_{s_2}^{v_3} - \varepsilon_{s_1}^{v_1}\varepsilon_{s_2}^{v_3} + \varepsilon_{s_1}^{v_1}\varepsilon_{s_2}^{v_2} - \varepsilon_{s_1}^{v_2}\varepsilon_{s_2}^{v_1}} \tag{12.34}$$

$$C_3^{s_1} = \frac{\varepsilon_{s_2}^{v_2} - \varepsilon_{s_2}^{v_1}}{\varepsilon_{s_1}^{v_2}\varepsilon_{s_2}^{v_3} - \varepsilon_{s_1}^{v_1}\varepsilon_{s_2}^{v_3} + \varepsilon_{s_1}^{v_1}\varepsilon_{s_2}^{v_2} - \varepsilon_{s_1}^{v_2}\varepsilon_{s_2}^{v_1}} \tag{12.35}$$

These expressions have the same denominator as those for the flux-control coefficients, equations 12.30–32, but now each numerator term contains one fewer elasticity than the denominator terms, because the concentration s_1 that occurs as a superscript on the left-hand side of the equation does not occur in the numerator elasticities. As before, the modulated enzyme is missing from all products, and one additional enzyme is also missing from each product. Each numerator term occurs twice in the three expressions, with opposite signs; for example, the term $\varepsilon_{s_2}^{v_3}$ in equation 12.33 is matched by the term $-\varepsilon_{s_2}^{v_3}$ in equation 12.34. This matching of numerator terms ensures that the summation relationship, equation 12.16, is obeyed.

12.6.3 Expression of summation and connectivity relationships in matrix form

Although I am avoiding matrix algebra as far as possible in this book, readers familiar with it may find it helpful to have the results of the previous section expressed so as to facilitate comparison with review articles that use the matrix formulation. This anyway becomes almost indispensable for taking metabolic control analysis beyond the most elementary level.

Consider, as an example, the following equation, in which I shall refer to the first matrix as the **C** matrix, the second as the **ε** matrix, and the right-hand side as the unit matrix:

$$\begin{bmatrix} C_1^J & C_2^J & C_3^J \\ C_1^{s_1} & C_2^{s_1} & C_3^{s_1} \\ C_1^{s_2} & C_2^{s_2} & C_3^{s_2} \end{bmatrix} \cdot \begin{bmatrix} 1 & -\varepsilon_{s_1}^{v_1} & 0 \\ 1 & -\varepsilon_{s_1}^{v_2} & -\varepsilon_{s_2}^{v_2} \\ 1 & 0 & -\varepsilon_{s_2}^{v_3} \end{bmatrix} = \begin{bmatrix} 1 & 0 & 0 \\ 0 & 1 & 0 \\ 0 & 0 & 1 \end{bmatrix} \tag{12.36}$$

Note first of all that the top row of the **C** matrix contains the three flux control coefficients, the second and third the concentration control coefficients for the two intermediates S_1 and S_2 respectively. The ε matrix contains a unit vector as first column, and the other entries contain all of the elasticities, of which two, $\varepsilon_{S_2}^{v_1}$ and $\varepsilon_{S_1}^{v_3}$, are replaced by zero because S_1 is assumed in equation 12.29 to have no effect on E_3 and S_2 is assumed to have no effect on E_1. The product of the first row of **C** and the first column of ε yields the top-left entry in the unit matrix, which is 1, and thus expresses the summation relationship for flux control coefficients (compare equation 12.15 or 12.20). In a similar way the summation relationships for concentration control coefficients are expressed by products of the other rows of **C** with the first column of ε. The connectivity relationship for flux control coefficients is expressed by the product of the top row of **C** with any column of ε apart from the first. The connectivity relationships for concentration control coefficients are expressed by all of the other possible products that have not been explicitly mentioned.

12.6.4 Connectivity relationship for a metabolite not involved in feedback

Every metabolite has at least two non-zero elasticities, because every metabolite affects the rates of the enzyme for which it is the substrate and the enzyme for which it is the product. However, metabolites that are not involved in feedback or feedforward effects and are not substrates or products of more than one enzyme will have *only* these two non-zero elasticities, and the connectivity relationship then assumes rather a simple form. For example, for S_1 in equation 12.29, we have

$$C_1^J \varepsilon_{S_1}^{v_1} = -C_2^J \varepsilon_{S_1}^{v_2} \tag{12.37}$$

which shows that the ratio of the flux control coefficients of two consecutive enzymes is equal to minus the reciprocal of the ratio of elasticities of the connecting metabolite (hence the name "connectivity relationship"):

$$\frac{C_1^J}{C_2^J} = -\frac{\varepsilon_{S_1}^{v_2}}{\varepsilon_{S_1}^{v_1}} \tag{12.38}$$

This relationship allows one to "walk" along a pathway relating control coefficients in pairs, and as control coefficients are in principle much more difficult to measure directly than elasticities, this is an important advantage.

12.6.5 The flux control coefficient of an enzyme for the flux through its own reaction

The increasingly algebraic expression of metabolic control analysis, and especially the increasing use of matrix algebra, can result in obscuring properties that are quite obvious if one takes care not to lose sight of the underlying chemistry. An example is provided by the degree of control exerted by an enzyme E_i on the flux through the reaction that it catalyses. If we ignore the complication of multiple enzymes in the system that catalyse the same reaction (that is to say we ignore the possibility of isoenzymes), then it is obvious that the rate v_i depends only on the activity of the enzyme itself as determined by its concentration and those of its own substrate S_{i-1} and product S_i as sensed through their non-zero elasticities, and those of any other metabolites with non-zero elasticities, which we can represent by an arbitrary intermediate S_k. Thus the same kind of thought experiment that led us to equation 12.17 will give the following expression:

$$\mathrm{d}J_i = \frac{\partial v_i}{\partial e_i}\mathrm{d}e_i + \frac{\partial v_i}{\partial s_{i-1}}\mathrm{d}s_{i-1} + \frac{\partial v_i}{\partial s_i}\mathrm{d}s_i + \frac{\partial v_i}{\partial s_k}\mathrm{d}s_k \qquad (12.39)$$

We can write the steady-state flux J_i through the step catalysed by E_i on the left-hand side of this expression rather than v_i because in the steady state they are identical. Dividing all terms by $v_i \mathrm{d}e_i/e_i$ (or $J_i \mathrm{d}e_i/e_i$) and multiplying where appropriate by fractions equivalent to unity (such as s_{i-1}/s_{i-1})

$$\frac{e_i\mathrm{d}J_i}{J_i\mathrm{d}e_i} = \frac{e_i}{v_i}\frac{\partial v_i}{\partial e_i} + \frac{s_{i-1}}{v_i}\frac{\partial v_i}{\partial s_{i-1}}\frac{e_i}{s_{i-1}}\frac{\mathrm{d}s_{i-1}}{\mathrm{d}e_i} + \frac{s_i}{v_i}\frac{\partial v_i}{\partial s_i}\frac{e_i}{s_i}\frac{\mathrm{d}s_i}{\mathrm{d}e_i} + \frac{s_k}{v_i}\frac{\partial v_i}{\partial s_k}\frac{e_i}{s_k}\frac{\mathrm{d}s_k}{\mathrm{d}e_i} \qquad (12.40)$$

The left-hand side of this equation is the definition of the flux control coefficient $C_i^{J_i}$, the first term on the right-hand side is the elasticity of E_i with respect to its own rate, usually assumed to be equal to unity, and each of the other terms can be recognized as an elasticity multiplied by a control coefficient. Making all the appropriate subtitutions, therefore, the equation may be written as follows:

$$C_i^{J_i} = 1 + \varepsilon_{s_{i-1}}^{v_i}C_i^{s_{i-1}} + \varepsilon_{s_i}^{v_i}C_i^{s_i} + \varepsilon_{s_k}^{v_i}C_i^{s_k} \qquad (12.41)$$

This equation now expresses the important idea that the flux control coefficient of any enzyme for the flux through its own reaction is completely determined by the sum of the products of its non-zero elasticities with the corresponding concentration control coefficients (Heinrich and Rapoport, 1974).

Although only three metabolites appear on the right-hand side of equation 12.41, representing the three *classes* of metabolites that commonly have non-zero elasticities, for any given enzyme the number may be more or less than three. It will not normally be less than two, however, because substrates and products always have non-zero elasticities: algebraically this must be true, and even numerically it will be unusual for a substrate or product elasticity to be negligible.

12.7 Response coefficients: the partitioned response

Although it is sometimes convenient for the algebra to treat changes in an enzyme activity as if they resulted from changes in the concentration of the enzyme, in fact the effects on the system are exactly the same regardless of how they were caused. The justification for this assertion lies in the treatment of external effectors on enzymes.

As an analogue of a control coefficient that expresses the dependence of a system variable such as flux on an internal parameter such as enzyme activity, one can define a *response coefficient* R_z^J to express the dependence of a system variable on an external parameter, such as the concentration z of an external effector Z:

$$R_z^J = \frac{\partial \ln J}{\partial \ln z} \tag{12.42}$$

An external effector such as Z can only produce a systemic effect by acting on one or more enzymes in the system. Thus it must have at least one non-zero elasticity $\varepsilon_z^{v_i}$, defined in exactly the same way as any other elasticity:

$$\varepsilon_z^{v_i} = \frac{\partial \ln v_i}{\partial \ln z} \tag{12.43}$$

We now consider how these two quantities are related to each other and to the flux control coefficient of the enzyme acted on by Z. Any effect of Z on the system can be counterbalanced by changing the concentration of the enzyme by an amount just sufficient to produce a net effect of zero. So we can write a zero change in rate as the sum of two effects:

$$\frac{\mathrm{d}v_i}{v_i} = \varepsilon_z^{v_i} \frac{\mathrm{d}z}{z} + \frac{\mathrm{d}e_i}{e_i} = 0 \tag{12.44}$$

and the corresponding zero change in flux is also the sum of two terms:

$$\frac{\mathrm{d}J}{J} = R_z^J \frac{\mathrm{d}z}{z} + C_i^J \frac{\mathrm{d}e_i}{e_i} = 0 \tag{12.45}$$

Dividing one equation by the other produces the expression for the *partitioned response*, which shows that the response coefficient is the product of the elasticity of the effector for the enzyme that it acts on and the control coefficient of that enzyme:

$$R_z^J = C_i^J \varepsilon_z^{v_i} \tag{12.46}$$

Although we have considered effects on fluxes here, exactly the same relationship applies to any variable of the system, so J in equation 12.46 can represent not only any flux but also any concentration.

The partitioned response not only explains why effects on enzyme activity can be treated as if they were changes in enzyme concentration; it also explains the difference between the definitions of a control coefficient represented by equations 12.12 and 12.13: these are apparently equivalent only because equation 12.7 was assumed to be true; there is an implicit elasticity of unity connecting the response coefficient defined by equation 12.13 with the control coefficient defined by equation 12.12. In general, any response coefficient can be written as the product of a control coefficient and an elasticity. A little thought should show that a relationship of this general kind must obviously apply: any effector can only act on a system variable by altering the activity of an enzyme, and transmission of this primary effect will be moderated according to the control coefficient of the enzyme in question.

This analysis shows that the infinitesimal response of a pathway to a signal depends only on the values of the elasticities with respect to the effector, not on the mechanism by which the effector acts. The mechanism does become relevant, however, for large effects, and then competitive and uncompetitive inhibitors, for example, may behave very differently (Section 12.3.4).

12.8   Control and regulation

Although the major ideas of metabolic control analysis date from 1973–1974, and have their roots in the work of Higgins (1965) a decade earlier, they were absorbed into the mainstream of thought about metabolic regulation rather slowly. Indeed, it took almost ten years before it began to be extended to new metabolic systems, such as respiration (Groen and co-workers, 1982) and gluconeogenesis (Groen and co-workers, 1983, 1986).

This slow acceptance of metabolic control analysis by the biochemical community is partly a consequence of a supposed disdain for the classic work on regulation [see Atkinson (1990) for example], with little use for such central concepts as feedback inhibition by end products (Yates and Pardee, 1956; Umbarger, 1956), or cooperative and allosteric interactions

(Monod, Changeux and Jacob, 1963; Monod, Wyman and Changeux, 1965; Koshland, Némethy and Filmer, 1966).

Part of the confusion has resulted from a lack of agreed definitions for certain crucial terms. "Control" is now accepted in the field to have the meaning attributed to it by Kacser and Burns (1973), but "regulation" continues to give difficulties. For some, regulation is little different from control, and Sauro (1990) for example took it to mean "some sort of response of metabolism to a change in an external influence"; for others it is quite different, having to do with the properties of regulatory enzymes in isolation. Hofmeyr and Cornish-Bowden (1991) argued that its use in biochemistry ought to be brought as close as possible to its use in everyday life. When we say that a domestic refrigerator is well regulated, for example, we mean that it is capable of maintaining a predetermined internal temperature constant in the face of large variations in heat flux that result from opening the door or variations in the external temperature. Metabolism is in almost exact analogy to this if one considers a well-regulated system to be one in which concentrations of internal metabolites (the "temperature") are maintained steady in the face of variations in metabolic flux. In economic terms, we usually regard a well-regulated economy as one in which the supply of goods is determined largely by the demand for them. Again, there are obvious metabolic analogies, and we should expect a well-regulated organism to be one in which the supply of precursors for protein synthesis is determined by the need for protein synthesis, and not solely by the supply of food. All of this is more difficult to achieve than it may appear, as it is becoming increasingly clear that metabolite concentrations respond much more readily to perturbations than fluxes do; for example, deletion of a gene from the yeast genome typically produces little or no change in growth or other metabolic fluxes, but significant changes in metabolite concentrations (see Cornish-Bowden and Cárdenas, 2001b).

Another important term that seems to mean one thing but actually means something else is "end product". It seems obvious that the end product in a metabolic system ought to be the sink into which the flux flows. But in ordinary use in the literature on metabolic regulation, for example Stadtman (1970), "end product" always refers to a metabolite such as threonine that is not excreted but is explicitly recognized as the starting point for other pathways. In virtually the entire experimental literature on metabolic regulation an end product is understood in this way; it never means a genuine end product of metabolism such as water or carbon dioxide.

It follows, therefore, that we cannot hope to understand the role of end-product inhibition in metabolic regulation unless we draw pathways as components of systems that explicitly recognize that there are steps after the release of "end product". Thus in their discussion of feedback inhibition

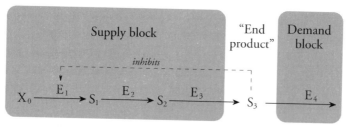

Figure 12.3. Regulatory structure of a biosynthetic pathway. The regulation of a typical biosynthetic pathway can most easily be rationalized by considering it to consist of a *supply block* of reactions that produce an *end product* at a rate that satisfies the requirements of a *demand block*, here shown as a single reaction. In most discussions of metabolic regulation the demand block is omitted and the end product is shown at the end of the process. However, this sort of representation loses the capacity to express regulation in terms of communication between the supply and demand blocks via the end product.

Kacser and Burns (1973) included a step after the formation of end product, though they did not explain the reason for doing so. Many textbooks, however, omit this step, and thereby render meaningful analysis of metabolic regulation impossible.

For integrating the classical regulatory concepts into metabolic control analysis (Hofmeyr and Cornish-Bowden, 1991, 2000), one can represent a pathway with feedback inhibition as a two-step pathway, with a supply block, consisting of all the reactions that lead to the end product, and a demand block, consisting of the reactions that consume it (Figure 12.3). If the supply block were the entire pathway, the end product would be an external parameter and any effect that it had on the flux would have to be treated in terms of a response coefficient $R_{s_3}^{J_s}$, but this response coefficient is conceptually the same as the block elasticity $\varepsilon_{s_3}^{v_{123}}$ that defines its effect on the supply flux considered as the local rate.

It follows from this kind of discussion that the boundaries of a system and the distinctions between internal and external parameters or between local and systemic properties cannot be regarded as absolute. To understand how an end product such as S_3 in Figure 12.3 can fulfil its regulatory role, it is not sufficient to regard it solely as an internal metabolite; we must also study sub-systems where it becomes an external parameter, so that we can ask questions like "if the supply block in Figure 12.3 were the complete system, what effect would s_3 have on the supply flux?".

Using this type of analysis, one can study how to achieve effective regulation of a pathway such as that of Figure 12.3 by demand, by which we mean not only that the flux responds sensitively to changes in demand, but also that the concentration of end product changes little when the flux changes. Control of flux by demand requires the supply elasticity, $\varepsilon_{s_3}^{v_{123}}$, in the complete system (the same as the response coefficient $R_{s_3}^{J_s}$ in the supply block considered in isolation) must be as large as possible compared with the demand elasticity $\varepsilon_{s_3}^{v_4}$. (Being an inhibitory elasticity the supply elasticity is negative, so "as large as possible" means "as far below zero as possible".) Effective homeostasis of S_3 requires the absolute sum (ignoring minus signs) of the two elasticities to be as large as possible. Both criteria thus favour making the supply elasticity large, but they pull in opposite directions for the demand elasticity, so some compromises are inevitable.

The results of a study of the importance of cooperativity in effective regulation (Hofmeyr and Cornish-Bowden, 1991) proved to be surprising, as they appeared at first to suggest that it was much less important than had been thought since the 1960s. However, it must be emphasized that this is an illusion: cooperativity is certainly as necessary for effective regulation as has been thought, but its role is somewhat different from what one might naively imagine.

Changing the degree of cooperativity in the feedback inhibition of E_1 by S_3 in Figure 12.3 in the range of Hill coefficients from 1 (no cooperativity) to 4 (approximately the maximum cooperativity observed in any single effector–enzyme interaction) has almost no effect on the control of flux by demand: the curves showing flux as a function of demand (expressed by the limiting rate V_4 of the demand block) show near-proportionality between flux and demand over a 25-fold range, regardless of the Hill coefficient. This must surprise anyone who thinks that cooperativity is essential for flux regulation. However, as has been emphasized already, flux regulation is only part of regulation, and it is of little use without concentration regulation: we should not be satisfied with a refrigerator that tolerated a wide range of heat fluxes but had no control over the internal temperature! When the concentration of end product is considered as well as the flux, the effect of the Hill coefficient becomes large: over the same 25-fold range of demand considered above, a Hill coefficient of 4 causes s_3 to be restricted to less than a threefold range, whereas with a Hill coefficient of 1 it varies more than tenfold in a demand range of only about threefold.

In summary, cooperativity of feedback interactions is indeed essential for effective regulation, but it is not sufficient to say that it allows effective regulation of flux by demand; one must say that it allows effective regulation of flux by demand while maintaining homeostasis.

12.9   Mechanisms of regulation

To a considerable extent metabolic control analysis takes the properties of individual enzymes as it finds them, regarding a mechanistic explanation of these properties as outside its domain. Moreover, we have already discussed the most important such mechanisms in Chapter 11 of this book, and there is no need to discuss them again here. In addition to these, however, there are also three important regulatory mechanisms that involve multiple enzymes, which demand fuller discussion: these are *channelling* of intermediates between enzymes, *interconvertible enzyme cascades* involving covalent modification, and the *amplification* of the effects of small changes in ATP concentration by adenylate kinase.

12.9.1   Metabolite channelling

Channelling involves the idea that the metabolite shared by two consecutive enzymes in a pathway may be directly transferred from one to the other, without being released into free solution, or at least without achieving equilibrium with the metabolite in free solution. There are various versions of this idea (see Scheme 1 of Ovádi, 1991), but the essentials are shown in Figure 12.4a, based on the mechanism suggested by Gutfreund (1965). This can be regarded as a combination of a perfect channel (Figure 12.4b) with an ordinary free-diffusion mechanism (Figure 12.4c).

Although there is at least one enzyme, the multifunctional enzyme tryptophan synthase, for which the evidence for channelling (of indole) is overwhelming and generally accepted (see Yanofsky, 1989), channelling in general remains controversial, at least for enzymes forming "dynamic" complexes, which are complexes that exist only transiently during the transfer of metabolite.

The enzymes malate dehydrogenase and citrate synthase illustrate the sort of difficulties that can arise when trying to establish whether channelling actually occurs. Lindbladh and co-workers (1994) made a "fusion protein" of these enzymes from yeast, using genetic techniques to produce a single protein containing both activities, and Shatalin and co-workers (1999) made a similar fusion protein for the same pair of enzymes from pigs. In both studies the transient lag time for product formation was shorter for the fusion protein than it was for the free enzymes, suggesting that the intermediate oxaloacetate was channelled between the active sites. To reach such a conclusion one needs to assume that the individual enzymes have exactly the same kinetic properties when fused as they have in the free state. However, when this was carefully checked for the fusion protein of the yeast enzymes, the kinetic parameters were found to differ from those of the free enzymes to a sufficient degree to account for the decreased lag time (Pettersson and co-workers, 2000), without requiring any channelling.

The controversy about whether channelling actually occurs in the systems where it is proposed is certainly important in relation to metabolic control,

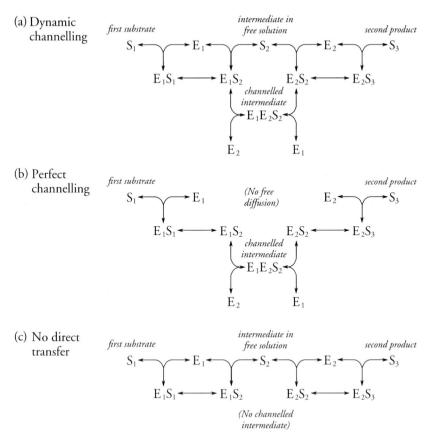

Figure 12.4. Metabolite channelling. If a metabolite S_2 is both the product of an enzyme E_1 and the substrate of an enzyme E_2, one can envisage that as well as being released into free solution it might be transferred directly from one enzyme to another by an encounter between the first enzyme–substrate complex E_1S_2 with the second enzyme E_2. *Dynamic channelling* (a) can be regarded as a combination of perfect channelling (b), in which no free intermediate is formed, with an ordinary free-diffusion mechanism (c) in which there is no direct transfer.

because if it does not occur it cannot have any relevance to control. Nonetheless, this is only part of the question, because it does not necessarily follow that channelling has any significant metabolic consequences even if it does occur. This point has been much less discussed, possibly because the advantages of channelling have been perceived to be obvious, as a result of confusion between the properties of a perfect channel and those of the sort of dynamic channelling mechanism that is the subject of the controversy.

It might seem intuitively obvious that even in a mechanism such as that of Figure 12.4a increasing the rates of the channelling steps at the expense of the free-diffusion steps must decrease the steady-state concentration of free intermediate, but this is an illusion. Although this may well decrease the rate at which S_2 is released from E_1S_2, it will also decrease the rate at which S_2 is taken up by E_2, and the net result may be in either direction. Whether there is a small net effect or not is ultimately a question of definition, as it is not easy to separate genuine effects of channelling from effects that could equally well arise from changes in the catalytic activity of the enzyme that have nothing to do with channelling; however, there is no reasonable doubt that even if an effect of channelling on intermediate concentrations exists in the steady state it is too small to fulfil a useful regulatory role (Cornish-Bowden and Cárdenas, 1993). For this reason there is no necessity to consider it further here.

12.9.2   Interconvertible enzyme cascades

Cooperativity of interactions with individual enzymes is an important way of making feedback inhibition more effective as a regulatory mechanism. It has a serious drawback, however, that prevents it from providing a universal way of increasing elasticities: the degree of cooperativity is limited in practice by the rarity of enzymes with Hill coefficents greater than 4; as the elasticity for an interaction cannot exceed the corresponding Hill coefficient, this means that individual enzyme–metabolite interactions do not normally result in elasticities greater than 4. This is too small for a device intended to operate as a switch: it implies the need for a threefold change in metabolite concentration to bring about a change from 10% to 90% of full activity (compare Section 11.1.2).

Much higher effective elasticities become possible when one comes to consider multi-enzyme systems of the kind illustrated in Figure 12.5, sometimes called "cascades".

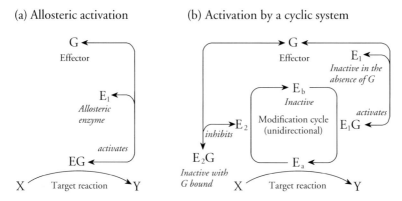

Figure 12.5. Interconvertible enzyme cascade. (a) In ordinary allosteric activation the activator G binds directly to an enzyme to increase its activity, but in an interconvertible enzyme system (b) it acts on the enzymes that catalyse the irreversible conversions between the inactive and active forms of a target enzyme. Additional substrates (not shown), such as ATP and water, are needed to make the interconversion reactions irreversible.

Interconvertible enzyme systems have been known for more than a quarter of a century, and many examples are known, including many involving protein kinases and phosphatases (E. G. Krebs and Beavo, 1979). Glutamine synthetase from *Escherichia coli* is inactivated by adenylylation and reactivated by deadenylylation (Chock, Rhee and Stadtman, 1980, 1990); it has been studied in detail and has served as the basis for extensive theoretical work (Chock and Stadtman, 1977; Stadtman and Chock, 1977, 1978).

In the context of this chapter, the essential point is that interconvertible enzyme systems can generate high sensitivity to signals, much higher than is possible for single enzymes (Goldbeter and Koshland, 1981, 1984). It is tempting to suppose that this high sensitivity is inherent in the structure of the cycle, but in reality if the kinetic parameters of the cycle reactions are assigned arbitrary values, the typical result is a system that generates *less* sensitivity than a single non-cooperative enzyme. Very high sensitivity results only if several conditions are satisfied (Cárdenas and Cornish-Bowden, 1989): the interactions of the effector with the modifier enzymes should be predominantly catalytic rather than specific (uncompetitive rather than competitive in the terminology of inhibition); the inactivation reaction should cease to operate at much lower concentrations of effector than are needed to activate the activating reaction; both modifier enzymes should operate close to saturation, a condition especially emphasized by Goldbeter and Koshland (1981, 1982, 1984) under the name "zero-order ultrasensitivity". If all these conditions are satisfied the sensitivity possible with the mechanism of Figure 12.5 is enormously greater than is possible for a single enzyme; even with severe constraints allowed for the kinetic parameters of the modifier enzymes one can easily obtain the equivalent of a Hill coefficient of 30, or even of 800 if one relaxes the constraints while still staying within the range of behaviour commonly observed with real enzymes (Cárdenas and Cornish-Bowden, 1989).

The first two of these conditions imply that experimenters should take special care to note what might appear to be insignificant kinetic properties of modifier enzymes. Even if the uncompetitive component of the inhibition of a modifier enzyme is an order of magnitude weaker than the competitive component it may still be essential to the effective working of the system. Likewise, if one observes that the phosphatase in a cycle is activated only at supposedly unphysiological concentrations of an effector ten times higher than those effective for inhibiting the kinase, this does not mean that the effector is irrelevant to the action of the phosphatase; it means that the whole system is well designed for generating high sensitivity.

As interconvertible enzyme systems can generate so much more sensitivity than individual cooperative enzymes one may wonder why they are not universally used in metabolic regulation. However, unlike individual cooperative enzymes interconvertible enzyme systems consume energy, because both modification reactions are assumed to be irreversible; this is possible only if they involve different co-substrates, for example the activation might be phosphorylation by ATP whereas the inactivation might be hydrolysis.

12.9.3   The metabolic role of adenylate kinase

The third mechanism that I shall consider here is rarely discussed in the context of metabolic control analysis, possibly because its existence has been known for so long (first suggested by H. A. Krebs, 1964) that it has been forgotten that it is a multienzyme regulation mechanism at all. This concerns the enzyme adenylate kinase (often called myokinase) which catalyses the interconversion of the three adenine nucleotides:

$$ATP + AMP = 2ADP \qquad\qquad (12.47)$$

Adenylate kinase is present with high catalytic activity in some tissues (such as muscle) where at first sight its reaction appears to have no metabolic function; certainly in many cells where the enzyme is found the long-term flux through the reaction is negligibly small. Why then is it present, and why at such high activity? It is hardly sufficient to say just that its role is to maintain the reaction at equilibrium, because the concentrations of ATP and ADP normally change so little that the reaction would hardly ever be far from equilibrium even if the enzyme concentration were much lower than it is. The answer appears to be that if the adenine nucleotides exist predominantly as ATP (as they do), and if equation 12.47 is always at equilibrium, not just approximately but exactly, then small changes in the balance between ATP and ADP will be translated into large relative changes in the concentration of AMP, so that enzymes that are specifically affected by AMP can respond with high sensitivity to the original small changes.

This idea is illustrated in Figure 12.6, the top part of which shows that if the three adenine nucleotides are maintained at a total concentration $[ATP] + [ADP] + [AMP] = 5\,mM$ with an equilibrium constant of $[ATP][AMP]/[ADP]^2 = 0.5$, then concentrations of ATP around $4\,mM$ correspond to such low concentrations of AMP that almost all of any variation in the ATP concentration is translated into an opposite and almost equal variation in the ADP concentration. At first sight it seems scarcely any different from what one would have if the AMP were not there. However, although its concentration in these conditions is small, its *fractional* variations are large compared with those of ATP and ADP. Any enzyme that binds AMP tightly (so that it can detect it even though its concentration is small

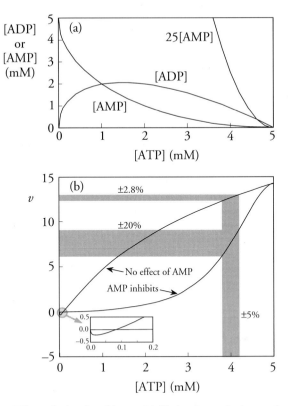

Figure 12.6. Effect of adenylate kinase. (a) If the three adenine nucleotides are always in equilibrium, small changes in ATP concentration around 4 mM result in large relative changes in the AMP concentration. (b) This allows an enzyme that binds AMP tightly to show a much larger response to ATP than it would have if there were no variations in AMP concentration. The curves in (b) were calculated assuming reversible Michaelis–Menten kinetics with respect to ATP and ADP, and linear competitive inhibition by AMP. If the individual responses were cooperative the net effects could be much larger than those shown. The small negative rates at low ATP concentrations (magnified in the inset) are due to the reverse reaction, which occurs (albeit to a negligible extent) when the [ATP]/[ADP] ratio is sufficiently small.

compared with those of ATP and ADP) can respond sensitively, as illustrated by Figure 12.6b. The presence of a high activity of adenylate kinase in a cell thus allows AMP to act as an amplifier of small signals: in the example of Figure 12.6, 5% variation in the ATP concentration produces only a 2.8% change in rate if the enzyme does not respond to AMP, but a 20% change in rate if it does; in effect, the mechanism has increased the effective elasticity with respect to ATP by a factor of more than seven, from about 0.56 (2.8/5) to about 4 (20/5). By itself this effect is not enough to explain, for example, the observation that glycolytic flux in insect flight muscles may increase 100-fold when the ATP concentration falls by 10% with a simultaneous 2.5-fold increase in AMP concentration (Sacktor and Wormser-Shavit, 1966; Sacktor and Hurlbut, 1966), but it certainly makes a major contribution to the response. Moreover, enzymes that respond to AMP normally do so cooperatively, whereas in Figure 12.6 only linear competitive inhibition was assumed, to avoid complicating the discussion by examining two different sorts of effect at the same time.

12.10   Computer modelling of metabolic systems

We have been mainly concerned in this chapter with the principles that govern the behaviour of multi-enzyme systems, and those, indeed, have been the main concerns of metabolic control analysis in general. Nonetheless, it is difficult to visualize the properties of systems of more than trivial size, and although methods exist that fulfil in multi-enzyme systems the role of the King–Altman method in one-enzyme kinetics (Hofmeyr, 1989) they require considerable study before one can arrive at the sort of intuitive understanding that the King–Altman method provides quite easily (see Section 4.7). In addition, experiments have always been an essential aid in efforts to understand enzyme behaviour, but non-trivial multi-enzyme experimentation is a formidable task, often virtually impossible.

For both of these reasons, the capacity to model multi-enzyme systems in the computer has been an important tool for supplementing theoretical analysis. At a time when as much as 30 minutes of central-processor time were needed to simulate 75 ms of glycolysis, high-level programming languages barely existed, and programming involved great effort to circumvent arbitrary restrictions like the inability of the computer to represent numbers greater than one, Garfinkel and Hess (1964) nonetheless succeeded in setting up a working model of glycolysis. This has been followed by many others, and it is now hardly necessary for the prospective simulator to be a programmer, as various programs designed for metabolic modelling are readily available (Mendes, 1997; Sauro, 1993, 2000; Voit and Ferreira, 2000), and more general mathematical programs can be used for the same purpose, as thoroughly discussed by Mulquiney and Kuchel (2003).

Nonetheless, even with the use of such programs the researcher needs to pay some attention to general principles that apply to the rate equations that it is appropriate to use. In marked contrast to studies of single enzymes, where it is usually possible to set up conditions that avoid the need to consider product inhibition or reversibility, models for multi-enzyme systems must always take account of effects of products, because there is no way to ensure that product concentrations are zero in the conditions of interest. It will usually also be necessary to take account of effects of other metabolites that are present in the system even though they are not the substrates or products of the enzyme for which a rate equation is to be written. It follows from all this that the rate equations needed for models are almost inevitably more complicated than those used for studying the same enzymes one at a time. In particular, most rate equations need to be entered in their full reversible forms. For a simple enzyme, therefore, the fundamental equation is not equation 2.15, the Michaelis–Menten equation, but equation 2.46, the reversible Michaelis–Menten equation. For enzymes with complicated kinetic behaviour this immediately raises a problem, as equations of the sort introduced in Chapter 11 are almost never normally written in reversible form, and even in their usual irreversible forms they are not only complicated but also include more parameters than one can normally expect to have reliable empirical values for.

The latter problem is fairly easily resolved, because the difficulty of distinguishing experimentally between models of cooperativity becomes a virtue in the context of metabolic modelling: if one cannot easily identify the correct equation to describe an enzyme's behaviour it will not make a large difference to the behaviour of a metabolic model if one uses the wrong one. In particular, the Hill equation (equation 11.3) can usually describe the behaviour of an enzyme in the range of interest just as accurately as kinetic versions of equations 11.37 or 11.42, and is thus often used for cooperative enzymes in metabolic models. However, this leaves the first difficulty, that all of these equations are almost never written as reversible equations, and in some contexts at least one needs to allow for reversibility in a reaction catalysed by a cooperative enzyme.

There is a need, therefore, for an empirical equation that satisfies the following properties: it must always define a rate in the direction dictated by the thermodynamic state, it should degenerate to the appropriate irreversible Hill equation when either substrate or product concentration is zero, and with Hill coefficient h equal to the number of interacting sites it should lead to the correct equation for maximum cooperativity for that number of sites. The first of these conditions means that it must be possible to express the rate as a thermodynamic term, positive or negative according to the direction the reaction needs to take to reach equilibrium, multiplied by a positive kinetic term:

$$v = \left(1 - \frac{p/a}{K}\right) \text{Pos}(a, p, \ldots) \tag{12.48}$$

In this equation K, the equilibrium constant, is the value of p/a at equilibrium, and $\text{Pos}(a, p, \ldots)$ is any positive function of a, p and any other relevant concentrations. Although we have not previously needed this equation in this book, all of the reversible equations that have appeared can be written in the same form. For example, equation 2.46 can be written as follows:

$$v = \frac{\left(1 - \frac{p/a}{k_A/k_P}\right) \cdot k_A e_0 a}{1 + \frac{a}{K_{mA}} + \frac{p}{K_{mP}}} \tag{12.49}$$

and as the equilibrium constant is k_A/k_P this clearly has the form of equation 12.48.

Returning to the Hill equation, a reversible form that satisfies the requirements listed above is as follows (Hofmeyr and Cornish-Bowden, 1997):

$$v = \frac{\frac{Va}{a_{0.5}} \cdot \left(1 - \frac{p/a}{K}\right) \cdot \left(\frac{a}{a_{0.5}} + \frac{p}{p_{0.5}}\right)^{b-1}}{1 + \left(\frac{a}{a_{0.5}} + \frac{p}{p_{0.5}}\right)^{b}} \tag{12.50}$$

in which V is the limiting rate of the forward reaction, $a_{0.5}$ is the substrate concentration that gives a rate of $0.5V$ in the absence of product, $p_{0.5}$ is the corresponding parameter for the reverse reaction, K is the equilibrium constant and b is the Hill coefficient.

This equation can readily be generalized to accommodate modifiers. For example, if it is written as follows:

$$v = \frac{\frac{Va}{a_{0.5}} \cdot \left(1 - \frac{p/a}{K}\right) \cdot \left(\frac{a}{a_{0.5}} + \frac{p}{p_{0.5}}\right)^{b-1}}{\frac{1 + (x/x_{0.5})^b}{1 + \beta(x/x_{0.5})^b} + \left(\frac{a}{a_{0.5}} + \frac{p}{p_{0.5}}\right)^{b}} \tag{12.51}$$

then x represents the concentration of a substance that can act either as an inhibitor or as an activator, depending on the value of β: if $\beta < 1$ then it is an inhibitor, and if $\beta > 1$ it is an activator.

These equations are far simpler than anyone could hope to derive from mechanistically realistic models of cooperativity, but they may still be more complicated than one might wish when modelling, and they contain parameters that may not be experimentally known, such as $p_{0.5}$. We need, therefore, to return to the general question of when it is safe to represent reactions in metabolic models with irreversible equations. Previous discussions (for example Hofmeyr and Cornish-Bowden, 1997) have tended to recommend that only an exit reaction into a metabolic sink should be treated as irreversible, and not all authors (for example Mendes, Kell and Westerhoff, 1992) would allow even this exception. On the other hand most published models (for example Heinrich and Rapoport, 1973; Bakker and co-workers, 1997) have in practice treated reactions with large equilibrium constants, such as that catalysed by pyruvate kinase, as irreversible. It thus appeared surprising that modifying the model of Bakker and co-workers (1997) slightly to allow for the reversibility of pyruvate kinase caused it to behave very differently, with a major redistribution of flux control (Eisenthal and Cornish-Bowden, 1998).

Subsequent investigation (Cornish-Bowden and Cárdenas, 2001a) clarified the issue: as discussed in Section 12.3.1, even if the negative term in the numerator of the rate expression is genuinely negligible this does not justify ignoring inhibitory effects of products. In retrospect this appears quite obvious, though it escaped recognition in more than 40 years of metabolic modelling, as reactions considered to be irreversible have almost always been treated as product-insensitive as well.

Problems

12.1 Consider an enzyme with rate given by the irreversible Hill equation (equation 11.3), that is to say $v = Va^h/(K_{0.5}^h + a^h)$, under conditions where the reaction is far from equilibrium and product inhibition is negligible. What is the value of the elasticity ε_a^v when the enzyme is half-saturated, with $a = K_{0.5}$? What are the limiting values that it approaches when a is very small or very large?

12.2 For the three-step pathway of equation 12.29, calculate the three flux control coefficients under conditions where the elasticities with respect to the two intermediates are as follows: $\varepsilon_{s_1}^{v_1} = -0.2$, $\varepsilon_{s_1}^{v_2} = 0.3$, $\varepsilon_{s_2}^{v_2} = -0.1$, $\varepsilon_{s_2}^{v_3} = 0.2$. (Assume that S_1 has no effect on E_3 and S_2 has no effect on E_1, in other words that $\varepsilon_{s_1}^{v_3} = \varepsilon_{s_2}^{v_1} = 0$.)

12.3 Manipulation of the activity of an enzyme E_i in a metabolic pathway reveals that it has a flux control coefficient of 0.15 under physiological conditions. Its elasticity towards the product S_i of its reaction is found to be −0.25 under the same conditions. The next enzyme E_j in the pathway cannot be directly manipulated, and so its control coefficients cannot be directly measured. However, studies with the purified enzyme indicate that it has an elasticity of 0.2 towards its substrate S_i under physiological conditions. Assuming that S_i has no significant interactions with any other enzymes in the pathway, estimate the flux control coefficient of E_j.

12.4 The "top-down" approach described by Brown, Hafner and Brand (1990) involves varying the activities of several enzymes in constant proportion and using the resulting variations in flux and metabolite concentrations to estimate control coefficients for the whole block of enzymes rather than for the individual enzymes within the block. Devise a "thought experiment" to deduce the relationship between any control coefficient of a block of enzymes and the corresponding control coefficients of the enzymes composing the block.

12.5 Much of modern biotechnology is based on the premise that identifying the enzymes that catalyse the rate-limiting steps in pathways leading to desirable products, cloning these enzymes and then overexpressing them in suitable organisms will allow greatly increased yields of the desirable products. What implications does the analysis of this chapter have for such a strategy?

Chapter 13

Fast Reactions

13.1 Limitations of steady-state measurements

13.1.1 The transient state

It should be obvious that experimental methods for investigating fast reactions, with half-times of much less than 1 s, must be different from those used for slower reactions, because in most of the usual methods it takes seconds or more to mix the reactants. Less obviously, the kinetic equations needed for the study of fast reactions are also different, because in most enzyme-catalysed reactions the steady state is established fast enough to be considered to exist throughout the period of investigation, provided that this period does not include the first second after mixing (Section 2.5). Consequently, most of the equations that have been discussed in this book are based on the steady-state assumption. By contrast, fast reactions are concerned, almost by definition, with the *transient state* (or *transient phase*) of a reaction before the establishment of a steady state and cannot be described by steady-state rate equations. This chapter deals with experimental and analytical aspects of this phase.

The differential equations that define simple chemical reactions, such as those considered in Chapter 1, are linear and have solutions that consist of exponential terms of the form $A \exp(-\lambda t)$: here t is the time, A is a constant known as the *amplitude*, and λ is a constant that is called the *frequency constant* in this chapter (but see Section 13.1.2). For example, the second term of equation 1.6 is an exponential term with frequency constant k and amplitude a_0. Such an exponential term is equal to the amplitude when t is zero, but decays towards zero as t increases, and eventually becomes negligible. As illustrated by this example, a frequency constant is a first-order or pseudo-first-order rate constant, but it is a more general term as it can also be applied to processes that are not of first order.

Enzyme-catalysed reactions are more complicated, because the differential equations that define them are not linear and do not have analytical solutions consisting of exponential terms. Nonetheless, it is usually possible, as will be discussed in this chapter, to set up experimental conditions that allow accurate linear approximations to the true differential equations to be used, and so the behaviour of exponential terms remains relevant.

13.1.2   The relaxation time

It is common, especially in discussions of the sort of methods described in Section 13.3.7, to replace the frequency constant λ by its reciprocal, usually written as τ and called the *relaxation time* (or *time constant*). Some authoritative texts, such as Hammes and Schimmel (1970) and Gutfreund (1995), switch arbitrarily from one convention to the other, using mainly frequency constants to discuss systems far from equilibrium and mainly relaxation times to discuss systems close to equilibrium. This is a potential source of confusion quite apart from the spurious contrast between systems close to and far from equilibrium that it implies. Even when relaxation times are used consistently they present problems, as their expressions are usually more complicated than those of the corresponding frequency constants: for this reason it is common in texts that use relaxation times to see equations with left-hand sides that consist of sums of reciprocals. As time is a somewhat less abstract concept for most people than frequency, there might be some advantage in expressing relationships in terms of relaxation times if the times in question had convenient physical meanings, but they do not: even for a simple first-order reaction the relaxation time is the time required for the amount of reactant to decrease by about 63% (more exactly, to decrease by a factor of e), and such a period can hardly appear less abstract than its reciprocal. It is noteworthy, for example, that when Hiromi (1979) provides a table (his Table 4.2) entitled "Physical meaning of the relaxation time, τ", the quantity actually tabulated is not τ but $1/\tau$.

A minor disadvantage of writing equations in terms of relaxation frequencies is that the quantity symbolized as λ has no universally accepted and recognized name. In simple reactions it is a first-order rate constant, but using a term such as "apparent first-order rate constant" is not only cumbersome, but also a potential source of confusion if applied to complicated examples far removed from first-order kinetics. For this reason in this book I shall call it the *frequency constant*.

13.1.3   "Slow" and "fast" steps in mechanisms

Before entering into detail on methods for analysing fast reactions it is useful to ask why we need to make transient-phase measurements at all: what can we learn from them that we cannot learn from the steady state?

Steady-state measurements have been useful for elucidating the mechanisms of enzyme-catalysed reactions, but they have the major disadvantage that, at best, the steady-state rate of a multi-step reaction provides information about the slowest step, and steady-state measurements do not normally provide information about the faster steps. Yet if the mechanism of an enzyme-catalysed reaction is to be understood it is necessary to have information about all steps.

Before taking the argument any further, however, we need to dispose of an apparent absurdity. In the steady state of any linear process all steps proceed at the same net rate (with the same difference between forward and reverse rates), so it appears meaningless to designate one of them as the "slowest step", and Northrop (2001), for example, calls this an "obvious misnomer". If one takes "slowest" to mean proceeding at the lowest rate, then Northrop is certainly correct, but if we take it to mean accounting for the greatest part of the total time then a different view is possible.

As a simple example, we may write equation 8.20 (or the middle part of equation 2.48, which is equivalent) in reciprocal form as follows:

$$\frac{1}{k_A} = \frac{1}{k_1} + \frac{K_1}{k_2} + \frac{K_1 K_2}{k_3} \tag{13.1}$$

where $K_1 = k_{-1}/k_1$ and $K_2 = k_{-2}/k_2$ are the equilibrium constants (written in the reverse direction) for steps 1 and 2. Generalizing this, we see that the reciprocal of the specificity constant is the sum of the reciprocals of the forward rate constants with each multiplied by the product of equilibrium constants back to the first step. This quantity, the specificity time (Section 2.3.4), can thus be regarded as the sum of a series of times. Although the reversibility of the steps means that the time taken by individual molecules to traverse the whole process may vary greatly, it remains true that there is an average time that can be considered the sum of the average amounts of time spent traversing the separate steps; indeed, Van Slyke and Cullen (1914) discussed the kinetics of urease in just these terms many years ago. There will always be a step that contributes as least as much to the total time as any other, and it seems no great abuse of language to call this the slowest step, and to refer to steps that contribute significantly less as faster steps.

Thus although all steps have the same rate in the steady state, it does not follow from this that an equal amount of time is required for every step, so when we describe some steps as slower than others we do not mean that they proceed more slowly but that they account for a higher proportion of the time consumed.

Northrop (1981) also called into question the whole idea of a rate-limiting step in enzyme mechanisms, and Ray (1983) made a thorough analysis that parallels in some respects the discussion of rate-limiting steps in metabolism that was initiated by Kacser and Burns (1973) and discussed in Section 12.5; indeed, Ray defined a *sensitivity index* for use in the study of isotope effects that has properties similar to those of flux control coefficients. It is perfectly possible (though not necessary) for one step to take much more time than all the others put together, and if this is true it is reasonable to call it the rate-limiting step. Even if it does not completely dominate the sum, the time for the slowest step will still usually be similar in magnitude to the time for the whole process, so it will be quite common for a single step to be roughly rate-limiting. Exceptions will arise if the whole process contains a large number of slow steps of similar times, but this is much less likely in mechanisms for individual enzymes than it is for metabolic systems: the latter may indeed contain a large number of components, and natural selection can be expected to have eliminated large variations in kinetic efficiency between them; but in chemical systems rate constants for the individual processes typically vary by orders of magnitude, and no natural selection can be invoked to suggest that the variations should be evened out. Considerations of this sort probably explain why the need to eliminate rate-limiting steps from discussions of metabolic regulation became evident around a decade before the corresponding question was seriously discussed for enzyme mechanisms.

It is probably worth adding that most of the objections to the term *rate-limiting* relate to the tendency to treat it as synonymous with the more objectionable term *rate-determining*, which is now found less often in chemical or biochemical writing than it once was (though it has by no means disappeared entirely). As long as one takes the idea of a limit quite literally, as an absolute boundary, then it is quite true that the smallest first-order or pesudo-first-order rate constant in a sequence sets a limit to the frequency of the whole process that cannot be exceeded. The same applies in the metabolic context: for any unbranched sequence of enzyme reactions the smallest limiting rate in the series sets a limit to the rate of the process as a whole.

13.1.4   Ambiguities in the steady-state analysis of systems with intermediate isomerization

As discussed in Chapter 7, the experimenter has considerable freedom to alter the relative rates of the various steps in a reaction, by varying the concentrations of the substrates. Consequently it is often possible to examine more than one step of a reaction despite this fundamental limitation of steady-state kinetics. However, isomerizations of intermediates along the reaction pathway cannot be separated in this way, as we may illustrate by reference to the simple three-step Michaelis–Menten mechanism (equation 2.47), for which the definitions of the Michaelis–Menten parameters were given in equation 2.48 for the forward reaction and in equation 2.49 for the reverse reaction. As there are six elementary rate constants in the mechanism, but only two Michaelis–Menten parameters for each direction of reaction, it follows that characterization of the steady state, or measurement of the Michaelis–Menten parameters, cannot provide enough information to specify all of the rate constants. As discussed in Section 8.7, measurements of primary isotope effects may allow this problem to be circumvented to some degree, but only with the introduction of new assumptions, and only for the simpler cases: if there are three steps instead of two between the binding of substrate and release of product no steady-state method can reveal the existence of all the steps, let alone give information about the magnitudes of the rate constants.

In general, as mentioned in Section 4.3, all of the intermediates in any part of a reaction mechanism that consists of a series of isomerizations of intermediates must be treated as a single species in steady-state kinetics. This is a severe limitation and provides the main justification for transient-state kinetics, which are subject to no such limitation.

The conveniently low rates observed in steady-state experiments are commonly achieved by working with small concentrations of enzyme. This may be an advantage if the enzyme is expensive or available only in small amounts, but it also means that all information about the enzyme is obtained at second hand, by observing its effects on reactants, and not by observing the enzyme itself. To observe the enzyme itself, one must use it in reagent quantities so that it can be detected by spectroscopic or other techniques. This usually results in such a high enzyme concentration that steady-state methods cannot be used.

The advantages of transient-state methods may seem to make steady-state kinetics obsolete, but there is no sign yet that steady-state methods are being superseded, and one may expect them to predominate for many years to come, in part because the theory of the steady state is simpler, and steady-state measurements require simpler equipment. In addition, the small amounts of enzyme needed for steady-state measurements allow them to be used for many enzymes for which transient-state experiments would be prohibitively expensive.

13.1.5 Ill-conditioning

As well as these practical considerations, the analysis of transient-state data can suffer from a numerical difficulty known as *ill-conditioning*. This means that, even in the absence of experimental error, it is possible to obtain convincing fits to the same experimental results with a wide range of constants

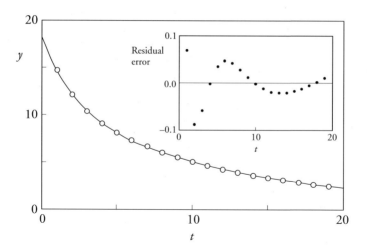

Figure 13.1. Ill-conditioned character of exponential functions. The points were calculated from $y = 5.1\exp(-0.769t) + 4.7\exp(-0.227t) + 9.3\exp(-0.0704t)$, the line from $y = 7.32\exp(-0.4625t) + 10.914\exp(-0.07776t)$. Although in principle the lack of fit can be made obvious by plotting the differences between observed and calculated y values against time (inset), this presupposes a very high degree of experimental precision.

and indeed of equations. This is illustrated in Figure 13.1, which shows a set of points and a line calculated from two different equations, both of the type commonly encountered in transient-state kinetics. Although a plot of residual errors, that is to say differences between observed and calculated points, shows an obvious systematic character (inset to Figure 13.1), this plot requires such an expanded ordinate scale that even a small amount of random error in the data would submerge all evidence of lack of fit. The practical implication is that it is often impossible to extract all of the extra information that is theoretically available from transient-state measurements unless the various processes have very different frequency constants, or the terms with similar frequency constants have very different amplitudes.

One must not exaggerate the importance of this example. To some degree the problems visible in Figure 13.1 are a consequence of choosing a linear time scale: if a logarithmic time scale were used, with evenly spaced values of $\ln t$, the different exponential terms would be better resolved. Ill-conditioning is anyway not a special problem of transient-state kinetics, but applies to all types of quantitative experiments for which there is a temptation to calculate more parameters from the observations than they can support. Its particular relevance to transient-state experiments relates to the claim that these contain much more information than steady-state experiments, because in practice much of the extra information may be difficult to extract. It is also striking in Figure 13.1 that even though the three-term equation is similar in form to the two-term equation there is no correspondence in values between the two sets of parameters: there is no sense in which one can say that inclusion of the third term has just added to the information already provided by the first two.

13.2 Product release before completion of the catalytic cycle

13.2.1 "Burst" kinetics

While studying the chymotrypsin-catalysed hydrolysis of nitrophenylethyl carbonate, Hartley and Kilby (1954) observed that the release of nitro-phenolate became almost linear after a short period, and that extrapolation of the straight lines back to the product axis gave positive intercepts (Figure 13.2). Because the substrate was not a specific one for the enzyme and was consequently poor, they had to work with high enzyme concentrations, and the magnitudes of the intercepts, which are known as *bursts* of product, were proportional to the enzyme concentration. This suggested a mechanism

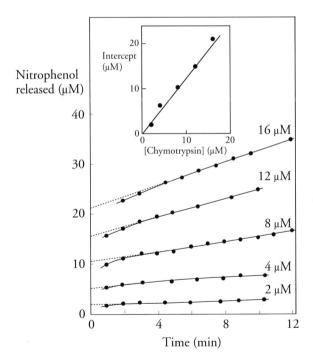

Figure 13.2. A "burst" of product release. Data of Hartley and Kilby (1954) for the chymotrypsin-catalysed hydrolysis of nitrophenylethyl carbonate are shown, and the curves are labelled with the enzyme concentrations. The intercepts obtained by extrapolating the straight portions of the progress curves back to zero time are proportional to and almost equal to these concentrations (inset).

in which the products were released in two steps, the nitrophenolate being released first:

$$E + A \xrightleftharpoons[k_{-1}]{k_1} EA \xrightarrow{k_2} EQ \xrightarrow{k_3} E + Q \qquad (13.2)$$

$$P$$

If the final step is rate-limiting, so that k_3 is small compared with k_1a, k_{-1} and k_2, then the enzyme exists almost entirely as EQ in the steady state. P can be released before EQ is formed, however, and so in the transient state P can be released much faster than at the steady-state rate. One might suppose that the amount of P released in the burst would be equal to the amount of enzyme, and not just proportional to it. However, this is accurately true only if k_3 is much smaller than the other rate constants; otherwise the burst is smaller than the stoicheiometric amount, as will now be shown, following a derivation based on that of Gutfreund (1955).

If a is large enough to be treated as a constant during the time period considered, and if k_1a is large compared with $(k_{-1} + k_2 + k_3)$, then shortly after mixing the system effectively simplifies to the following form:

$$EA \xrightleftharpoons[Q]{k_2} \quad \xrightarrow{P} \quad EQ \qquad (13.3)$$

$$\xleftarrow{k_3}$$

because the reaction $E + A \rightarrow EA$ can be regarded as instantaneous and irreversible, and the concentration of free enzyme becomes negligible. This is then a simple reversible first-order reaction (compare Section 1.4), with the following solution:

$$[EA] = \frac{e_0\{k_3 + k_2 \exp[-(k_2 + k_3)t]\}}{k_2 + k_3} \qquad (13.4)$$

$$[EQ] = \frac{k_2 e_0\{1 - \exp[-(k_2 + k_3)t]\}}{k_2 + k_3} \qquad (13.5)$$

Expressions for the rates of release of the two products are easily obtained by multiplying the first of these two equations by k_2 and the second by k_3:

$$\frac{dp}{dt} = k_2[EA] = \frac{k_2 e_0\{k_3 + k_2 \exp[-(k_2 + k_3)t]\}}{k_2 + k_3} \qquad (13.6)$$

$$\frac{dq}{dt} = k_3[EQ] = \frac{k_2 k_3 e_0\{1 - \exp[-(k_2 + k_3)t]\}}{k_2 + k_3} \qquad (13.7)$$

In the steady state, when t is large, the exponential term is negligible and the two rates are equivalent:

$$\frac{dp}{dt} = \frac{dq}{dt} = \frac{k_2 k_3 e_0}{k_2 + k_3} \tag{13.8}$$

In the transient phase, however, dp/dt is initially much larger than dq/dt, so that whereas P displays a burst, Q displays a *lag* when the linear parts of the progress curves are extrapolated back to zero time. The magnitude of the burst can be calculated by integrating equation 13.6 and introducing the condition $p = 0$ when $t = 0$:

$$p = \frac{k_2 k_3 e_0 t}{k_2 + k_3} + \frac{k_2^2 e_0 \{1 - \exp[-(k_2 + k_3)t]\}}{(k_2 + k_3)^2} \tag{13.9}$$

The steady-state part of the progress curve is obtained by considering the same equation after the transient has decayed to zero:

$$p = \frac{k_2 k_3 e_0 t}{k_2 + k_3} + \frac{k_2^2 e_0}{(k_2 + k_3)^2} \tag{13.10}$$

This is the equation for a straight line, and the intercept on the p axis gives π, the magnitude of the burst:

$$\pi = \frac{k_2^2 e_0}{(k_2 + k_3)^2} = \frac{e_0}{(1 + k_3/k_2)^2} \tag{13.11}$$

Thus the burst in P is not equal to the enzyme concentration but approximates to it if k_2 is large compared with k_3. The equation implies that the burst can never exceed the enzyme concentration, but reality is more complicated, because the substrate concentration is not truly constant and decreases throughout the steady-state phase. The steady-state portion of the progress curve is therefore not exactly straight, and if the rate decreases appreciably during the period examined extrapolation to the axis can cause the magnitude of the burst to be overestimated. This type of error can be avoided by ensuring that there are no perceptible deviations from linearity during the steady-state phase.

13.2.2 Active site titration

The discovery of burst kinetics led to an important method for titrating enzymes. It is generally difficult to obtain an accurate measure of the molarity of an enzyme: rate assays provide concentrations in activity units such as nkat/ml, which are adequate for comparative purposes, but do not provide true concentrations unless they have been calibrated in some way; most other assays are really protein assays and are therefore unspecific unless the enzyme is known to be pure and fully active. However, equation 13.11 shows that, if a substrate can be found for which k_3 is either very small or zero, then the burst π is both well-defined and equal to the concentration of active sites. The substrates of chymotrypsin that were examined originally, p-nitrophenylethyl carbonate and p-nitrophenyl acetate, had inconveniently large k_3 values, but subsequently Schonbaum, Zerner and Bender (1961) found that under suitable conditions trans-cinnamoylimidazole gave excellent results. At pH 5.5 this compound reacts rapidly with chymotrypsin to give imidazole and trans-cinnamoylchymotrypsin, but no further reaction readily occurs, k_3 being close to zero. So measurement of the amount of imidazole released by a solution of chymotrypsin provides a measure of the amount of enzyme.

Active-site titration by means of burst measurements differ from rate assays in being relatively insensitive to changes in the rate constants: a rate assay demands precisely defined pH, temperature, buffer composition and other conditions if it is to be reproducible, but the magnitude of a burst is unaffected by relatively large changes in k_2, such as might result from chemical modification of the enzyme, unless k_2 is decreased to the point where it is comparable in magnitude to k_3. As measured by this technique, therefore, chemical modification alters the molarity of an enzyme either to zero or not at all. For this reason, enzyme titration has also been called an *all-or-none assay* (Koshland, Strumeyer and Ray, 1962).

A recent discussion of enzyme titration methods (including single-turnover kinetic experiments, discussed below in Section 13.4.2) is given by Brocklehurst, Resmini and Topham (2001).

13.3   Experimental techniques

13.3.1   Classes of method

Steady-state experiments are usually carried out with a timescale of several minutes, at least. They have not required the development of special equipment, because, in principle, any method that permits the analysis of a reaction mixture at equilibrium can be adapted to allow analysis during the course of reaction. In the study of fast reactions, however, the short time periods involved have required specially designed instruments and methods that are not just obvious adaptations of those used for steady-state experiments.

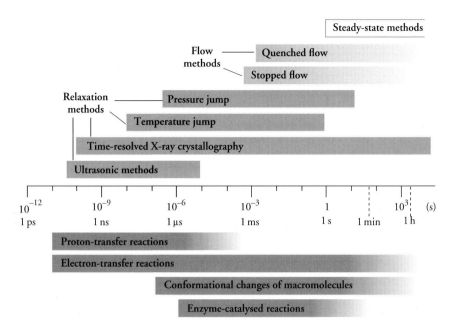

Figure 13.3. Time domains of methods and processes. The upper part of the Figure shows the time ranges in which various methods discussed in the text can be used. The lower part shows the typical ranges in which processes of interest occur.

The typical frequency constants of processes important for the understanding of enzyme-catalysed reactions range from around 10^{11} s^{-1} for the fastest proton- or electron-transfer reactions to less than 1 s^{-1} for the slower specific enzyme-catalysed reactions, or much slower, taking minutes or even hours, for reactions with unspecific substrates, as shown schematically in the lower half of Figure 13.3. However, as shown at the top of the figure, ordinary steady-state methods cannot be applied on a time scale of less than seconds (and with difficulty even then). An important group of rapid-mixing methods (Section 13.3.2–4) allow this range to be extended to the millisecond scale, but this is still too slow for many processes. Flash photolysis (Section 13.3.5) and relaxation methods (Section 13.3.7) bring all but the fastest chemical steps within range.

Other methods in addition to those shown in Figure 13.3 are useful in some circumstances. *Flash photolysis* and *pulse radiolysis* are techniques for generating unstable short-lived intermediates rapidly and then observing their subsequent reactions, and when they can be used the range accessible to measurement extends to processes as fast as 10^{13} s^{-1}. Even though there are few biological systems to which they can be directly applied, the information that they have given about simple chemical reactions contributes to our understanding of the chemical steps that occur in enzyme-catalysed reactions.

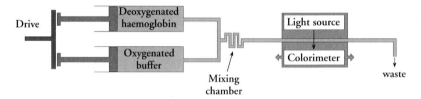

Figure 13.4. Continuous-flow method (schematic). By locating the detection system at different positions along the long observation tube one can examine mixtures that have aged for different times. This apparatus requires large amounts of protein, but does not need a rapidly responding detection system.

13.3.2 Continuous flow

For processes with frequency constants of the order of $1000\,\mathrm{s}^{-1}$ or less, the principal techniques used are rapid-mixing methods known collectively as *flow methods*. They are all derived ultimately from the *continuous-flow* method devised by Hartridge and Roughton (1923) for measuring the rate of combination of oxygen with haemoglobin. In this method the reaction was initiated by forcibly mixing the two reagents, reduced haemoglobin and oxygenated buffer, so that the mixture was made to move rapidly down a tube of 1 m in length. To prevent laminar flow of incompletely mixed components through the tube, it was essential to include a mixing chamber in the system to create turbulence and ensure complete and instaneous mixing. As long as the flow rate is constant, the mixture observed at any point along the tube in such an experiment has a constant age determined by the flow rate and the distance from the mixing chamber. So, by making measurements at several points along the tube one can obtain a progress curve for the early stages of reaction.

The principle of the apparatus is shown in Figure 13.4. This is not intended as a realistic representation of the original apparatus, but is drawn so as to emphasize the relationship of the method to the stopped-flow method, which will be considered shortly.

The continuous-flow method required large amounts of materials, which in practice limited its use to the study of haemoglobin. Nonetheless, the experiments of Hartridge and Roughton (1923) are among the most instructive in the history of biochemistry, worthy of study by all experimentalists, whether they are interested in reaction kinetics or not, because they illustrate how scientific ingenuity can overcome seemingly impossible obstacles. They were done before automatic devices for measuring light intensity became commercially available, and the equipment that did exist required several seconds of manual adjustment for making each measurement. Although it might seem self-evident that processes occurring on a time scale of milliseconds could not be studied with such equipment, Hartridge and Roughton showed that they could.

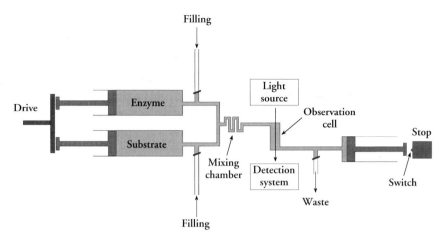

Figure 13.5. Essentials of the stopped-flow apparatus.

13.3.3   Stopped flow

Various major improvements to the design of rapid-mixing apparatus by Millikan (1936), Chance (1940, 1951), Gibson and Milnes (1964) and others led to the development of the *stopped-flow* method, which has become the most widely used method for studying fast reactions. The essentials of the apparatus are shown in Figure 13.5, and consist of the following: (a) two drive syringes containing the reacting species, (b) a mixing device, (c) an observation cell, (d) a stopping syringe, and (e) a detecting and recording system capable of responding sufficiently rapidly. The reaction is started by pushing the plungers of the two drive syringes simultaneously. This causes the two reactants to mix, and the mixture is forced through the observation cell and into the stopping syringe. A short movement of the plunger of the stopping syringe brings it to a mechanical stop, which prevents further mixing and simultaneously activates the detection and recording system. The time that inevitably elapses between the first mixing of reactants and the arrival of the mixture in the observation cell is of the order of 1 ms, and is called the *dead time* of the apparatus.

In its usual form, the stopped-flow method requires a spectrophotometer for following the course of the reaction. This makes it particularly useful for studying reactions that produce a large change in absorbance at a convenient wavelength, such as a dehydrogenase-catalysed reaction in which the oxidized form of NAD is reduced. The method is not, however, restricted to such cases, because other detection systems can be used. For example, many enzyme-catalysed reactions are accompanied by the release or uptake of protons, which can be detected optically by including a pH indicator in the reaction mixture, an approach that dates back to the continuous-flow

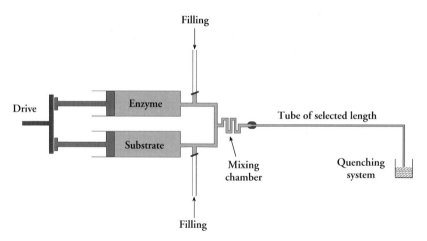

Figure 13.6. The quenched-flow method. The age of the system after mixing is varied by varying the length of the tube connecting the mixing chamber to the quenching system.

method (Brinkman, Margaria and Roughton, 1934). In other cases one may exploit changes of fluorescence during the reaction (see for example Hastings and Gibson, 1963).

13.3.4   Quenched flow

There are sometimes doubts about the chemical nature of the events observed spectroscopically in the stopped-flow method (Porter, 1967). These can in principle be overcome by using the *quenched-flow* method (Figure 13.6). In this method the reaction is stopped ("quenched") shortly after mixing, for example by a second mixing with a denaturing agent, such as trichloroacetic acid, that rapidly destroys enzyme activity; another method is to cool the mixture rapidly to a temperature at which the reaction rate is negligible. By varying the time between the initial mixing and the subsequent quenching, one can obtain a series of samples that can be analysed by chemical or other means, from which a record of the chemical progress of the reaction can be reconstructed.

The quenched-flow method requires much larger amounts of enzyme and other reagents than the stopped-flow method, because each run yields only one point on the time course, whereas each stopped-flow run yields a complete time course. It is often therefore appropriate to apply the quenched-flow method only after preliminary stopped-flow experiments have established the proper questions to be asked. Consider, for example, the data shown in Figure 13.7, which were obtained by Eady, Lowe and Thorneley (1978) in studies of the dependence on $MgATP^{2-}$ of the reaction catalysed by nitrogenase. This enzyme is responsible for the biological fixation of molecular

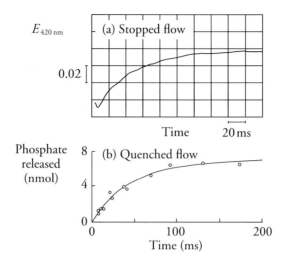

Figure 13.7. Comparison of stopped-flow and quenched-flow data for the reaction catalysed by nitrogenase from *Klebsiella pneumoniae* (Eady, Lowe and Thorneley, 1978). The stopped-flow trace (a) records the electron transfer between the two components of nitrogenase, whereas the quenched-flow observations (b) measure the rate of ATP hydrolysis as the rate of appearance of inorganic phosphate. The equality of the frequency constants shows that the two processes are coupled.

nitrogen, and consists of two proteins, known as the iron protein and the molybdenum–iron protein. $MgATP^{2-}$ is required for the transfer of electrons from the iron protein to the molybdenum–iron protein, and is hydrolysed to $MgADP^-$ during the reaction. The oxidation of the iron protein can be observed directly in the stopped-flow spectrophotometer at 420 nm. When the reaction is initiated by mixing with $MgATP^{2-}$ it shows a single relaxation with a frequency constant of $24 \pm 2 \ s^{-1}$ (Figure 13.7a). By itself, this observation does not establish that hydrolysis of $MgATP^{2-}$ and electron transfer are directly coupled; instead, $MgATP^{2-}$ might merely be an activator of the iron protein. To resolve this question the rate of hydrolysis had to be directly measured, and this was done by measuring the production of inorganic phosphate by the quenched-flow method. This process proved to have a frequency constant of $23 \pm 2 \ s^{-1}$ (Figure 13.7b), indistinguishable from the value measured in the stopped-flow method, and confirming that the two reactions were synchronous.

In the earliest versions of the quenched-flow method, the time between mixing and quenching was varied by varying the physical design of the apparatus, that is, by varying the flow rate and the length of tube between the two mixing devices. In practice there were severe restrictions on the time scales that could be used and the method was impractical for general use. Many of the problems were overcome in the *pulsed quenched-flow* method, which was described by Fersht and Jakes (1975). In this arrangement, the reaction is initiated exactly as in a stopped-flow experiment, and a second set of syringes is used for quenching. These are actuated automatically after a preset time has elapsed after the initial mixing. The period between mixing and quenching is controlled electronically and does not depend on the physical dimensions of the apparatus. As there is no need for long tubes, this system is much more economical of reagents than the conventional quenched-flow method.

More detailed information about flow methods may be found in books devoted specifically to rapid reactions (for example Hiromi, 1979), or with a strong emphasis on them (for example Fersht, 1999). More elaborate methods include rapid scanning spectrophotometry (Hollaway and White, 1975), for which the apparatus needed is now available commercially, though mainly likely to be found in specialized laboratories. However, it offers considerable advantages over the simpler types of stopped-flow equipment as complete spectra of the reaction mixture can be observed during the transient. This has allowed, for example, detailed study of the intermediates produced during the reaction of myeloperoxidase with hydrogen peroxide (Marquez, Huang and Dunford, 1994).

13.3.5   Flash photolysis

Mixing of reagents cannot be done efficiently in less than about 0.2 ms, and stopping the flow of a mixture through an apparatus requires about 0.5 ms. (Although one can conceive of stopping the process more abruptly, in practice shock waves would be created that would generate artefactual transients in the detection system.) Quenching, either chemically or by cooling, also requires finite time. There is therefore a lower limit of about 0.5 ms to the dead time that it is possible to achieve with flow methods, and it is unlikely that improved design will decrease this appreciably. Consequently processes that are virtually complete within 0.5 ms cannot be observed by flow methods. This is a severe restriction for the enzymologist, because most enzyme-catalysed reactions contain some such processes, and there are some for which the complete catalytic cycle requires less than 1 ms. Fersht (1999), for example, lists seven enzymes with catalytic constants of the order of 10^3 s^{-1} or more.

Avoiding the dead time for mixing at the start of a reaction is possible if one uses *flash photolysis* (Norrish and Porter, 1949), a technique that uses a high-energy pulse of radiation to produce the reactive form of a reactant when it is already mixed with the other reactants. This approach is limited by the need to have a suitable photosensitive precursor of any reactant one needs to study, but for major biochemical substances like ATP this is no longer a problem, since the introduction of "caged ATP" (Goldman and co-workers, 1982), an ester of the terminal phospho group of ATP with 2-nitrophenylethanol.

X-ray crystallography was long regarded as being far too slow to have kinetic applications, but this picture has changed considerably in the past fifteen years as a consequence of developments in several fields. Photolysis now allows reaction intermediates to be generated inside crystals from unreactive precursors, but it would be of limited usefulness without additional methods to collect X-ray data rapidly, and to perform the subsequent analysis. Short exposure times of less than a second are made possible by the use of synchrotron radiation, which is extremely intense but polychromatic, in contrast to the much weaker monochromatic radiation used in conventional X-ray crystallography. Polychromatic radiation was used in Laue's original X-ray diffraction experiments (Friedrich, Knipping and Laue, 1913), but it was rapidly supplanted by monochromatic radiation, which gave results that were far easier to analyse. This revival after many years of a method that appeared obsolete almost immediately after its first use required yet another development, namely the enormous increase in computing power that has been evident in the past few decades, which has made it almost routine to analyse data that previously seemed impossibly complicated.

Although synchrotron radiation is inherently polychromatic it is possible to generate it with a relatively narrow range of wavelengths, and monochromators exist that could in principle be used to narrow the range still further to allow use of essentially monochromatic radiation. In practice this is not done, however, because it has compensating disadvantages. Use of monochromatic radiation requires rotation of the crystal by about $0.2°$ during the radiation, interfering with observations of processes occurring on short time scales. However, polychromatic radiation allows the entire set of X-ray reflections to be excited instantaneously without rotation, and this provides the major reason for using it.

As an example, we may consider a study of isocitrate dehydrogenase by Stoddard and co-workers (1998). To prevent premature binding of the substrate, isocitrate, it was diffused into crystals of the enzyme in the form of a photolabile nitrophenyl derivative. The crystals were then exposed to X-rays for 10 ms, photolysed to release the isocitrate and exposed again to X-rays. Subsequent analysis of the crystallographic data allowed analysis of substrate binding, and further experiments were done with similarly labelled intermediates in the reaction. Outlined like this in a few words such experiments may appear straightforward, but in reality a high degree of expertise and care is needed at all stages, from the design and execution of the appropriate chemical modifications to the final analysis of the crystallographic data, as discussed in a recent review by Stoddard (2001).

Experiments designed to decrease the dependence on trapping, which incurs the danger of observing structures that are not necessarily those of real interest, have been applied to myoglobin (Ren and co-workers, 2001; Srajer and co-workers, 2001). These are less far advanced than the studies of isocitrate dehydrogenase just mentioned, but they show considerable promise.

13.3.6   Magnetic resonance methods

Magnetic resonance methods, especially *nuclear magnetic resonance*, have a time scale of applicability similar to that of the temperature-jump method. They are increasingly being used for the study of protein structure, and have applications also to the measurement of rate constants in systems at equilibrium, especially rate constants for dissociation of ligands, for example from enzyme–inhibitor complexes. When a paramagnetic metal ion such as Mn^{2+}, Fe^{2+} or Gd^{3+} interacts with an enzyme, either as a natural physiological component of the reaction or, more often, as a substitute for a physiological ion such as Mg^{2+}, its electron paramagnetic resonance can also be used, instead of or as a supplement to nuclear resonance. A recent discussion of the use of methods of this kind for determining rate constants is given by Monasterio (2001).

13.3.7   Relaxation methods

The problem of mixing time and other considerations of this sort led to the development of *relaxation* methods (Eigen, 1954) for studying very fast processes. These methods do not require mixing of reactants during the period of observation, although, as described below, it can be useful to combine them with the stopped-flow method. In a standard relaxation method a mixture at equilibrium is subjected to a *perturbation* that alters the equilibrium constant, and one then observes the system proceeding to the new equilibrium, a process known as *relaxation*. (In a sense, of course, mixing reactants is itself a perturbation, and one can consider any chemical reaction as a relaxation. In practice, however, experiments in which the only perturbation is mixing of reactants are not usually regarded as relaxation studies.) Various different relaxation methods exist, of which the most familiar is the *temperature-jump* method, with a perturbation consisting of an increase in temperature brought about by passing a large electric discharge through the reaction mixture: in this way one can easily produce an increase in temperature of about 10 °C in about 1 μs. Another kind of perturbation is a pressure jump: this is useful for probing volume changes that may occur during the catalytic process, but is more difficult to use because the changes in kinetic constants that occur after a pressure change are typically much smaller than those that result from the same input of energy in the form of a temperature change (see Section 1.7.4).

The perturbation produced by an electric discharge is not instantaneously converted into a change in temperature uniformly distributed over the whole reaction volume. Instead, it takes at least 1 μs, and considerable care in the design of the apparatus is needed to ensure that all parts of the reaction mixture are heated uniformly. Thus one can only regard the heating as instantaneous if one confines attention to processes that occur more than 1 μs after the beginning of the perturbation. It is not likely that appreciably shorter times can ever be achieved with irreversible perturbations, but much faster processes can be studied with *sinusoidal* perturbations. Ultrasonic waves, for example, with frequencies as high as 10^{11} s^{-1}, produce local fluctuations in temperature and pressure as they propagate through a medium. These fluctuations produce oscillations in the values of all the rate constants of the system, and study of the absorption of ultrasonic energy by a reaction mixture yields information about these rate constants (Hammes and Schimmel, 1970). Enzyme systems are generally too complicated for direct application of this approach, but the study of simple systems has nonetheless provided information valuable for the enzymologist: for example, the work of Burke, Hammes and Lewis (1965) on poly-L-glutamate showed that a major conformational change of a macromolecule, the helix–coil transition, can occur at a rate of 10^5–10^7 s^{-1}. Obviously, conformational changes in enzymes do not have to occur in the same range of rates, but similar rates must be possible, and so the need for a fast conformational change does not provide any objection to a proposed mechanism for enzymic catalysis.

One disadvantage of observing the relaxation of a system to equilibrium is that the equilibrium concentrations of the transient species of particular importance in the catalytic process may be too small to detect. This difficulty can be overcome by combining the stopped-flow and temperature-jump methods, and commercial stopped-flow spectrophotometers now include a temperature-jump capability. The reactants are mixed as in a conventional stopped-flow experiment, and are subsequently subjected to a temperature jump after a steady state has been attained. Relaxation to the new steady state characteristic of the higher temperature is then observed. This sort of experiment allows the observation of processes in the early phase of reaction that are too fast for the conventional stopped-flow method, because the reactants are already mixed when the temperature jump occurs.

In the standard kind of temperature-jump apparatus, in which the temperature change is produced by an electrical discharge, there is no provision to prevent the system from cooling after the heating has occurred. This is the reason for the upper limit of applicability of about 1 s shown in Figure 13.3.

However, if precautions are taken to avoid this problem the period can be increased indefinitely. For example, Buc, Ricard and Meunier (1977) used the temperature-jump method to study relaxations of wheat-germ hexokinase lasting as long as 40 min.

13.4 Transient-state kinetics

13.4.1 Systems far from equilibrium

In Section 2.5, I examined the validity of the steady-state assumption by deriving an equation for the kinetics of the two-step Michaelis–Menten mechanism without assuming a steady state. This derivation was only possible, however, because the substrate concentration was treated as constant, which was clearly not exactly correct. Solution of the differential equations is unfortunately impossible for nearly all mechanisms of enzyme catalysis unless some assumptions are made; approximations must always be introduced, therefore, if any analysis is to be possible. In transient-state experiments one usually tries to set up conditions such that the mechanism approximates to a sequence of first-order steps, because this is the most general sort of mechanism that has an exact solution.

As discussed in Chapter 4, virtually all steady-state systems in enzyme mechanisms can be analysed in terms of a single method, that of King and Altman (1956). This approach can be adapted to the analysis of the transient-state kinetics of systems far from equilibrium, as set out in detail by Pettersson (1978), but it is simpler to examine some particular cases, starting with the following two-step mechanism, to illustrate how sequences of first-order steps can be analysed:

$$X_0 \underset{k_{-1}}{\overset{k_1}{\rightleftharpoons}} X_1 \underset{k_{-2}}{\overset{k_2}{\rightleftharpoons}} X_2 \tag{13.12}$$

The system is defined by a *conservation equation* and three rate equations. The conservation equation is as follows:

$$x_0 + x_1 + x_2 = x_{tot} \tag{13.13}$$

and ensures that the requirements of stoicheiometry are satisfied, by requiring the sum of the three concentrations to be a constant, x_{tot}. The three rate equations are as follows:

$$\frac{dx_0}{dt} = -k_1 x_0 + k_{-1} x_1 \tag{13.14}$$

$$\frac{dx_1}{dt} = k_1 x_0 - (k_{-1} + k_2)x_1 + k_{-2}x_2 \tag{13.15}$$

$$\frac{dx_2}{dt} = k_2 x_1 - k_{-2}x_2 \tag{13.16}$$

Any one of these three equations is redundant, as their sum is simply the first derivative of the conservation equation, equation 13.13:

$$\frac{dx_0}{dt} + \frac{dx_1}{dt} + \frac{dx_2}{dt} = 0 \tag{13.17}$$

To solve the system, therefore, we can take it as defined by equations 13.13–15, ignoring equation 13.16.

Solution of a set of three differential equations in three unknown concentrations is most easily achieved by eliminating two of the concentrations to produce a single differential equation in one unknown. First x_2 can be eliminated by using equation 13.13 to express it in terms of the other two concentrations:

$$x_2 = x_{\text{tot}} - x_0 - x_1 \tag{13.18}$$

and substituting this in equation 13.15:

$$\begin{aligned}
\frac{dx_1}{dt} &= k_1 x_0 - (k_{-1} + k_2)x_1 + k_{-2}(x_{\text{tot}} - x_0 - x_1) \\
&= (k_1 - k_{-2})x_0 - (k_{-1} + k_2 + k_{-2})x_1 + k_{-2}x_{\text{tot}}
\end{aligned} \tag{13.19}$$

Differentiation of equation 13.14 yields

$$\begin{aligned}
\frac{d^2 x_0}{dt^2} &= -k_1 \frac{dx_0}{dt} + k_{-1} \frac{dx_1}{dt} \\
&= -k_1 \frac{dx_0}{dt} + k_{-1}(k_1 - k_{-2})x_0 - k_{-1}(k_{-1} + k_2 + k_{-2})x_1 + k_{-1}k_{-2}x_{\text{tot}}
\end{aligned} \tag{13.20}$$

Next x_1 is eliminated by rearranging equation 13.14 into an expression for $k_{-1}x_1$ in terms of x_0,

$$k_{-1}x_1 = k_1 x_0 + \frac{dx_0}{dt} \tag{13.21}$$

which can be substituted into equation 13.20:

$$\frac{d^2 x_0}{dt^2} = -k_1 \frac{dx_0}{dt} + k_{-1}(k_1 - k_{-2})x_0 - (k_{-1} + k_2 + k_{-2})\left(k_1 x_0 + \frac{dx_0}{dt}\right) + k_{-1}k_{-2}x_{\text{tot}}$$

$$\tag{13.22}$$

If this is rearranged,

$$\frac{d^2 x_0}{dt^2} + (k_1 + k_{-1} + k_2 + k_{-2})\frac{dx_0}{dt} + (k_{-1}k_{-2} + k_1 k_2 + k_1 k_{-2})x_0 = k_{-1}k_{-2}x_{tot}$$

(13.23)

it is seen to have the following standard form:

$$\frac{d^2 x_0}{dt^2} + P\frac{dx_0}{dt} + Qx_0 = R \qquad (13.24)$$

and to have the following solution:

$$x_0 = x_{0\infty} + A_{01}\exp(-\lambda_1 t) + A_{02}\exp(-\lambda_2 t) \qquad (13.25)$$

in which $x_{0\infty}$ is the value of x_0 at equilibrium, A_{01} and A_{02} are constants of integration that give the amplitudes of the two transients, and λ_1 and λ_2 are the corresponding frequency constants. As in most elementary accounts of relaxation kinetics, I shall concentrate on the information content of the frequencies, which are simpler to treat mathematically than the amplitudes. Nonetheless, amplitudes provide a potentially rich source of additional information. Because relaxation methods involve input of energy, usually in the form of heat, every relaxation amplitude depends on the thermodynamic properties of the system, such as the enthalpy of reaction ΔH. Consequently, amplitude measurements can provide more accurate information about these thermodynamic properties than is available from ordinary measurements. Thusius (1973) describes how this is done for the simple case of formation of a 1:1 complex, and gives references to other sources of information.

The values of the frequency constants in equation 13.25 are as follows:

$$\lambda_1 = 0.5[P + (P^2 - 4Q)^{0.5}] \qquad (13.26)$$

$$\lambda_2 = 0.5[P - (P^2 - 4Q)^{0.5}] \qquad (13.27)$$

The solutions for the other two concentrations x_1 and x_2 have the same form as equation 13.25, with the same pair of frequency constants as those in equations 13.26–27 but with different amplitudes.

If $k_{-1}k_2$ is small compared with $(k_{-1}k_{-2} + k_1 k_2 + k_1 k_{-2})$, the expressions for the frequency constants simplify to $(k_1 + k_{-1})$ and $(k_2 + k_{-2})$, not necessarily in that order, as λ_1 is always the larger and λ_2 is always the smaller. This simplification is not always permissible, but the expressions for the sum and product of the frequency constants always take fairly simple forms:

$$\lambda_1 + \lambda_2 = k_1 + k_{-1} + k_2 + k_{-2} = P \tag{13.28}$$

$$\lambda_1 \lambda_2 = k_{-1}k_{-2} + k_1k_2 + k_1k_{-2} = Q \tag{13.29}$$

This example illustrates several points that apply more generally. Any mechanism that consists of a sequence of n steps that are first-order in both directions can be solved exactly (Matsen and Franklin, 1950). The solution for the concentration of any reactant or intermediate consists of a sum of $(n + 1)$ terms, the first being its value at equilibrium and the others consisting of n transients, with frequencies that are the same for all concentrations and amplitudes that are characteristic of the particular concentration. In favourable cases the frequency constants for some or all of the transients can be associated with particular steps in the mechanism; when this is true the frequency constant is equal to the sum of the forward and reverse rate constants for the step concerned.

Another general point, not illustrated by the above analysis, is that reactants that are separated from the rest of the mechanism by irreversible steps have simpler relaxation spectra than other reactants, because some of their amplitudes are zero. Consider for example the following five-step mechanism, in which two of the steps are irreversible:

$$X_0 \rightleftarrows X_1 \rightarrow X_2 \rightleftarrows X_3 \rightarrow X_4 \rightleftarrows X_5 \tag{13.30}$$

In principle, each reactant should have five frequency constants, and this is indeed what ought to be found for X_4 and X_5. But X_2 and X_3 are isolated from the last step by the irreversible fourth step; they therefore have one zero amplitude each, and only four frequency constants. X_0 and X_1 are isolated from the rest of the mechanism by the irreversible second step; they therefore have three zero amplitudes each, and only two frequency constants. Regardless of the presence of irreversible steps, the total number of relaxations observed for any concentration cannot exceed the number predicted by the mechanism, but is often less, either because processes with similar frequency constants are not resolved, or because some of the amplitudes are too small to be detected.

All mechanisms for enzyme catalysis include at least one second-order step, but any such step can be made to follow pseudo-first-order kinetics with respect to time, by ensuring that one of the two reactants involved is in large excess over the other. It follows that at least one of the observed frequency constants contains a pseudo-first-order rate constant and thus its expression includes a concentration dependence, thus allowing measured frequency constants to be assigned to particular steps. Consider, for example, the following mechanism, which represents half of a substituted-enzyme mechanism (Section 7.2.2) studied in the absence of the second substrate:

$$
E + A \underset{k_{-1}}{\overset{k_1}{\rightleftarrows}} EA \xrightarrow{k_2} E' + P \tag{13.31}
$$

For this mechanism equations 13.28–29 take the following form:

$$
\lambda_1 + \lambda_2 = k_1 a + k_{-1} + k_2 \tag{13.32}
$$

$$
\lambda_1 \lambda_2 = k_1 k_2 a \tag{13.33}
$$

So a plot of the sum of the two frequency constants against a yields a straight line of slope k_1 and intercept $(k_{-1} + k_2)$ on the ordinate, and a plot of their product against a yields a straight line through the origin with slope $k_1 k_2$. All three rate constants can thus be calculated from measurements of the frequency constants.

13.4.2   Simplification of complicated mechanisms

Although a system of n unimolecular steps can in principle be analysed exactly, regardless of the value of n, it is in practice difficult to resolve exponential processes unless they are well separated on the time axis. Consequently the number of transients detected may well be less than the number present. The degree of separation necessary for resolution depends on the amplitudes, but one can make some useful generalizations. If two processes have amplitudes of opposite sign they are relatively easy to resolve, even if the frequency constants are within a factor of 2. The reason for this is fairly obvious: if a signal appears and then disappears at least two transients must be involved. If two neighbouring transients have amplitudes of the same sign it is much more difficult to resolve them, because the decay curve is monotonic and unless the faster process has a much larger amplitude than the slower one its presence may pass unnoticed (compare Figure 13.1, above).

Slow relaxations are in general easier to measure than fast ones, because they can be observed in a time scale in which all of the faster ones have decayed to zero. In principle, therefore, one can examine the faster processes by subtracting out the contributions of the slower ones. Consider, for example, the following equation:

$$
x = x_\infty + A_1 \exp(-\lambda_1 t) + A_2 \exp(-\lambda_2 t) \tag{13.34}
$$

and assume that λ_1 is smaller than λ_2 by a factor of at least 10. One can evaluate x_∞ by allowing the reaction to proceed to equilibrium, and then determine A_2 and λ_2 by making measurements over a period from about $0.5/\lambda_2$ to about $5/\lambda_2$. These three constants then allow $x_\infty + A_2 \exp(-\lambda_2 t)$ to be calculated at any time, and by doing this in the early part of the progress curve and subtracting the result from the observed value of x one obtains data for a single relaxation $A_1 \exp(-\lambda_1 t)$. In this process, known as *peeling*, the errors accumulate as one proceeds: any inaccuracy in x_∞ contributes to the errors in A_2 and λ_2, and any inaccuracies in x_∞, A_2 and λ_2 contribute to the errors in A_1 and λ_1. So although in principle any number of relaxations can be resolved by this method, in practice the faster processes are much less well defined than the slower ones, and it is advisable to create experimental conditions in which the number of relaxations is as small as possible. There was a simple example of this in Section 13.2.1: although the three-step mechanism used to explain the burst of product release should in principle give rise to two relaxations, the number was decreased to one by using such a high substrate concentration that the first relaxation could be treated as instantaneous.

Considered from the point of view of the enzyme, enzyme-catalysed reactions are usually cyclic; that is, the first reactant, the free enzyme, is also the final product. This does not prevent solution of the differential equations (provided, as before, that every step is first-order or pseudo-first-order), but it does lead to more complicated transient kinetics than one obtains with non-cyclic reactions, because a cyclic system relaxes to a steady state rather than to equilibrium. It is therefore useful to simplify matters by eliminating the cyclic character from the reaction. There are various ways of doing this, of which conceptually the simplest is to choose a substrate for which the steady-state rate is so small that it can be ignored. In effect, this is what one does in using an active-site titrant (Section 13.2.2). It has the important disadvantage, however, that it usually means studying the enzyme with an unnatural substrate.

A different approach is to carry out a *single-turnover experiment*, in which the rate is limited by substrate, not by enzyme, in conditions with $e_0 \gg a_0$. The Michaelis–Menten mechanism can then be written as follows:

$$ A \underset{k_{-1}}{\overset{k_1 e}{\rightleftharpoons}} EA \underset{k_{-2}e}{\overset{k_2}{\rightleftharpoons}} P \qquad (13.35) $$

This has the form of equation 13.12, because the reaction must stop when the substrate is consumed, and so no recycling of enzyme can take place. This approach has been comparatively little used in recent years, but it remains potentially valuable for determining the true active-site molarities of enzymes (or other catalysts, such as catalytic antibodies) without an assumption about

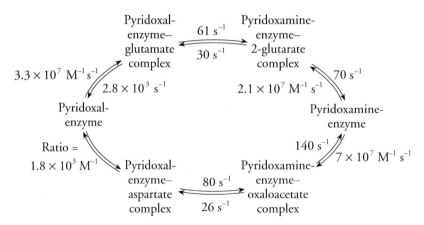

Figure 13.8. The mechanism of the reaction catalysed by glutamate–aspartate transaminase, with the rate constants assigned by Hammes and Fasella (1962).

purity, and it remains valid in the presence of complications such as the capacity of some protein molecules to bind reactants without catalysing any reaction (Topham and co-workers, 2000; Brocklehurst, Resmini and Topham, 2001). The essential idea is that the usual steady-state methods readily provide values of parameters such as $K_m/V = 1/k_Ae_0$ that include the enzyme concentration e_0 as a factor, whereas pre-steady-state methods, such as the plots described at the end of Section 13.4.1, provide the corresponding quantities without the factor e_0; by dividing one by the other, therefore, one can calculate the true enzyme molarity (Bender and co-workers, 1966; Reiner, 1969).

In reactions with more than one substrate (other than hydrolytic reactions), one can prevent enzyme recycling by omitting one substrate from the reaction mixture. This is especially useful for enzymes that follow a substituted-enzyme mechanism, because some chemical reaction occurs, and is potentially measurable, even in incomplete reaction mixtures. An early example of this approach was the study of glutamate–aspartate transaminase by Hammes and Fasella (1962). By studying the partial reactions of this enzyme, mainly by the temperature-jump method, they were able to assign values to 10 of the 12 rate constants that occur in the mechanism (Figure 13.8). Transaminases form a particularly attractive class of enzymes for such studies on account of the easily monitored spectral changes in the coenzyme, pyridoxal phosphate, that occur during the reaction.

Treatment of enzymes that follow a ternary-complex mechanism is less straightforward, because a complete reaction mixture is normally required before any chemical change can occur. Nonetheless Pettersson (1976) has provided a rigorous treatment of the transient kinetics of ternary-complex reactions, and applied it to resolve some ambiguities in the reaction catalysed by alcohol dehydrogenase from horse liver (Kvassman and Pettersson, 1976). Single-turnover experiments can be carried out with ternary-complex enzymes by keeping one substrate (not both) at a much lower concentration than that of the enzyme.

One of the attractive features of fast-reaction kinetics is that it can often provide conceptually simple information about mechanisms without any of the algebraic complications that can hardly be avoided in steady-state work. An obvious example of this was the original burst experiment of Hartley and Kilby (1954), in which the order of release of products was established with a high degree of certainty by the observation that one product (the first) was released in a burst (Section 13.2.1). Strictly, one ought to show that the second product does not also show a burst, because in principle both products could be released in a burst if the last step of the reaction were a rate-limiting isomerization of the free enzyme to regenerate its original form.

Similarly, one can deduce the order of addition of substrates in a ternary-complex mechanism by varying the combinations of reagents in the syringes in a stopped-flow experiment. If there is a compulsory order of addition, the trace seen when the enzyme is premixed with the substrate that binds first is likely to be simpler than that seen when the enzyme is premixed with the substrate that binds second: premixing with the first substrate means that the first complex is already formed when the syringes are actuated, but premixing with the second substrate achieves nothing because no reaction can take place until the enzyme is exposed to the first substrate.

13.4.3 Systems close to equilibrium

In the temperature-jump apparatus the perturbation of the equilibrium constant is not usually large enough to create a state that is far from equilibrium. Analysis of the relaxation kinetics is therefore fairly simple, with no requirement for all higher-order steps to be made pseudo-first-order, because it is the terms in products of concentrations that render the differential equations insoluble in strict terms, and these can be neglected in systems close to equilibrium. A simple binding reaction illustrates this point:

$$\text{E} \quad + \quad \text{A} \quad \underset{k_{-1}}{\overset{k_1}{\rightleftharpoons}} \quad \text{EA} \qquad (13.36)$$
$$e_\infty + \Delta e \quad a_\infty + \Delta a \qquad\qquad x_\infty + \Delta x$$

If e_∞, a_∞ and x_∞ are the equilibrium concentrations of E, A and EA respectively, and all of these and also the rate constants k_1 and k_{-1} are defined as those that apply at the higher temperature, that is to say the constants that define the system *after* the perturbation, then the concentrations at any instant t can be represented as $(e_\infty + \Delta e)$, $(a_\infty + \Delta a)$ and $(x_\infty + \Delta x)$ respectively. Then the rate is given by the following expression:

$$\frac{dx}{dt} = k_1(e_\infty + \Delta e)(a_\infty + \Delta a) - k_{-1}(x_\infty + \Delta x) \qquad (13.37)$$

However, $d\Delta x/dt$ is the same as dx/dt, and, by the stoicheiometry of the reaction, $\Delta e = \Delta a = -\Delta x$, so

$$\frac{d\Delta x}{dt} = k_1(e_\infty - \Delta x)(a_\infty - \Delta x) - k_{-1}(x_\infty + \Delta x)$$
$$= k_1 e_\infty a_\infty - k_{-1}x_\infty - [k_1(e_\infty + a_\infty) + k_{-1}]\Delta x + k_1(\Delta x)^2 \qquad (13.38)$$

As it stands this is a non-linear differential equation that has no analytical solution. But the term that makes it non-linear is $k_1(\Delta x)^2$, and if the system is close to equilibrium, as assumed at the beginning, this term can be neglected. Moreover, the net rate at equilibrium is zero, as in any equilibrium, and so $k_1 e_\infty a_\infty - k_{-1}x_\infty = 0$, with the result that the first two terms of equation 13.38 vanish and it simplifies to a linear differential equation:

$$\frac{d\Delta x}{dt} = - [k_1(e_\infty + a_\infty) + k_{-1}]\Delta x \qquad (13.39)$$

This can be solved directly by separating the variables and integrating (compare equation 1.1 in Section 1.2.1, which has the same form), with the following result:

$$\Delta x = \Delta x_0 \exp\{-[k_1(e_\infty + a_\infty) + k_{-1}]t\} \qquad (13.40)$$

in which Δx_0 is the magnitude of the perturbation when $t = 0$. Thus, provided that the initial perturbation is small, the relaxation of a single-step reaction is described by a single transient with a frequency constant λ given by

$$\lambda = k_1(e_\infty + a_\infty) + k_{-1} \qquad (13.41)$$

As e_∞, a_∞ and k_{-1}/k_1 can normally be measured independently, measurement of λ permits individual values to be assigned to k_1 and k_{-1}.

A similar analysis can be applied to any mechanism close to equilibrium, without regard to whether the individual steps are first-order or not, because the non-linear terms of the type $k_1(\Delta x)^2$ can always be neglected. In general, a mechanism with n steps has a solution with n transients, though this number may be decreased by thermodynamic constraints on the allowed values for the rate constants. In addition, the number of transients observed experimentally is often less than the theoretical number, because of failure to detect transients that have small amplitudes or are poorly resolved from neighbouring ones.

In Section 13.3.7 we saw that it is often convenient to observe a system relaxing to a steady state rather than to equilibrium. Such a system can be analysed in the same sort of way as one close to equilibrium. Consider, for example, the Michaelis–Menten mechanism:

$$
\underset{e_{ss} + \Delta e \qquad a_0}{E \; + \; A} \; \underset{k_{-1}}{\overset{k_1}{\rightleftharpoons}} \; \underset{x_{ss} + \Delta x}{EA} \; \overset{k_2}{\longrightarrow} \; E + P \qquad (13.42)
$$

If this is perturbed from a steady state, so that the rate constants are no longer those that defined the original steady state, it will relax to a new steady state at the following rate:

$$
\frac{d\Delta x}{dt} = k_1 (e_{ss} + \Delta e) a_0 - (k_{-1} + k_2)(x_{ss} + \Delta x) \qquad (13.43)
$$

After introducing the steady-state condition $k_1 e_{ss} a_0 = (k_{-1} + k_2) x_{ss}$, and the stoicheiometric requirement $\Delta e = -\Delta x$, this simplifies as follows:

$$
\frac{d\Delta x}{dt} = -(k_1 a_0 + k_{-1} + k_2)\Delta x \qquad (13.44)
$$

which is readily integrable to give a solution that consists of a single transient:

$$
\Delta x = \Delta x_0 \exp[-(k_1 a_0 + k_{-1} + k_2)t] \qquad (13.45)
$$

A more advanced and more detailed account of relaxation kinetics of enzymes is given by Hammes and Schimmel (1970).

Problems

13.1 An enzyme of molecular mass 50 kDa is studied in a single-turnover experiment at a substrate concentration of 1 μM. Measurement of the rate of product appearance at an enzyme concentration of 1 mg/ml reveals two transients, with frequency constants of 220 and 10 s^{-1}. At 5 mg/ml enzyme the corresponding values are 360 and 32 s^{-1}. Assuming the simplest mechanism consistent with these observations, estimate the values of the rate constants.

13.2 The following data can be expressed by an equation of the form $y = A + B \exp(-\lambda_1 t) + C \exp(-\lambda_2 t)$. Assuming that λ_1 is more than

$200\,s^{-1}$, estimate the values of A, C and λ_2, and use the results to estimate B and λ_1.

t (ms)	y	t (ms)	y	t (ms)	y
1	78	7	40	25	20
2	68	8	37	30	19
3	60	9	34	35	18
4	53	10	32	40	17
5	48	15	25	45	17
6	43	20	22	50	17

13.3 In steady-state studies of an enzyme, a competitive inhibitor was found to have $K_i = 20\,\mu M$, and this value was interpreted as a true equilibrium constant. The Michaelis constant for the substrate was found to be 0.5 mM. Subsequently the following stopped-flow experiments were carried out: in experiment (a), one syringe of the apparatus contained 50 μM enzyme, and the other contained 0.4 mM inhibitor and 10 mM substrate; in experiment (b), one syringe of the apparatus contained 0.4 mM inhibitor and 50 μM enzyme, and the other contained 10 mM substrate. The transient phase of the reaction had a frequency constant of $130\,s^{-1}$ in experiment (a) but $67\,s^{-1}$ in experiment (b). Estimate the *on* and *off* rate constants for binding of the inhibitor to the free enzyme. Do the results of the stopped-flow experiments support the original interpretation of K_i as a true equilibrium constant?

Chapter 14

Estimation of Kinetic Constants

14.1 The effect of experimental error on kinetic analysis

Most enzyme kinetic experiments carried out since the 1930s have been analysed by means of linear plots of the sort discussed in Sections 2.6.2–4, and of these by far the most popular has been and remains the double-reciprocal plot of reciprocal rate against reciprocal substrate concentration (Section 2.6.2). The appearance of small computers in every laboratory has led to an increase in the number of experiments analysed with computer programs, but graphical methods remain in widespread use, most definitely for illustrating results but to a considerable extent also for the actual analysis. In any case computer-based methods constitute an advance only if they are used with some understanding of the underlying calculations and the assumptions implicit in them.

It might seem that all valid plots ought to give the same results, apart from slight subjective variations derived from different ideas about where to draw the best lines, and that computation would also give the same results apart from greater consistency because of eliminating this subjective element. If experiments could be done with perfect accuracy this would indeed be the expectation, but in reality experimental error is always present and causes different methods to give different parameter values, because different methods handle the error in different ways.

This is illustrated by some sample calculations in Table 14.1. The top half of the Table shows the effect on $1/v$ and a/v of assuming the same additive error of $+0.1$ in each of five observations spanning the range from $K_m/5$ to $5K_m$. Notice that although the errors in the v values are all exactly the same, the resulting errors in $1/v$ vary more than 20-fold, and those in a/v almost twofold. (This large difference explains why the error bars that were shown in Figure 2.5, on page 43, varied greatly in length, whereas those in Figure 2.6, on page 44, varied much less.) Moreover, as the bottom half of the

Table 14.1. Effect of experimental error on transformed observations. The upper half of the table illustrates how an error of $+0.1$ in the value of a rate v leads to an error in the reciprocal rate $1/v$ that varies very greatly in magnitude with the true value of v, whereas it produces an error in the ratio a/v that is much more weakly dependent on the true value. The lower half of the table shows, in contrast, that an error of $+5\%$ has similar effects on the errors in the two derived quantities.

a	v_{true}	ε	v	$\dfrac{1}{v_{true}}$	$\dfrac{1}{v}$	$\dfrac{1}{v} - \dfrac{1}{v_{true}}$	$\dfrac{a}{v_{true}}$	$\dfrac{a}{v}$	$\dfrac{a}{v} - \dfrac{a}{v_{true}}$
0.2	1	+0.1	1.1	1.00000	0.90909	−0.09091	0.20000	0.18182	−0.01818
0.5	2	+0.1	2.1	0.50000	0.47619	−0.02381	0.25000	0.23810	−0.01190
1	3	+0.1	3.1	0.33333	0.32258	−0.01075	0.33333	0.33333	−0.01075
2	4	+0.1	4.1	0.25000	0.24390	−0.00610	0.50000	0.48780	−0.01220
5	5	+0.1	5.1	0.20000	0.19608	−0.00392	1.00000	0.98039	−0.01961
0.2	1	+5%	1.05	1.00000	0.95238	−0.04762	0.20000	0.19048	−0.00952
0.5	2	+5%	2.10	0.50000	0.47619	−0.02381	0.25000	0.23810	−0.01190
1	3	+5%	3.15	0.33333	0.31746	−0.01587	0.33333	0.31746	−0.01587
2	4	+5%	4.20	0.25000	0.23810	−0.01190	0.50000	0.47619	−0.02381
5	5	+5%	5.25	0.20000	0.19048	−0.00952	1.00000	0.95238	−0.04762

Table shows, the results are not even qualitatively the same when the original errors in v are a constant proportion of the true values.

For simplicity all of the errors in Table 14.1 are calculated as positive values, though in reality some would be positive and some negative. If the calculation had been done with negative errors the results would have been similar to those shown (apart from sign changes), but not identical, because the non-linear transformations of v into $1/v$ and a/v have asymmetrical effects in the two directions: that is why the error bars in Figures 2.5–7 are asymmetric (all of them, though the asymmetry is only noticeable for some of the longer ones).

Even these complications could, with some effort, be overcome, if real experimental errors were systematic (all of the same size and in the same direction) and of known magnitude, as in Table 14.1. In reality, however, they are of unknown magnitudes and they vary at random from one observation to another.

Thus transformation of the original observations for more convenient plotting produces non-linear distortions of any errors present in the original data, and these distortions make it difficult to handle these errors properly. The distortions may be large (as for $1/v$ in the upper half of Table 14.1), or they may be reasonably small (as for a/v in the upper half of Table 14.1); they may be of similar magnitudes for different transformations (as in the lower half of Table 14.1), or they may be of quite different magnitudes for the same pair of transformations (as in the upper half of Table 14.1).

All of this means that estimating kinetic parameters from graphs is not as straightforward as it may appear at first sight. Many authors (including this one in an earlier book: Cornish-Bowden, 1976c) have drawn the conclusion that graphs should be avoided and that serious parameter estimation should always rely on computation. This conclusion is made especially appealing by the fact that nearly all of the methods to be described in this chapter are suitable for expression as computer programs, and many are far too laborious to try to apply any other way. It is mistaken, nonetheless, because it overlooks two points: first, the human eye is much better at recognizing peculiar or unexpected behaviour than any existing computer program; second, all computer programs embody some assumptions about the properties of the experimental errors underlying the data, and if these are not appropriate the results from a computer program will not be any better than what one can expect from a poorly executed graph. It is obvious, moreover, that many users of computer programs have little understanding of these implicit assumptions. For example, one may read inconsistent statements in many papers, to the effect that the observations were found to have approximately uniform percentage errors and were accordingly given equal weight: as we shall see below, equal weight is appropriate for observations with uniform errors in absolute magnitude, not when expressed as a percentage.

Blind use of a computer program written by someone else is rarely a safe procedure. It cannot convert a badly designed experiment into a well-designed one, and it cannot convert poor data into precise information. Nonetheless, it is hardly realistic in the twenty-first century to suggest that all computer users should be programmers; in any case, several good programs for fitting enzyme kinetic data are now available. Provided that they are used with some attention to their underlying theory, they can be a valuable tool. This chapter provides an introduction to the theory, but it is dealt with in more detail in another book (Cornish-Bowden, 1995b), which includes a program for implementing the methods described on the IBM PC and compatible computers.

Before leaving the topic of graphs we should note that their function goes beyond analysing data: they also have the essential role of illustrating the results of analysis. If readers of a paper see, as they do with ever increasing frequency, parameter values accompanied by no indication about how they were obtained beyond an uninformative statement that the data "were fitted to the hyperbolic Michaelis–Menten equation" (Portaro and co-workers, 2000), or the parameters "were calculated fitting the data using the GRAFIT program" (Stabile, Curti and Vanoni, 2000), they have no way of judging whether the fitting was done with appropriate attention to weighting and testing for systematic error; hence they have no basis for putting any trust in the values quoted. Accompanying the quantitative results by at least one graph to illustrate the quality and extent of the data should be regarded as essential, but, increasingly, it is not. On the contrary, pressure from editors to omit what they see as inessential information is producing a trend in the opposite direction: more and more papers not only present no primary data but present no secondary data either. For example, the parameters in a recent study of sialyltransferase (Nash and co-workers, 2000) were obtained by analysis of progress curves, but not only were no progress curves (the primary data) shown; no rates derived from them (the secondary data) were shown either.

14.2   Least-squares fit to the Michaelis–Menten equation

14.2.1   Introduction of error in the Michaelis–Menten equation

The Michaelis–Menten equation is usually written as in equation 2.15, reproduced here:

$$v = \frac{Va}{K_\mathrm{m} + a} \tag{14.1}$$

or in some equivalent way. Written like this it is incomplete, however, as it ignores the effect of experimental error, and as the whole point of statistical estimation of parameters is to minimize the effects of experimental error omitting it from the equation is a recipe for searching in the dark.

To allow for the effect of experimental error on the observations, equation 14.1 can be modified in various ways, of which two correspond to simple assumptions about the errors. If we suppose that all observations have the same *coefficient of variation*, which means that the errors are approximately uniform when expressed in percent, then a *multiplicative* error term $(1 + e_i)$ is appropriate:

$$v_i = \frac{Va_i(1 + e_i)}{K_m + a_i} \tag{14.2}$$

but if we think that they have the same *standard deviation*, which means that the errors are approximately uniform when expressed in units of rate (for example in mM/s), then an *additive* error term e_i' is appropriate:

$$v_i = \frac{Va_i}{K_m + a_i} + e_i' \tag{14.3}$$

In both equations the subscripts i are to show that we shall be dealing in this chapter with the ith of a sample of n observations, not with a single isolated observation.

The values of e_i' in equation 14.3 are not the same as those of e_i in equation 14.2: that is why we need different symbols. It is not essential to write the equations in this way: we could use equation 14.2 to study a uniform standard deviation, or equation 14.3 to study a uniform coefficient of variation, but that would make the analysis needlessly difficult. The advantage of using equation 14.2 for uniform coefficient of variation and equation 14.3 for uniform standard deviation is that we then have an e_i or e_i' that behaves simply without correction factors. In this chapter we shall take equation 14.2 as the standard form, because it corresponds better to the results that have been obtained in the few investigations that have been made of experimental error in enzyme kinetics (for example Storer, Darlison and Cornish-Bowden, 1975; Askelöf, Korsfeldt and Mannervik, 1976; Mannervik, Jakobson and Warholm, 1986). I shall not go through the algebra that results from using equation 14.3, but will give some of the results from such algebra when appropriate. As a bonus, both the algebra and the arithmetic are simpler when we assume a constant coefficient of variation; however, this is *not* a valid reason for preferring it, and the choice should be based on the underlying error structure of the data, not on convenience.

Expressing the Michaelis–Menten equation as equation 14.2 illustrates why linear transformations such as those in Section 2.6 can give unsatisfactory results. As they were obtained from equation 2.15 (the same as equation 14.1) by perfectly correct algebra, it is difficult to see at first why they can be other than correct. Once it is realized that it is not the algebra that is at fault but the starting point, and that a more complete expression of the Michaelis–Menten equation, as in equation 14.2, cannot be recast as the equation for a straight line, the difficulty disappears.

14.2.2   Estimation of the Michaelis–Menten parameters

Estimating V, V/K_m and K_m as accurately as possible is a question of finding the values that make the deviations e_i as small as possible. If there are more than one or two observations it is not in general possible to find values that make all the e_i zero. Instead, one can make the average of all the e_i^2 as small as possible. We use e_i^2 here rather than e_i to avoid complications due to the occurrence of both positive and negative values. (In principle one might achieve a similar effect by simply ignoring the signs of the e_i, but this leads to such difficult algebra that it is not usually done.) Thus we can define the *best-fit* values of the parameters as those that jointly minimize the *sum of squares* of deviations, SS, defined as follows:

$$SS = \sum_{i=1}^{n} e_i^2 \tag{14.4}$$

The summation is made over all observations, as indicated, but as summation limits are normally obvious in statistical calculations they can usually be omitted without danger of ambiguity, and in the rest of this chapter they will not be shown explicitly. Rearrangement of equation 14.2 allows e_i to be expressed in terms of v_i, a_i, K_m and V:

$$e_i = \frac{K_m v_i}{V a_i} + \frac{v_i}{V} - 1 \tag{14.5}$$

and substituted into equation 14.4:

$$SS = \sum \left(\frac{A v_i}{a_i} + B v_i - 1 \right)^2 \tag{14.6}$$

in which $A = K_m/V$ and $B = 1/V$ are the ordinate intercept and slope respectively of a plot of a_i/v_i against a_i (Section 2.6.3 and Figure 2.6). Although the parameter values that make SS a minimum can be found by partially differentiating equation 14.6 with respect to V and K_m (or V and V/K_m, or V/K_m

and K_m, according to taste) and setting both derivatives to zero, the same conclusion may be reached by a shorter and simpler route by solving for A and B first. Partial differentiation of equation 14.6 with respect to A and B gives the following pair of partial derivatives:

$$\frac{\partial SS}{\partial A} = \sum \left[\frac{2v_i}{a_i} \left(\frac{2Av_i}{a_i} + Bv_i - 1 \right) \right] \tag{14.7}$$

$$\frac{\partial SS}{\partial B} = \sum \left[2v_i \left(\frac{2Av_i}{a_i} + Bv_i - 1 \right) \right] \tag{14.8}$$

Defining $\hat{A}$ and $\hat{B}$ as the values of A and B that make SS a minimum, both expressions can be set to zero and rearranged into a pair of simultaneous equations in $\hat{A}$ and $\hat{B}$:

$$\hat{A} \sum \frac{v_i^2}{a_i^2} + \hat{B} \sum \frac{v_i^2}{a_i} = \sum \frac{v_i}{a_i} \tag{14.9}$$

$$\hat{A} \sum \frac{v_i^2}{a_i} + \hat{B} \sum v_i^2 = \sum v_i \tag{14.10}$$

which have the following solution:

$$\hat{A} = \frac{\sum v_i^2 \sum \dfrac{v_i}{a_i} - \sum \dfrac{v_i^2}{a_i} \sum v_i}{\sum \dfrac{v_i^2}{a_i^2} \sum v_i^2 - \left(\sum \dfrac{v_i^2}{a_i} \right)^2} \tag{14.11}$$

$$\hat{B} = \frac{\sum \dfrac{v_i^2}{a_i^2} \sum v_i - \sum \dfrac{v_i^2}{a_i} \sum \dfrac{v_i}{a_i}}{\sum \dfrac{v_i^2}{a_i^2} \sum v_i^2 - \left(\sum \dfrac{v_i^2}{a_i} \right)^2} \tag{14.12}$$

Substitution of the definitions of the Michaelis–Menten parameters in terms of A and B into these two equations converts them into expressions for the best-fit values, $\hat{V}$, $\widehat{V/K_m}$ and $\hat{K}_m$:

$$\hat{V} = \frac{1}{\hat{B}} = \frac{\sum \dfrac{v_i^2}{a_i^2} \sum v_i^2 - \left(\sum \dfrac{v_i^2}{a_i} \right)^2}{\sum \dfrac{v_i^2}{a_i^2} \sum v_i - \sum \dfrac{v_i^2}{a_i} \sum \dfrac{v_i}{a_i}} \tag{14.13}$$

$$\widehat{V/K}_{\mathrm{m}} = \frac{1}{\hat{A}} = \frac{\sum \dfrac{v_i^2}{a_i^2} \sum v_i^2 - \left(\sum \dfrac{v_i^2}{a_i}\right)^2}{\sum v_i^2 \sum \dfrac{v_i}{a_i} - \sum \dfrac{v_i^2}{a_i} \sum v_i} \qquad (14.14)$$

$$\hat{K}_{\mathrm{m}} = \frac{\hat{A}}{\hat{B}} = \frac{\sum v_i^2 \sum \dfrac{v_i}{a_i} - \sum \dfrac{v_i^2}{a_i} \sum v_i}{\sum \dfrac{v_i^2}{a_i^2} \sum v_i - \sum \dfrac{v_i^2}{a_i} \sum \dfrac{v_i}{a_i}} \qquad (14.15)$$

(For $\widehat{V/K}_{\mathrm{m}}$ the circumflex indicating that this is a best-fit estimate qualifies the whole fraction, which is considered as a single parameter.) This result, which was first given by Johansen and Lumry (1961), is exact; no further refinement is necessary to minimize SS as defined by equation 14.6.

14.2.3   Corresponding results for a uniform standard deviation in the rates

If we take equation 14.3 as a starting point instead of equation 14.2, substituting e_i' into equation 14.4 instead of e_i in the definition of SS, this is equivalent to assuming that the original rates are subject to a constant standard deviation instead of a constant coefficient of variation. The analysis is more difficult, because with this definition there are no analytical expressions for the best-fit parameters. Instead they have to be obtained by a series of approximations (see Cornish-Bowden, 1995b). The derivation will not be given here, but the results are as follows:

$$\hat{V} = \frac{\sum \dfrac{\hat{v}_i^3 v_i}{a_i^2} \sum \hat{v}_i^3 v_i - \left(\sum \dfrac{\hat{v}_i^3 v_i}{a_i}\right)^2}{\sum \dfrac{\hat{v}_i^3 v_i}{a_i^2} \sum \hat{v}_i^3 - \sum \dfrac{\hat{v}_i^3 v_i}{a_i} \sum \dfrac{\hat{v}_i^3}{a_i}} \qquad (14.16)$$

$$\widehat{V/K}_{\mathrm{m}} = \frac{\sum \dfrac{\hat{v}_i^3 v_i}{a_i^2} \sum \hat{v}_i^3 v_i - \left(\sum \dfrac{\hat{v}_i^3 v_i}{a_i}\right)^2}{\sum \hat{v}_i^3 v_i \sum \dfrac{\hat{v}_i^3}{a_i} - \sum \dfrac{\hat{v}_i^3 v_i}{a_i} \sum \hat{v}_i^3} \qquad (14.17)$$

$$\hat{K}_{\mathrm{m}} = \frac{\sum \hat{v}_i^3 v_i \sum \dfrac{\hat{v}_i^3}{a_i} - \sum \dfrac{\hat{v}_i^3 v_i}{a_i} \sum \hat{v}_i^3}{\sum \dfrac{\hat{v}_i^3 v_i}{a_i^2} \sum \hat{v}_i^3 - \sum \dfrac{\hat{v}_i^3 v_i}{a_i} \sum \dfrac{\hat{v}_i^3}{a_i}} \qquad (14.18)$$

These cannot be used directly, because they involve the calculated rates $\hat{v}_i$, which are unknown until the parameters have been calculated. To get around this problem, we start by assuming that the measured rates are reasonably accurate estimates of the calculated rates, calculating preliminary estimates of the Michaelis–Menten parameters from equations 14.16–18 with $\hat{v}_i$ replaced by v_i throughout. These preliminary estimates can be used to calculate preliminary estimates of the calculated rates, which can be used to get better parameter estimates, which can be used to get better calculated rates... and so on, until the results are self-consistent.

As this calculation involves approximations, there is no unique way of doing it, and a different approach, described by Wilkinson (1961), is often referred to in the literature. This converges to exactly the same numerical results and requires about the same amount of computation: there is thus no strong reason to prefer one rather than the other. Slightly different formulae can also be found in the literature that give slightly different results from those from Wilkinson's method because they are not quite correct: for example, in an earlier book (Cornish-Bowden, 1976c) I gave expressions equivalent to equations 14.16–18 with $\hat{v}_i^3 v_i$ replaced throughout by $\hat{v}_i^2 v_i^2$; Cleland (1967) gave ones equivalent to equations 14.16–18 with $\hat{v}_i^3 v_i$ replaced throughout by $\hat{v}_i^4$.

14.3 Statistical aspects of the direct linear plot

14.3.1 Comparison between classical and distribution-free statistics

The least-squares approach to data fitting is the most convenient general method, and the most widely used (apart from drawing lines on graphs by eye), but it is not necessarily the best. To demonstrate that the least-squares solution to a problem is the "best" solution, we must not only define what we mean by "best"; we must also make a whole battery of assumptions about the nature of the underlying experimental error, as follows:

1. The random errors in the measurements are distributed according to the normal (Gaussian) curve of error.
2. Only one variable, the so-called *dependent variable*, is subject to error. In enzyme kinetic measurements this is generally taken to be the rate.
3. The proper weights are known. In the context of this chapter this means knowing whether equation 14.2 or 14.3 or some other equation

is the correct starting point for introducing errors into the kinetic equation.

4. The errors are uncorrelated, which means that the magnitude or sign of any particular error implies nothing about the magnitude or sign of any other error.

5. Systematic error can be ignored, which means that the distribution curve for each error has a mean of zero. In more biochemical terms, it means that we are sure we are fitting the right equation, that we are not, for example, fitting the Michaelis–Menten equation to data generated by a sigmoid dependence of rate on substrate concentration.

In practice we know little or nothing about the truth or otherwise of any of these, and in consequence classical least-squares data analysis is built on a much more fragile theoretical foundation than is commonly realized, even though in general it is best to base conclusions on as few unproved assumptions as possible. An alternative branch of statistics, known as *distribution-free* or *non-parametric* statistics, offers hope of escape, because it dispenses with all of the above assumptions apart from the last, which is retained in a weaker form: in the absence of other information any error is assumed to be as likely to be positive as to be negative.

Perhaps the simplest idea to follow from this reduction in the number of assumptions is that the preferred estimator of the mean of a set of values is not the sample mean but the sample *median*, or the middle value when they are arranged in order: for example the mean of the numbers 1, 2, 3, 5 and 20 is $(1 + 2 + 3 + 5 + 20)/5 = 31/5 = 6.2$, and the median is 3. Notice that a single atypical value has a much greater effect on the mean than it does on the median. The mean is the archetype of a least-squares estimator, whereas the median is the archetype of a distribution-free estimator.

The advantages of using distribution-free methods can be considerable, because they generally give much more reliable information when the assumptions listed above are false. On the other hand nothing comes without a price, and here the price is that if all the classical assumptions are true one will have somewhat poorer estimates than one could get from least-squares analysis. However, in general the price is small compared with the potential benefits. Not the least of these benefits is that the theory underlying distribution-free analysis is much easier to understand than classical statistical theory. As no distributions are assumed, no distribution theory is needed, and indeed little theory at all, beyond that needed for understanding coin-tossing experiments, as the idea that every error is as likely to be positive as to be negative is similar to the hypothesis that a coin is as likely to fall on one face as on the other.

14.3.2   Application to the direct linear plot

The direct linear plot (Section 2.6.5) was conceived as a way of introducing distribution-free ideas into enzyme kinetics, at the same time greatly simplifying the procedures and concepts. For any non-duplicate pair of observations (a_i, v_i) and (a_j, v_j), there is a unique pair of values of the parameters V and K_m that satisfy both observations exactly when substituted in the Michaelis–Menten equation:

$$V_{ij} = \frac{a_i - a_j}{\dfrac{a_i}{v_i} - \dfrac{a_j}{v_j}} \tag{14.19}$$

$$K_{m,ij} = \frac{v_j - v_i}{\dfrac{v_i}{a_i} - \dfrac{v_j}{a_j}} \tag{14.20}$$

These parameter values define the coordinates of the point of intersection of the two lines representing the two observations in the direct linear plot. Altogether, n observations provide a maximum of $n(n-1)/2$ such pairs of values. (This is the maximum value rather than the actual value because the pairs of values obtained from duplicate observations with $a_i = a_j$ are meaningless and must be omitted from consideration.)

The best-fit estimate of K_m can now be taken as the median of the set of values $K_{m,ij}$, and the best-fit estimate of V as the median of the set of values V_{ij}. If there is an odd number of values, the median is the middle one when they are arranged in rank order; if there is an even number, it is the mean of the middle two. I shall use the symbols K_m^* and V^* (rather than, say, $\hat{K}_m$ and $\hat{V}$) for these estimates to emphasize that they are not least-squares estimates. There are two main reasons for defining them as medians for analysing the direct linear plot, in preference to a more familiar estimator like the arithmetic mean: first, the estimates $K_{m,ij}$ and V_{ij} are automatically ranked when the plot is drawn, as illustrated in Figure 14.1, and so no calculation is required to find the medians; second, the median, unlike other types of average, is insensitive to extreme values, which inevitably occur quite often in the direct linear plot, because some of the lines may be nearly parallel.

The main advantage of this approach over the method of least squares is that it requires no calculation, and it incorporates no abstract or difficult idea such as the normal distribution of errors. However, it also has some important statistical advantages (Cornish-Bowden and Eisenthal, 1974), which will be outlined in the next two sections.

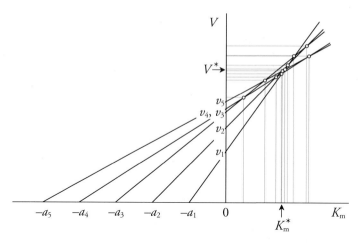

Figure 14.1. Determination of median estimates from the direct linear plot. Each pair of lines intersects at a point whose coordinates give an estimate of K_m and an estimate of V, as marked by the projections on the two axes. As these intersections are automatically ordered by the plot, determination of the medians is a simple matter of counting to find the middle estimate of each series, or halfway between the middle pair if (as in this example) there is an even number of intersection points. If there were any intersections outside the first quadrant, giving rise to negative parameter estimates, these would require special attention (see Figure 14.2). If such intersections are numerous, or if the arrangement of intersections shows a clear and reproducible pattern, one should investigate the possibility that the data do not fit the Michaelis–Menten equation.

14.3.3   Lack of need for weighting

The distribution-free analysis of the direct linear plot is only marginally inferior to the method of least squares when all of the least-squares assumptions (listed in Section 14.3.1) are true. In other circumstances it can be much superior, because its dependence on fewer and weaker assumptions makes it insensitive to departures from the expected error structure. For example, the least-squares estimates of the Michaelis–Menten parameters given by equations 14.16–18 are by no means the same as or only trivially different from those given by equations 14.13–15; if one uses equations 14.16–18 when one ought to be using equations 14.13–15, or vice versa, one will lose all the optimum properties of the method of least squares and obtain poor estimates of the parameters. What this means in practice is that to use the method of least squares with confidence it is essential to know the proper weights that apply to the observations, to know whether they are uniform in coefficient of variation or uniform in standard deviation, or some combination of the two, but in practice one almost never knows this. The distribution-free approach dispenses with any need to know the proper weights, and although it may give marginally worse results than the correctly weighted least-squares method it will normally do much better than the least-squares method with incorrect weights.

14.3.4   Insensitivity to outliers

An *outlier*, or "wild observation", is an observation with a much larger error than expected from the error distribution of the majority of observations. If an experiment contains one, it can have a drastic effect on least-squares parameter estimation, but almost none on distribution-free estimation. The reason for this is quite simple, and derives from the ordinary properties of means and medians: if an octogenarian walks into a room full of children, the mean age of the people in the room may easily be doubled, but the median will hardly change. Median estimation of parameters as described in Section 14.3.2 is not as insensitive as this, because any wild observation contributes to $(n - 1)$ wild lines in the plot, but it is still much more resistant to outliers than least-squares estimation.

To summarize this section and the preceding one, we may say that under ideal conditions the least-squares method always performs a little better than any distribution-free alternative, but it is so sensitive to departures from ideal conditions that in practice it can easily perform much worse.

14.3.5   Handling of negative parameter estimates

In experiments with rather large errors in the rates, the direct linear plot as originally described [as shown in Figures 2.8 (page 47) and 14.1] has a small negative bias, leading to V^* and K_m^* values that are not distributed about the corresponding true values but about values that are too small (Cornish-Bowden and Eisenthal, 1978). This bias derives from the occurrence of intersection points in the third quadrant, that is to say points with $K_{m,ij}$ and V_{ij} both negative. To understand how to correct the bias it is helpful to consider why intersections should ever occur outside the first quadrant, and Figure 14.2 shows typical combinations of a and v values that lead to intersections in the first, second and third quadrants; intersection in the fourth quadrant is not shown as this is impossible unless the data contain negative v values.

Intersections in the second and third quadrants arise from different causes, and they require different treatment. Intersections in the second quadrant (Figure 14.2b) can occur because the enzyme shows substrate inhibition and the observations come from the part of the curve where v decreases with increasing a. If this is the correct explanation it should be confirmed by other intersections in the second quadrant from other observations in a similar range of a values. However, if there is no evidence for systematic departure from Michaelis–Menten kinetics, an isolated intersection in the second quadrant is

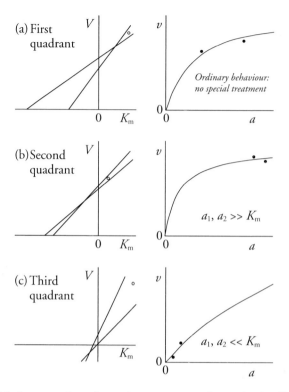

Figure 14.2. Interpretation of negative parameter estimates in the direct linear plot. In each pair of plots, the one on the left shows a pair of observations in (K_m, V) coordinates and the one on the right shows the same observations in (a, v) coordinates, with a line drawn according to the parameter values indicated by an open circle (o) on the left. The graphs are schematic rather than exact: *they are not drawn to any consistent scale*. (a) Experimental error will ordinarily occur but the corresponding intersection point occurs in the first quadrant and both parameter estimates are positive. (b) If both a values are larger than K_m experimental error may produce an intersection in the second quadrant if the v values are incorrectly ranked. (c) If both a values are smaller than K_m experimental error may produce an intersection in the third quadrant if the v/a values are incorrectly ranked.

most likely to occur when both a values are large compared with K_m, so that the two true v values are similar enough to one another for experimental error to rank the observed values in the wrong order. This implies that the true value of V is similar to the two observed v values, and that the true value of K_m is much smaller than the two a values. As this is also implied by the values of V_{ij} and $K_{m,ij}$ obtained by treating the intersection in the second quadrant at face value, it follows that there is no reason to give such intersections any special treatment.

Intersections in the third quadrant (Figure 14.2c) have a different explanation. They can also be a symptom of systematic error, implying that the plot of v against a is sigmoid and not a rectangular hyperbola, but again, this can be confirmed or ruled out by examining other intersections from observations at similar a values. If systematic error is not confirmed, the most likely cause is that the two a values are low enough in relation to the true K_m for experimental error to have caused the two v/a values to be incorrectly ranked. This implies that the true K_m is large compared with the smaller a value and the true V is large compared with the smaller v value: this is the *opposite* of what is implied by the values of V_{ij} and $K_{m,ij}$ obtained by treating the intersection at face value, as both of these are negative in the third quadrant. This is the reason for the slight bias if the direct linear plot intersections are all taken at face value, and it may be corrected by treating intersections in the third quadrant as giving large *positive* values of *both* V_{ij} and $K_{m,ij}$. Actual numerical values do not have to be assumed; it is sufficient to say that they are large and positive. This is because numerical values of the extreme members of a sample are unnecessary for determining the median.

This remedy in effect treats a negative number as being beyond infinity rather than below zero. Although many people may feel uneasy about doing this, they should realize that extrapolating past an infinite rate is what every user of the double-reciprocal plot does every time the intercept on the abscissa axis is interpreted as $-1/K_m$ (Figure 2.5, page 43).

None of these complications affects the alternative form of the direct linear plot illustrated in Figure 2.9 (page 48), because with this plot all smooth displacements of the lines brought about by errors of reasonable magnitude result in smooth displacements of the intersections: they move smoothly across either of the two axes and never flip from one quadrant to another by way of infinity. Because of this no special rules are needed for interpreting intersections outside the first quadrant of this plot. From the computational point of view this means that instead of determining the median values of V and K_m as described above one should first determine the median values of $1/V$ and K_m/V and then calculate any desired parameters from them. For least-squares analysis, one can calculate the parameters in any order one wishes, and the result is the same: for example, comparing equations 14.13–15 one may see that $\widehat{V/K_m}$ is the same whether it is calculated from equation 14.14 or by dividing the expression for $\hat{V}$ in equation 14.13 by that for $\hat{K}_m$ in equation 14.15: $\widehat{V/K_m} \equiv \hat{V}/\hat{K}_m$. The corresponding relationship does *not* apply to median estimates, because these are not analytical functions of the observations.

14.4   Precision of estimated kinetic parameters

I have tried to keep mathematical complications to a minimum in this chapter, and will not describe how formulae for assessing the precision of parameter estimates are derived, but will only give some results that may be useful (though they should be used with caution, as calculations made without understanding are a source of many misconceptions). Explanations of how they are arrived at may be found elsewhere (Cornish-Bowden, 1995b).

To avoid dealing separately with equations 14.13–15 and 14.16–18, it is convenient to begin by noting that all of the sets of expressions for the best-fit Michaelis–Menten parameters are special cases of the following pair of expressions for a *weighted* best fit to the double-reciprocal plot:

$$\hat{V} = \frac{\sum \dfrac{w_i}{a_i^2} \sum w_i - \left(\sum \dfrac{w_i}{a_i}\right)^2}{\sum \dfrac{w_i}{a_i^2} \sum \dfrac{w_i}{v_i} - \sum \dfrac{w_i}{a_i} \sum \dfrac{w_i}{a_i v_i}} \qquad (14.21)$$

$$\widehat{V/K_m} = \frac{\sum \dfrac{w_i}{a_i^2} \sum w_i - \left(\sum \dfrac{w_i}{a_i}\right)^2}{\sum w_i \sum \dfrac{w_i}{v_i a_i} - \sum \dfrac{w_i}{a_i} \sum \dfrac{w_i}{v_i}} \qquad (14.22)$$

$$\hat{K}_m = \frac{\sum w_i \sum \dfrac{w_i}{v_i a_i} - \sum \dfrac{w_i}{a_i} \sum \dfrac{w_i}{v_i}}{\sum \dfrac{w_i}{a_i^2} \sum \dfrac{w_i}{v_i} - \sum \dfrac{w_i}{a_i} \sum \dfrac{w_i}{a_i v_i}} \qquad (14.23)$$

In these equations the weights w_i are the appropriate weights to apply to values of $1/v_i$ for two different hypotheses about the weighting appropriate for v_i: if the rates are assumed to be uniform in coefficient of variation the weight for $1/v_i$ is $w_i = v_i^2$ and equations 14.21–23 are identical to equations 14.13–15; if they are assumed to be uniform in standard deviation the weight for $1/v_i$ is $w_i = \hat{v}_i^3 v_i$ and they are identical to equations 14.16–18.

There are various calculations that one can do to assess the quality of a fit, but two questions stand out: how well does the calculated line agree with the observed points, and how precisely are the parameter values defined? The minimum sum of squares provides a starting point for answering the first question. In terms of the weighted fit it may be defined as follows:

$$\hat{SS} = \sum w_i \left(\frac{1}{v_i} - \frac{1}{\hat{v}_i}\right)^2 \qquad (14.24)$$

which is equivalent to equation 14.6 if one substitutes $w_i = v_i^2$ and calculates $\hat{v}_i$ with the best-fit parameter values. As it stands, however, the sum of squares is not satisfactory because it is a sum rather than an average. The obvious remedy is to convert it to an average by dividing by n, the number of observations. However, that is too simple, because it seriously overestimates the precision if n is small: the reason may be seen by considering the extreme case of $n = 2$, when $\widehat{SS}$ will always be zero, as it is always possible to fit two observations exactly with the Michaelis–Menten equation (or another two-parameter equation), even though there is no reason to suppose that the observations are exact. Although this makes it obvious that a correction is needed, it is less obvious how large the correction ought to be; here I shall just state without proof that an unbiassed estimate of the experimental error can be obtained by dividing $\widehat{SS}$ by $n - 2$:

$$\sigma_{\exp}^2 = \frac{\widehat{SS}}{n - 2} \tag{14.25}$$

This quantity is called the *experimental variance*. It provides a route to assessing the precision of the estimates of the Michaelis–Menten parameters, because their variances may be expressed in terms of it:

$$\sigma^2(\hat{V}) = \frac{\hat{V}^4 \sigma_{\exp}^2 \sum w_i/a_i^2}{\sum w_i/a_i^2 \sum w_i - \left(\sum w_i/a_i\right)^2} \tag{14.26}$$

$$\sigma^2\left(\hat{V}/\hat{K}_{\mathrm{m}}\right) = \frac{\left(\hat{V}/\hat{K}_{\mathrm{m}}\right)^4 \sigma_{\exp}^2 \sum w_i}{\sum w_i/a_i^2 \sum w_i - \left(\sum w_i/a_i\right)^2} \tag{14.27}$$

$$\sigma^2(\hat{K}_{\mathrm{m}}) = \frac{\hat{V}^2 \sigma_{\exp}^2 \left(\sum w_i + 2\hat{K}_{\mathrm{m}} \sum w_i/a_i + \hat{K}_{\mathrm{m}}^2 \sum w_i/a_i^2\right)}{\sum w_i/a_i^2 \sum w_i - \left(\sum w_i/a_i\right)^2} \tag{14.28}$$

Notice that even though the relationship between the three parameter estimates is straightforward, $\widehat{V/K_{\mathrm{m}}} \equiv \hat{V}/\hat{K}_{\mathrm{m}}$, the relationships among their three variances are not obvious at all.

One minor complication is that when the rates are uniform in standard deviation the values of w_i appropriate for inserting in equations 14.24–28 are *not* the same as those needed in equations 14.21–23: for *calculating* the best-fit parameters we should use $w_i = \hat{v}_i^3 v_i$, as already stated, but once they are known we should replace these weights by $w_i = \hat{v}_i^2 v_i^2$ for use in equations 14.24–28. The reasons for this change are subtle (Cornish-Bowden, 1982) and not important, because the results are little changed if one simply ignores it and uses the same weights throughout. No corresponding complication applies to the assumption of uniform coefficient of variation in the rates, when the same weights $w_i = v_i^2$ are appropriate throughout.

Variance estimates do not have the same dimensions as the parameters to which they refer, but the squares of those dimensions. The square root of a variance has the same dimensions as the parameter to which it refers, however, and is often called the *standard error*, its value being quoted in the following style: $V = 10.7 \pm 1.2\,\mathrm{mM\,s^{-1}}$, for example, where $10.7\,\mathrm{mM\,s^{-1}}$ is the estimated value and $1.2\,\mathrm{mM\,s^{-1}}$ is the square root of the estimated variance. Note that there is no point, and much absurdity, in recording a result any more precisely than this (except in a worked example where the interest is more in the arithmetic than in the result), for example as $V = 10.7137 \pm 1.2077\,\mathrm{mM\,s^{-1}}$, though one sometimes sees such expressions in the literature. As a general rule, there is rarely any reason to express a standard error with more than two significant figures, and the corresponding parameter value should not be expressed to more decimal places than its standard error.

Once we have calculated a standard error, what does it mean? It means that (a) if all the assumptions listed in Section 14.3.1 are true, including in particular that we have fitted the true model and there is no systematic error, (b) if we regard the particular experiment we have carried out as one of a universe of such experiments that we might have carried out in exactly the same way and in exactly the same conditions, and (c) if the number of observations n in each experiment is infinite, then the true value of the parameter will lie within one estimated standard deviation of the calculated parameter value in about 68% of this universe of conceivable experiments. If this seems rather an abstract meaning derived from a large amount of supposition, that is unfortunate: wishing that a standard deviation meant something more concrete will not make it true! The third supposition is clearly false in all real experiments, as we never have an infinite number of observations, but fortunately the bias due to the use of a less-than-infinite value of n can be corrected, according to a principle discovered by W. S. Gosset, a brewer who published theoretical work in statistics under the name Student. By means of his theory (Student, 1908) one can readily convert a standard error into a confidence limit at any level of confidence, not just at 68%. As a *rough* guide, one may take twice the standard error as a confidence limit for 95%, and three times the standard error as a confidence limit for 99%. This still leaves us with a definition of a confidence limit that is more abstract than we might like, and it still depends on many assumptions of which some are almost certainly false. There is little we can do about the abstractness, but biochemists who are unhappy about so many assumptions can decrease them by using distribution-free estimates.

The theory needed for calculating distribution-free confidence limits of Michaelis–Menten parameters (Cornish-Bowden, Porter and Trager, 1978) is not particularly difficult (in fact it is easier to understand than that which underlies the calculation of standard errors) but unfortunately setting it out clearly would require much more space than would be appropriate for a book on enzyme kinetics. I have, however, described it elsewhere (Cornish-Bowden, 1995b).

Equations 14.26–28 have an important consequence for the relative precision with which the three Michaelis–Menten parameters can be estimated, but this consequence is obscured by the different dimensions of the three parameters. However, dividing the variances by the squares of the corresponding parameter estimates, they become:

$$\frac{\sigma^2(\hat{V})}{\hat{V}^2} = \frac{\hat{V}^2 \sigma_{\text{exp}}^2 \sum \frac{w_i}{a_i^2}}{\sum \frac{w_i}{a_i^2} \sum w_i - \left(\sum \frac{w_i}{a_i}\right)^2} \tag{14.29}$$

$$\frac{\sigma^2(\widehat{V/K_\text{m}})}{(\widehat{V/K_\text{m}})^2} = \frac{(\widehat{V/K_\text{m}})^2 \sigma_{\text{exp}}^2 \sum w_i}{\sum \frac{w_i}{a_i^2} \sum w_i - \left(\sum \frac{w_i}{a_i}\right)^2} \tag{14.30}$$

$$\frac{\sigma^2(\hat{K}_\text{m})}{\hat{K}_\text{m}^2} = \frac{\sigma_{\text{exp}}^2 \left[\hat{V}^2 \sum w_i + 2\hat{V} \cdot \widehat{V/K_\text{m}} \sum \frac{w_i}{a_i} + (\widehat{V/K_\text{m}})^2 \sum \frac{w_i}{a_i^2}\right]}{\sum \frac{w_i}{a_i^2} \sum w_i - \left(\sum \frac{w_i}{a_i}\right)^2}$$

$$= \frac{\sigma^2(\hat{V})}{\hat{V}^2} - \frac{2\text{cov}(\hat{V}, \widehat{V/K_\text{m}})}{\hat{V} \cdot \widehat{V/K_\text{m}}} + \frac{\sigma^2(\widehat{V/K_\text{m}})}{(\widehat{V/K_\text{m}})^2} \tag{14.31}$$

in which the quantity $\text{cov}(\hat{V}, \widehat{V/K_\text{m}})$ is called the *covariance* of the two parameters, and has the following value:

$$\text{cov}(\hat{V}, \widehat{V/K_\text{m}}) = -\frac{(\hat{V} \cdot \widehat{V/K_\text{m}}) \sigma_{\text{exp}}^2 \sum \frac{w_i}{a_i}}{\sum \frac{w_i}{a_i^2} \sum w_i - \left(\sum \frac{w_i}{a_i}\right)^2} \tag{14.32}$$

As this expression must be negative in all normal circumstances (that is to say for positive values of the weights, parameter estimates, rates and substrate concentrations) it is clear that the middle term in equation 14.31 implies subtracting a negative value from the sum of two positive values (or adding a positive value to this sum). This in turn means that K_m is in all normal circumstances less well defined than either V or V/K_m.

This argument may appear excessively mathematical, but the qualitative result can be obtained less rigorously and more simply. In principle V can be estimated from a single measurement of the rate at a very high substrate concentration, and V/K_m can in principle be estimated from a single measurement at a very low substrate concentration. However, estimating K_m requires a minimum of two measurements — in the simplest case the two just mentioned — and will be subject to the errors in both of these together with additional errors derived from the process of combining them.

The conclusion from this analysis has far-reaching implications for many aspects of enzyme kinetics, even if one route for reaching it was obscure and difficult, and the other was weak. One implication is that it makes it easy to recognize when reported parameters and their precision estimates have been obtained with an invalid program, because, except in exceptional circumstances equations 14.31–32 require them to obey the following inequality:

$$\frac{\sigma^2(\hat{K}_m)}{\hat{K}_m^2} \geq \frac{\sigma^2(\hat{V})}{\hat{V}^2}, \frac{\sigma^2(\widehat{V/K_m})}{(\widehat{V/K_m})^2} \tag{14.33}$$

More simply, this means that the coefficient of variation of K_m (the standard error expressed as a percentage of the parameter estimate) should be larger than both of the corresponding values for V and V/K_m. If all three parameter values are listed in a table, then inclusion of precision estimates for K_m and V but not V/K_m should also raise suspicions, because there is no reason for a properly written computer program not to calculate and report all three.

As the primary parameters were taken in Section 14.2 as $1/V$ and K_m/V one might suspect that the lower precision of the K_m estimate was just a consequence of treating it as a derived parameter, but that is not so. One can do the initial fitting directly with K_m and V as primary parameters, as is done in the method of Wilkinson (1961), but, as pointed out in Section 14.2.3, the resulting parameter estimates are exactly the same as they are if the method described in this book is used. One would then estimate the precision of V/K_m from those of K_m and V by means of an equation similar to equation 14.31, which has the following form:

$$\frac{\sigma^2(\widehat{V/K_m})}{(\widehat{V/K_m})^2} = \frac{\sigma^2(\hat{K}_m)}{\hat{K}_m^2} - \frac{2\mathrm{cov}(\hat{V},\hat{K}_m)}{\hat{V}\hat{K}_m} + \frac{\sigma^2(\hat{V})}{\hat{V}^2} \tag{14.34}$$

However, in contrast to equation 14.31, in which the middle term implies subtracting a negative value, the middle term here implies subtracting a large positive value, because estimates of K_m and V always have a high positive correlation. Thus the conclusion shown in equation 14.33 is entirely consistent with equation 14.34, and the general relationship does not derive from an arbitrary choice about the order in which the parameters are calculated.

14.5   Residual plots and their uses

After data have been fitted to an equation, it is always worth checking whether the assumptions made in the analysis were reasonable. Has one fitted the right equation, or would a more complicated one have given a more credible explanation of the observations, or would a simpler one (with fewer parameters) have done just as well? Has one made reasonable statistical assumptions — for example, is the coefficient of variation constant, as assumed in Section 14.2.2, or would a constant standard deviation (Section 14.2.3) be a better approximation, or does the experimental error vary in a more complicated way than either of these imply?

Much the easiest way of attacking these questions is to examine the *residuals* after fitting, the differences between the observed rates v and the corresponding rates $\hat{v}$ calculated from the best-fit equation. I shall not attempt an exhaustive treatment in this section, but will rather indicate the sorts of uses to which residuals can be put.

If the observed v values truly have uniform coefficient of variation, the simple differences $(v - \hat{v})$ should tend to increase in absolute magnitude as the calculated rate $\hat{v}$ increases, but the *relative* differences $(v/\hat{v} - 1)$ should be scattered in a parallel band about zero. On the other hand, if the observed rates really have constant standard deviation, the simple differences should be scattered in a parallel band about zero and the relative differences should tend to decrease in absolute magnitude as $\hat{v}$ increases. Figure 14.3 shows examples of such *residual plots*. They are easy to execute, regardless of how complicated is the equation that has been fitted, and they may provide valuable information that is not readily available in any other way.

It is advisable to have as many points as possible in a residual plot, preferably more than 30, if it is being used to decide on the correct weighting strategy. This is to avoid being unduly influenced by the aberrant behaviour of one or two observations, as there is always a large amount of scatter. Nonetheless, even a plot with as few as five to ten points can be useful as long as one takes the interpretation as a suggestion rather than as definite information.

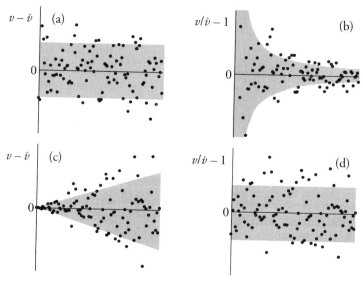

Figure 14.3. Residual plots for assessing the correctness of a weighting scheme. (a,b) The upper pair of plots show the expected residual plots when all the observed rates have the same standard deviation: (a) if the simple difference $v - \hat{v}$ is plotted against $\hat{v}$ the points are scattered in a parallel band about the abscissa axis, but (b) if the relative difference $v/\hat{v} - 1$ is plotted the band is trumpet-shaped with the opening to the left. (c,d) The lower pair of plots show the expected residual plots when all the observed rates have the same coefficient of variation: (c) now the plot of $v - \hat{v}$ produces a wedge-shaped band, opening to the right, whereas (d) the plot of $v/\hat{v} - 1$ produces a parallel band.

Residual plots are also valuable for recognizing if the deviations from the fitted equation are due more to systematic error (failure to fit the right equation) than to random scatter. For this purpose the interpretation can be quite clear even if there are only a few points. For example, the systematic trend in the residual plot shown as an inset to Figure 13.1 (page 322) would be unmistakeable even if there were only six or seven points rather than 19. That was an artificially constructed example, but a real one is illustrated in Figure 14.4: notice that without the residual plot the fit to the wrong model appears excellent to the eye, but with the residual plot the systematic nature of the deviations is obvious.

In this example the systematic scatter of the points in Figure 14.4a is quite obvious, and could easily be detected by a computer program, provided that it made the appropriate tests — a major proviso in practice as many of the programs in common use today do not automatically check for systematic trends in the residuals, even ones that can easily be deduced from non-random runs of signs. Residual plots anyway reveal to the practised eye non-random behaviour that would be unlikely to be detected by a computer program. In their kinetic study of lactoylglutathione lyase and hydroxyacylglutathione

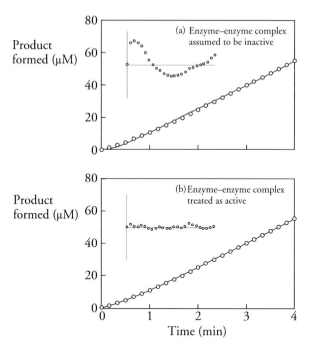

Figure 14.4. Model discrimination with the aid of residual plots. Two models for the time course of product release in the reaction catalysed by phosphoribulo-kinase (Lebreton and Gontero, 1999) are almost indistinguishable when compared as ordinary plots, but for the model treating complexed phosphoribulokinase as inactive (a) the residual plot shows clear evidence of systematic error, whereas for the model assuming the complex to be active (b) it does not.

hydrolase, Martins and co-workers (2001a) give several examples of residual plots with non-random trends that would not be detected by any method as simple as testing runs of signs, and would be difficult to detect even by more elaborate automated methods.

Authors rarely supply residual plots in published work, so it is useful for the critical reader to be able to visualize what they would look like if they were drawn. High precision is not important for this purpose and it may often be sufficient to redraw a published plot as a residual plot by estimating the coordinates by eye. Even quicker and simpler, one can emphasize the residual error in a published plot by looking along the plotted line with the eye close to the paper. This works best if the line drawn is straight, but it is also possible with curves if the curvature is slight.

It may often be impossible to draw any firm conclusions from one plot alone, as the number of observations may be too small, but if several lines are drawn in a plot, representing different but related experimental conditions, or if several related plots are presented in the same paper, one may be able to judge them together. If all indicate the same sort of systematic trend, even if each one of them has only four or five points, there is a high probability that systematic error is present. A single line with five observations showing a U-shaped distribution of residuals (for example a sign pattern such as $+ + - - +$), need not have any special importance, but five such examples in the same figure demand further investigation. Examples of figures displaying obvious systematic errors that have passed unnoticed by the authors may be found in a few minutes by leafing through any recent issue of any journal that contains kinetic plots.

If a residual plot indicates that a more complicated equation is desirable one must then decide which more complicated equation to try. Again, residual plots should be helpful. Suppose, for example, that experiments on the effect of an inhibitor have been interpreted in terms of competitive inhibition even though a small but non-negligible uncompetitive component is present. The incorrect equation may well then give acceptable results at low substrate concentrations but inacceptable results at high substrate concentrations, because as the substrate concentration increases competitive effects become less important whereas uncompetitive effects become more important. So a plot of residuals against substrate concentration (with points for all inhibitor concentrations superimposed on the same plot) should show a definite systematic trend that would not be evident from a plot of the same residuals against calculated rate.

As a summary, Figure 14.5 is a gallery of different kinds of residual plot that may be used to diagnose different kinds of faults in an experiment or its analysis. Plot (a) shows the ideal result (similar to that in Figure 14.3a and 14.3d), whereas plot (b) shows evidence of inappropriate weighting (similar to Figure 14.3c, and similar in idea to Figure 14.3b, though different in detail). Plot (c) shows obvious systematic error, whereas plot (d) may show systematic error, but is mainly symptomatic of a poorly designed experiment incapable of giving a clear indication of the proper equation. Plot (e) shows evidence of excessive rounding of numerical values too early in the computation (Cárdenas and Cornish-Bowden, 1993), but has a broader general importance in that it illustrates how residual plots draw attention to oddities in the error distribution that normally pass completely unnoticed in any other kind of data analysis. Plot (f) shows that systematic error may be more complicated than a simple deviation from linearity and yet still be obvious in a residual plot. Plot (g) illustrates the effect of a single outlier.

Plot (h) is more realistic than the others, as it illustrates that deviations from expected behaviour may be quite small, and that more than one type of deviation may appear in the same plot: although some systematic error appears to be present it is no greater in magnitude than the obvious random scatter, and so it is suggestive rather than conclusive; likewise there is a suggestion of an outlier, but it does not stand out as obviously as in plot (g). Moreover, the

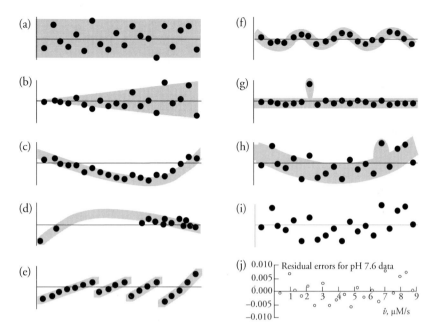

Figure 14.5. Selection of residual plots illustrating various kinds of behaviour. (a) Random scatter after fitting the correct model with appropriate weights; (b) inappropriate weighting; (c) systematic error; (d) inadequate experimental design; (e) effects of excessive rounding before fitting; (f) more complicated systematic effects; (g) effect of a single outlier; (h) various effects superimposed; (i) no shading, no labelling, axes as faint as possible while remaining visible; (j) excessive labelling, with data insufficiently prominent.

shading in the Figure — an interpretation of data, not actual data — strongly influences the interpretation likely to be placed on the plot by an observer. In plot (i), exactly the same points shown without shading and without labelling provide a more neutral view, with a weaker suggestion of systematic error and a much weaker suggestion of an outlier. In general, therefore, although shading is useful for indicating an interpretation to readers, the initial interpretation should ideally be made without shading and without distracting information of any kind. Compare plots (h) and (i) with plot (j), which again shows the same points but now burdened with so much extraneous information of the sort commonly included in published graphs that the points themselves are almost lost in a sea of irrelevance and the non-linear trend has become almost imperceptible. In a residual plot the ideal amount of labelling is no labelling at all — no scales, no numbers, no title, no variable names: all these, if needed, can be adequately communicated in the legend. The axes remain useful, but should be as imperceptible as possible while remaining visible, as in plot (i).

These last comments refer particularly to residual plots. Some of them do not apply to other kinds of graph, for which, in particular, relegating important information to the legend is not a good idea. Some points do also apply to other graphs, however, and in any plot the major emphasis should be on the data plotted and not, for example, on prominent grids. Tufte (1990), for example, recommends that if ready-made graph paper with prominent gridlines is used one should draw the plot on the *back* of the paper: they will show through the paper clearly enough to be used for drawing the plot, but faintly enough to be ignored afterwards. More generally, all three of his books (Tufte, 1983, 1990, 1997) are indispensable reading for anyone with a serious interest in the graphical display of information.

Problems

14.1 Determine the least-squares and distribution-free estimates of the parameters of the Michaelis–Menten equation for the following set of data:

a (mM)	v (mM min^{-1})	a (mM)	v (mM min^{-1})
1	0.219	6	0.525
2	0.343	7	0.512
3	0.411	8	0.535
4	0.470	9	0.525
5	0.490	10	0.540

For both sets of estimates, plot residuals against a. Is any trend apparent? If so, what experiments would you carry out to decide whether the trend is real and not an artefact of random error? Can any conclusions be drawn about the weights appropriate for least-squares analysis?

14.2 What would be the parameter estimates for the data in Problem 14.1 if the value of v for $a = 1$ mM were 0.159 mM min^{-1} instead of 0.219 mM min^{-1}? Discuss.

14.3 Suppose that some data have been fitted by the least-squares method using both of the weighting assumptions discussed in the text, both a constant coefficient of variation (Section 14.2.2) and a constant standard deviation (Section 14.2.3). If residual plots suggested not only that the simple error decreased with increasing rate, but also that the relative error increased in the same conditions, one might consider a compromise weighting scheme in which weights $w_i = v^3$ were defined for use in equations 14.21–23. Write down expressions for the Michaelis–Menten parameters in which these substitutions have been made, and check the result for dimensional consistency.

14.4 Intuitively one might think it reasonable that the coefficient of variation of any measured quantity might be approximately constant if the quantity were large, but that it would not be possible to measure a zero value exactly, and so at small values the tendency would be towards a constant standard deviation. Is this what the compromise weighting scheme suggested in Problem 14.3 implies?

References

S. J. Abbott, S. R. Jones, S. A. Weinmann, F. M. Bockhoff, F. W. McLafferty and J. R. Knowles (1979) "Chiral [^{16}O,^{17}O,^{18}O]phosphate monoesters: asymmetric synthesis and stereochemical analysis of [1(R)-^{16}O,^{17}O,^{18}O]phospho-(S)-propane-1,2-diol" *Journal of the American Chemical Society* **101**, 4323–4332 [Section 8.1]

G. S. Adair (1925) "The hemoglobin system. VI. The oxygen dissociation curve of hemoglobin" *Journal of Biological Chemistry* **63**, 529–545 [Section 11.2.4]

G. R. Ainslie, Jr., J. P. Shill and K. E. Neet (1972) "Transients and cooperativity. A slow transition model for relating transients and cooperative kinetics of enzymes" *Journal of Biological Chemistry* **247**, 7088–7096 [Section 11.8]

R. A. Alberty (1953) "The effect of enzyme concentration on the apparent equilibrium constant for an enzyme-catalyzed reaction" *Journal of the American Chemical Society* **75**, 1928–1932 [Section 7.3.1]

R. A. Alberty (1958) "On the determination of rate constants for coenzyme mechanisms" *Journal of the American Chemical Society* **80**, 1777–1782 [Section 7.3.1]

R. A. Alberty (2003) *Thermodynamics of Biochemical Reactions* (especially pages 1–17), Wiley–Interscience, Hoboken, New Jersey [Section 3.4]

R. A. Alberty and A. Cornish-Bowden (1993) "The pH dependence of the apparent equilibrium constant, K', of a biochemical reaction" *Trends in Biochemical Sciences* **18**, 288–291 [Section 9.4.3]

R. A. Alberty and B. M. Koerber (1957) "Studies of the enzyme fumarase. VII. Series solutions of integrated rate equations for irreversible and reversible Michaelis–Menten mechanisms" *Journal of the American Chemical Society* **79**, 6379–6382 [Section 2.9.4]

R. A. Alberty and R. N. Goldberg (1992) "Calculation of thermodynamic formation properties for the ATP series at specified pH and pMg" *Biochemistry* **31**, 10610–10615 [Section 3.4]

W. J. Albery and J. R. Knowles (1976) "Free-energy profile of the reaction catalyzed by triosephosphate isomerase" *Biochemistry* **15**, 5627–5631 [Section 8.7]

W. J. Albery and J. R. Knowles (1987) "Energetics of enzyme catalysis. I. Isotopic experiments, enzyme interconversion, and oversaturation" *Journal of Theoretical Biology* **124**, 137–171 [Section 8.5.2]

C. C. Allende, H. Chaimovich, M. Gatica and J. E. Allende (1970) "The aminoacyl transfer ribonucleic acid synthetases. II. Properties of an adenosine triphosphate-threonyl transfer ribonucleic acid synthetase complex" *Journal of Biological Chemistry* **245**, 93–101 [Section 7.8]

J. E. Allende, C. C. Allende, M. Gatica and M. Matamala (1964) "Isolation of threonyl adenylate-enzyme complex" *Biochemical and Biophysical Research Communications* **16**, 342–346 [Section 7.8]

S. R. Anderson and G. Weber (1965) "Multiplicity of binding by lactate dehydrogenases" *Biochemistry* **4**, 1948–1957 [Section 11.6.3]

B. H. Anderton and B. R. Rabin (1970) "Alkylation studies on a reactive histidine in pig heart malate dehydrogenase" *European Journal of Biochemistry* **15**, 568–573 [Section 6.3]

S. Arrhenius (1889) "Über die Reaktionsgeschwindigkeit bei der Inversion von Rohrzucker durch Säuren" *Zeitschrift für Physikalische Chemie* **4**, 226–248 [Section 1.7.1]

P. Askelöf, M. Korsfeldt and B. Mannervik (1976) "Error structure of enzyme kinetic experiments: implications for weighting in regression analysis of experimental data" *European Journal of Biochemistry* **69**, 61–67 [Section 14.2.1]

M. Z. Atassi and T. Manshouri (1993) "Design of peptide enzymes (pepzymes): surface-simulation synthetic peptides that mimic the chymotrypsin and trypsin active sites exhibit the activity and specificity of the respective enzyme" *Proceedings of the National Academy of Sciences U.S.A.* **90**, 8282–8286 [Problem 2.8]

D. E. Atkinson (1977) *Cellular Energy Metabolism and its Regulation*, pages 116–118 [Section 12.3.4], 275–282 [Section 2.7.1], Academic Press, New York

D. E. Atkinson (1990) "What should a theory of metabolic control offer to the experimenter?" pages 3–11 in *Control of Metabolic Processes* (edited by A. Cornish-Bowden and M. L. Cárdenas), Plenum Press, New York [Section 12.8]

B. M. Bakker, P. A. M. Michels, F. R. Opperdoes and H. V. Westerhoff (1997) "Glycolysis in bloodstream form *Trypanosoma brucei* can be understood in terms of the kinetics of the glycolytic enzymes" *Journal of Biological Chemistry* **272**, 3207–3215 [Section 12.10]

J. Balzarini, M.-J. Pérez-Pérez, A. San-Félix, M.-J. Camarasa, J. C. Bathurst, P. J. Barr and E. De Clercq (1992) "Kinetics of inhibition of human immunodeficiency virus type 1 (HIV-1) reverse transcriptase by the novel HIV-1-specific nucleoside analogue [2′,5′-bis-O-(*tert*-butyldimethylsilyl)-β-D-ribofuranosyl]-3′-spiro-5″-(4″-amino-1″,2″-oxathiole-2″,2″-dioxide)thymine (TSAO-T)" *Journal of Biological Chemistry* **267**, 11831–11838 [Section 5.4]

R. P. Bell (1973) *The Proton in Chemistry* (2nd edition), pages 226–296, Chapman and Hall, London [Section 8.6.1]

M. L. Bender, F. J. Kézdy and C. R. Gunter (1964) "The anatomy of an enzymatic catalysis: α-chymotrypsin" *Journal of the American Chemical Society* **86**, 3714–3721 [Section 10.3]

M. L. Bender, M. L. Begué-Cantón, R. L. Blakeley, L. J. Brubacher, J. Feder, C. R. Gunter, F. J. Kézdy, J. V. Killheffer, Jr., T. H. Marshall, C. G. Miller, R. W. Roeske and J. K. Stoops (1966) "The determination of the concentration of hydrolytic enzyme solutions: α-chymotrypsin, trypsin, papain, elastase, subtilisin, and acetylcholinesterase" *Journal of the American Chemical Society* **88**, 5890–5913 [Section 13.4.2]

M. Berthelot (1860) *Chimie Organique Fondé sur la Synthèse*, volume II, pages 655–656, Mallet–Bachelier, Paris [Section 2.1]

J. J. Berzelius (1836) "Quelques idées sur une nouvelle force agissant dans les combinaisons des corps organiques" *Annales de Chimie et de Physique* **61**, 146–151 [Section 1.6]

R. J. Beynon and J. Easterby (1996) *Buffer Solutions: the Basics*, Oxford University Press, Oxford [Section 9.4.5]

G. M. Blackburn, A. S. Kang, G. A. Kingsbury and D. R. Burton (1989) "Catalytic antibodies" *Biochemical Journal* **262**, 381–390 [Section 2.10.4]

D. Blangy, H. Buc and J. Monod (1968) "Kinetics of the allosteric interactions of phosphofructokinase from *Escherichia coli*" *Journal of Molecular Biology* **31**, 13–35 [Section 11.5.4]

E. A. Boeker (1982) "Initial rates. A new plot" *Biochemical Journal* **203**, 117–123 [Section 2.9.3]

E. A. Boeker (1984) "Integrated rate equations for enzyme-catalysed first-order and second-order reactions" *Biochemical Journal* **223**, 15–22 [Section 2.9.4]

E. A. Boeker (1985) "Integrated rate equations for irreversible enzyme-catalysed first-order and second-order reactions" *Biochemical Journal* **226**, 29–35 [Section 2.9.4]

M. R. Boocock and J. R. Coggins (1983) "Kinetics of 5-*enol*pyruvylshikimate-3-phosphate synthase inhibition by glyphosate" *FEBS Letters* **154**, 127–133 [Section 12.3.4]

J. Botts and M. Morales (1954) "Analytical description of the effects of modifiers and of enzyme multivalency upon the steady-state catalyzed reaction rate" *Transactions of the Faraday Society* **49**, 696–707 [Section 5.7.3]

T. R. C. Boyde (1980) *Foundation Stones in Biochemistry*, Voile et Aviron, Hong Kong

P. D. Boyer (1959) "Uses and limitations of measurements of rates of isotope exchange and incorporation in catalyzed reactions" *Archives of Biochemistry and Biophysics* **82**, 387–410 [Section 8.2]

R. B. Brandt, J. E. Laux and S. W. Yates (1987) "Calculation of inhibitor K_i and inhibitor type from the concentration of inhibitor for 50% inhibition for Michaelis–Menten enzymes" *Biochemical Medicine and Metabolic Biology* **37**, 344–349 [Section 5.5]

G. E. Briggs and J. B. S. Haldane (1925) "A note on the kinetics of enzyme action" *Biochemical Journal* **19**, 338–339; reprinted in pages 267–268 of Friedmann (1981) [Section 2.3.1]

R. Brinkman, R. Margaria and F. J. W. Roughton (1934) "The kinetics of the carbon dioxide–carbonic acid reaction" *Philosophical Transactions of the Royal Society of London, Series A* **232**, 65–97 [Section 13.3.3]

H. G. Britton (1966) "The concept and use of flux measurements in enzyme studies: a theoretical analysis" *Archives of Biochemistry and Biophysics* **117**, 167–183 [Section 8.5.1]

H. G. Britton (1973) "Methods of determining rate constants in single-substrate-single-product enzyme reactions. Use of induced transport: limitations of product inhibition" *Biochemical Journal* **133**, 255–261 [Section 8.5.2]

H. G. Britton (1994) "Flux ratios, induced transport and tracer perturbation" *Biochemical Journal* **302**, 965–966 [Section 8.5.2]

H. G. Britton and J. B. Clarke (1968) "The mechanism of the phosphoglucomutase reaction. Studies on rabbit muscle phosphoglucomutase with flux techniques" *Biochemical Journal* **110**, 161–183 [Section 8.5.3]

K. Brocklehurst (1994) "A sound basis for pH-dependent kinetic studies on enzymes" *Protein Engineering* **7**, 291–299 [Section 9.6]

K. Brocklehurst (1996) "pH-dependent kinetics" pages 175–190 in *Enzymology Labfax* (edited by P. C. Engel), Bios Scientific Publishers, Oxford [Section 9.6]

K. Brocklehurst (2002) "Electrochemical assays: the pH-stat" pages 157–170 in *Enzyme Assays* (2nd edition, edited by R. Eisenthal and M. J. Danson), Oxford University Press, Oxford [Section 3.1.1]

K. Brocklehurst, M. Resmini and C. M. Topham (2001) "Kinetic and titration methods for determination of active site contents of enzyme and catalytic antibody preparations" *Methods, a Companion to Methods in Enzymology* **24**, 153–167 [Sections 13.2.2, 13.4.2]

K. Brocklehurst and C. M. Topham (1993) "Some classical errors in the kinetic analysis of enzyme reactions" *Biochemical Journal* **295**, 898–899 [Section 4.6 (footnote)]

P. F. Brode III, C. R. Erwin, D. S. Rauch, B. L. Barnett, J. M. Armpriester, E. S. F. Wang and D. N. Rubingh (1996) "Subtilisin BPN′ variants: increased hydrolytic activity on surface-bound substrates via decreased surface activity" *Biochemistry* **25**, 3162–3169 [Section 2.9.3]

J. N. Brønsted (1923) "Einige Bemerkungen über den Begriff der Säuren und Basen" *Recueil des Travaux Chimiques des Pays-Bas* **42**, 718–728 [Section 9.2]

A. J. Brown (1892) "Influence of oxygen and concentration on alcohol fermentation" *Journal of the Chemical Society (Transactions)* **61**, 369–385 [Section 2.1]

A. J. Brown (1902) "Enzyme action" *Journal of the Chemical Society (Transactions)* **81**, 373–388; reprinted in pages 241–256 of Friedmann (1981) [Section 2.1]

G. C. Brown, R. Hafner and M. D. Brand (1990) "A 'top-down' approach to the determination of control coefficients in metabolic control theory" *European Journal of Biochemistry* **188**, 321–325 [Problem 12.4]

J. Buc, J. Ricard and J.-C. Meunier (1977) "Enzyme memory. 2. Kinetics and thermodynamics of the slow conformation changes of wheat-germ hexokinase L_I" *European Journal of Biochemistry* **80**, 593–601 [Section 13.3.7]

E. Buchner (1897) "Alkoholische Gährung ohne Hefezellen" *Berichte der Deutschen Chemischen Gesellschaft* **30**, 117–124; reprinted in pages 17–24 of Cornish-Bowden (1997); English translation in pages 93–101 of Boyde (1980), and (by H. C. Friedmann) in pages 25–31 of Cornish-Bowden (1997) [Section 2.1]

J. J. Burke, G. G. Hammes and T. B. Lewis (1965) "Ultrasonic attenuation measurements in poly-L-glutamic acid solutions" *Journal of Chemical Physics* **42**, 3520–3525 [Section 13.3.7]

M. L. Cárdenas (1995) *"Glucokinase": its Regulation and Role in Liver Metabolism*, page 30 [Section 2.6.1], pages 41–80 [Section 11.8], R.G. Landes, Austin, Texas

M. L. Cárdenas (2001) "The competition plot: a kinetic method to assess whether an enzyme that catalyzes multiple reactions does so at a unique site" *Methods, a Companion to Methods in Enzymology* **24**, 175–180 [Section 5.6.2]

M. L. Cárdenas and A. Cornish-Bowden (1989) "Characteristics necessary for an interconvertible enzyme cascade to give a highly sensitive response to an effector" *Biochemical Journal* **257**, 339–345 [Section 12.9.2]

M. L. Cárdenas and A. Cornish-Bowden (1993) "Rounding error: an unexpected fault in the output from a recording spectrophotometer" *Biochemical Journal* **292**, 37–40 [Sections 3.1.2, 14.5]

M. L. Cárdenas, E. Rabajille and H. Niemeyer (1978) "Maintenance of the monomeric structure of glucokinase under reacting conditions" *Archives of Biochemistry and Biophysics* **190**, 142–148 [Section 11.8]

M. L. Cárdenas, E. Rabajille and H. Niemeyer (1984a) "Fructose: a good substrate for rat-liver 'glucokinase' (hexokinase D)" *Biochemical Journal* **222**, 363–370 [Section 11.8]

M. L. Cárdenas, E. Rabajille and H. Niemeyer (1984b) "Suppression of kinetic cooperativity of hexokinase D ('glucokinase') by competitive inhibitors" *European Journal of Biochemistry* **145**, 163–171 [Section 11.8]

H. Cedar and J. H. Schwartz (1969) "The asparagine synthetase of *Escherichia coli*. II. Studies on mechanism" *Journal of Biological Chemistry* **244**, 4122–4127 [Section 8.4]

F. Cedrone, A. Ménez and E. Quéméneur (2000) "Tailoring new enzyme functions by rational redesign" *Current Opinion in Structural Biology* **10**, 405–410 [Section 2.10.2]

S. Cha (1968) "A simple method for derivation of rate equations for enzyme-catalyzed reactions under the rapid equilibrium assumption or combined assumptions of equilibrium and steady state" *Journal of Biological Chemistry* **243**, 820–825 [Section 4.6]

W. W.-C. Chan (1995) "Combination plots as graphical tools in the study of enzyme inhibition" *Biochemical Journal* **311**, 981–985 [Section 5.3.2]

B. Chance (1940) "The accelerated flow method for rapid reactions" *Journal of the Franklin Institute* **229**, 455–476 [Section 13.3.3]

B. Chance (1951) "Enzyme–substrate compounds" *Advances in Enzymology* **12**, 153–190 [Section 13.3.3]

Y.-C. Cheng and W. H. Prusoff (1973) "Relationship between the inhibition constant (K_I) and the concentration of inhibitor which causes 50 per cent inhibition (I_{50}) of an enzymatic reaction" *Biochemical Pharmacology* **22**, 3099–3108 [Section 5.5]

C. Chevillard, M. L. Cárdenas and A. Cornish-Bowden (1993) "The competition plot: a simple test of whether two reactions occur at the same active site" *Biochemical Journal* **289**, 599–604 [Section 5.6.2]

R. E. Childs and W. G. Bardsley (1975) "Time-dependent inhibition of enzymes by active-site-directed reagents. A theoretical treatment of the kinetics of affinity labelling" *Journal of Theoretical Biology* **53**, 381–394 [Problem 6.2]

R. S. Chittock, J.-M. Hawronskyj, J. Holah and C. W. Wharton (1998) "Kinetic aspects of ATP amplification reactions" *Analytical Biochemistry* **255**, 120–126 [Section 3.1.4]

P. B. Chock and E. R. Stadtman (1977) "Superiority of interconvertible enzyme cascades in metabolite regulation: analysis of multicyclic systems" *Proceedings of the National Academy of Sciences U.S.A.* **74**, 2766–2770 [Section 12.9.2]

P. B. Chock, S. G. Rhee and E. R. Stadtman (1980) "Interconvertible enzyme cascades in cellular regulation" *Annual Review of Biochemistry* **49**, 813–843 [Section 12.9.2]

P. B. Chock, S. G. Rhee and E. R. Stadtman (1990) "Metabolic control by the cyclic cascade mechanism: a study of *E. coli* glutamine synthetase" pages 183–194 in *Control of Metabolic Processes* (edited by A. Cornish-Bowden and M. L. Cárdenas), Plenum Press, New York [Section 12.9.2]

K. Chou, S. Jiang, W. Liu and C. Fee (1979) "Graph theory of enzyme kinetics" *Scientia Sinica* **22**, 341–358 [Section 4.5]

T.-C. Chou and P. Talalay (1977) "A simple generalized equation for the analysis of multiple inhibitions of Michaelis–Menten kinetic systems" *Journal of Biological Chemistry* **252**, 6438–6442 [Section 5.4]

B. Cigić and R. H. Pain (1999) "Location of the binding site for chloride ion activation of cathepsin C" *European Journal of Biochemistry* **264**, 944–951 [Section 5.7.2]

J. S. Clegg and S. A. Jackson (1988) "Glycolysis in permeabilized L-929 cells" *Biochemical Journal* **255**, 335–344 [Section 12.1.3]

W. W. Cleland (1963) "The kinetics of enzyme-catalyzed reactions with two or more substrates or products. I. Nomenclature and rate equations" *Biochimica et Biophysica Acta* **67**, 104–137 [Sections 7.2.4, 7.3.1, 7.3.4]

W. W. Cleland (1967) "The statistical analysis of enzyme kinetic data" *Advances in Enzymology* **29**, 1–32 [Section 14.2.3]

R. A. Cook and D. E. Koshland, Jr. (1970) "Positive and negative cooperativity in yeast glyceraldehyde 3-phosphate dehydrogenase" *Biochemistry* **9**, 3337–3342 [Section 11.2.5]

M. J. Corey and E. Corey (1996) "On the failure of de novo-designed peptides as biocatalysts" *Proceedings of the National Academy of Sciences U.S.A.* **93**, 11428–11434 [Problem 2.8]

D. R. Corey and M. A. Phillips (1994) "Cyclic peptides as proteases: a reevaluation" *Proceedings of the National Academy of Sciences U.S.A.* **91**, 4106–4109 [Problem 2.8]

A. Cornish-Bowden (1974) "A simple graphical method for determining the inhibition constants of mixed, uncompetitive and non-competitive inhibitors" *Biochemical Journal* **137**, 143–144 [Section 5.3.1]

A. Cornish-Bowden (1975) "The use of the direct linear plot for determining initial velocities" *Biochemical Journal* **149**, 305–312 [Sections 2.9.3, 3.1.2]

A. Cornish-Bowden (1976a) "Estimation of the dissociation constants of enzyme–substrate complexes from steady-state measurements" *Biochemical Journal* **153**, 455–461 [Sections 2.7.1, 9.6]

A. Cornish-Bowden (1976b) "Algebraic methods for deriving steady-state rate equations: practical difficulties with mechanisms that contain repeated rates" *Biochemical Journal* **159**, 167 [Section 4.8]

A. Cornish-Bowden (1976c) *Principles of Enzyme Kinetics*, Butterworths, London [Sections 14.1, 14.2.3, Problem 2.3]

A. Cornish-Bowden (1977) "An automatic method for deriving steady-state rate equations" *Biochemical Journal* **165**, 55–59 [Section 4.8]

A. Cornish-Bowden (1979) "Validity of a 'steady-state' treatment of inactivation kinetics" *European Journal of Biochemistry* **93**, 383–385 [Problem 6.2]

A. Cornish-Bowden (1982) "Weighting of linear plots in enzyme kinetics" *Journal of Molecular Science* **2**, 107–112 [Section 14.4]

A. Cornish-Bowden (1986) "Why is uncompetitive inhibition so rare?" *FEBS Letters* **203**, 3–6 [Section 12.3.4]

A. Cornish-Bowden (1987) "The time dimension in steady-state kinetics: a simplified representation of control coefficients" *Biochemical Education* **15**, 144–146 [Section 2.3.4]

A. Cornish-Bowden (1989) "Non-equilibrium isotope-exchange methods for investigating enzyme mechanisms" *Current Topics in Cellular Regulation* **30**, 143–169 [Section 8.5.1]

A. Cornish-Bowden (1995a) "Metabolic control analysis in theory and practice" *Advances in Molecular Cell Biology* **11**, 21–64 [Section 12.2]

A. Cornish-Bowden (1995b) *Analysis of Enzyme Kinetic Data*, Oxford University Press, Oxford [Sections 14.1, 14.2.3, 14.4]

A. Cornish-Bowden (1995c) "Kinetics of multi-enzyme systems" pages 121–136 in *Biotechnology: a Comprehensive Treatise* (2nd edition, edited by H.-J. Rehm and G. Reed), volume 9, VCH, Weinheim [Problem 12.5]

A. Cornish-Bowden (editor, 1997) *New Beer in an Old Bottle*, Universitat de València, Valencia, Spain [Section 2.1]

A. Cornish-Bowden (2002) "Enthalpy–entropy compensation: a phantom phenomenon" *Journal of Biosciences* **27**, 121–126 [Section 10.4]

A. Cornish-Bowden and M. L. Cárdenas (1987) "Co-operativity in monomeric enzymes" *Journal of Theoretical Biology* **124**, 1–23 [Section 11.8]

A. Cornish-Bowden and M. L. Cárdenas (editors, 1990) *Control of Metabolic Processes*, Plenum Press, New York [Section 12.2]

A. Cornish-Bowden and M. L. Cárdenas (1993) "Channelling can affect concentrations of metabolic intermediates at constant net flux: artefact or reality?" *European Journal of Biochemistry* **213**, 87–92 [Section 12.9.1]

A. Cornish-Bowden and M. L. Cárdenas (editors, 2000) *Technological and Medical Implications of Metabolic Control Analysis*, Kluwer, Dordrecht [Section 12.2]

A. Cornish-Bowden and M. L. Cárdenas (2001a) "Information transfer in metabolic pathways: effects of irreversible steps in computer models" *European Journal of Biochemistry* **268**, 6616–6624 [Sections 12.3.1, 12.10]

A. Cornish-Bowden and M. L. Cárdenas (2001b) "Silent genes given voice" *Nature* **409**, 571–572 [Section 12.8]

A. Cornish-Bowden and R. Eisenthal (1974) "Statistical considerations in the estimation of enzyme kinetic parameters by the direct linear plot and other methods" *Biochemical Journal* **139**, 721–730 [Section 14.3.2]

A. Cornish-Bowden and R. Eisenthal (1978) "Estimation of Michaelis constant and maximum velocity from the direct linear plot" *Biochimica et Biophysica Acta* **523**, 268–272 [Sections 2.6.5, 14.3.5]

A. J. Cornish-Bowden and J. R. Knowles (1969) "The pH-dependence of pepsin-catalysed reactions" *Biochemical Journal* **113**, 353–362 [Section 9.5]

A. Cornish-Bowden and D. E. Koshland, Jr. (1970) "A general method for the quantitative determination of saturation curves" *Biochemistry* **9**, 3325–3336 [Section 11.2.5]

A. Cornish-Bowden and D. E. Koshland, Jr. (1975) "Diagnostic uses of Hill (logit and Nernst) plots" *Journal of Molecular Biology* **95**, 202–212 [Section 11.2.5]

A. Cornish-Bowden, W. R. Porter and W. F. Trager (1978) "Evaluation of distribution-free confidence limits for enzyme kinetic parameters" *Journal of Theoretical Biology* **74**, 163–175 [Section 14.4]

A. Cornish-Bowden and J. T. Wong (1978) "Evaluation of rate constants for enzyme-catalysed reactions by the jackknife technique: application to liver alcohol dehydrogenase" *Biochemical Journal* **175**, 969–976 [Section 7.3.4]

A. Cortés, M. Cascante, M. L. Cárdenas and A. Cornish-Bowden (2001) "Relationships between inhibition constants, inhibitor concentrations for 50% inhibition and types of inhibition: new ways of analysing data" *Biochemical Journal* **357**, 263–268 [Section 5.5]

D. H. Craig, T. Barna, P. C. E. Moddy, N. C. Bruce, S. K. Chapman, A. W. Munro and N. S. Scrutton (2001) "Effects of environment on flavin reactivity in morphinone reductase: analysis of enzymes displaying differential charge near the N-1 atom and C-2 carbonyl region of the active-site flavin" *Biochemical Journal* **359**, 315–323 [Section 10.4]

K. Dalziel (1957) "Initial steady-state velocities in the evaluation of enzyme–coenzyme–substrate reaction mechanisms" *Acta Chemica Scandinavica* **11**, 1706–1723 [Section 7.3.1, Problem 7.7]

K. Dalziel (1969) "The interpretation of kinetic data for enzyme-catalysed reactions involving three substrates" *Biochemical Journal* **114**, 547–556 [Section 7.8]

R. M. Daniel, M. J. Danson and R. Eisenthal (2000) "The temperature optima of enzymes: a new perspective on an old phenomenon" *Trends in Biochemical Sciences* **26**, 223–225 [Section 10.2]

M. P. Deutscher (1967) "Rat liver glutamyl ribonucleic acid synthetase. I. Purification and evidence for separate enzymes for glutamic acid and glutamine" *Journal of Biological Chemistry* **242**, 1123–1131 [Section 3.2]

C. J. Dickenson and F. M. Dickinson (1978) "Inhibition by ethanol, acetaldehyde and trifluoroethanol of reactions catalysed by yeast and horse liver alcohol dehydrogenases" *Biochemical Journal* **171**, 613–627 [Section 5.7.3]

H. B. F. Dixon (1973) "Shapes of curves of pH-dependence of reactions" *Biochemical Journal* **131**, 149–154 [Section 9.6]

H. B. F. Dixon (1976) "The unreliability of estimates of group dissociation constants" *Biochemical Journal* **153**, 627–629 [Section 9.3.3]

H. B. F. Dixon (1979) "Derivation of molecular pK values from pH-dependences" *Biochemical Journal* **177**, 249–250 [Section 9.3.3]

M. Dixon (1953a) "The effect of pH on the affinities of enzymes for substrates and inhibitors" *Biochemical Journal* **55**, 161–170 [Section 9.4.4]

M. Dixon (1953b) "The determination of enzyme inhibitor constants" *Biochemical Journal* **55**, 170–171 [Section 5.3.1]

M. Dixon (1972) "The graphical determination of K_m and K_i" *Biochemical Journal* **129**, 197–202 [Section 6.1]

M. Dixon and E. C. Webb (1979) *Enzymes* (3rd edition), pages 323–331, Longman, London and Academic Press, New York [Section 7.8]

J. A. Doudna and T. R. Cech (2002) "The chemical repertoire of natural ribozymes" *Nature* **418**, 222–228 [Section 2.10.3]

M. Doudoroff, H. A. Barker and W. Z. Hassid (1947) "Studies with bacterial sucrose phosphorylase. I. The mechanism of action of sucrose phosphorylase as a glucose-transferring enzyme (transglucosidase)" *Journal of Biological Chemistry* **168**, 725–732 [Sections 7.2.2, 8.4]

C. Doumeng and S. Maroux (1979) "Aminotripeptidase, a cytosol enzyme from rabbit intestinal mucosa" *Biochemical Journal* **177**, 801–808 [Problem 5.5]

J. E. Dowd and D. S. Riggs (1965) "A comparison of estimates of Michaelis–Menten kinetic constants from various linear transformations" *Journal of Biological Chemistry* **240**, 863–869 [Sections 2.6.2, 2.6.4]

R. G. Duggleby (1995) "Analysis of enzyme progress curves by nonlinear regression" *Methods in Enzymology* **249**, 61–90 [Section 2.9.4]

R. G. Duggleby (2001) "Quantitative analysis of the time courses of enzyme-catalyzed reactions" *Methods, a Companion to Methods in Enzymology* **24**, 168–174 [Section 2.9.4]

G. S. Eadie (1942) "The inhibition of cholinesterase by physostigmine and prostigmine" *Journal of Biological Chemistry* **146**, 85–93 [Section 2.6.4]

R. R. Eady, D. J. Lowe and R. N. F. Thorneley (1978) "Nitrogenase of *Klebsiella pneumoniae*: a pre-steady state burst of ATP hydrolysis is coupled to electron transfer between the component proteins" *FEBS Letters* **95**, 211–213 [Section 13.3.4]

J. S. Easterby (1981) "A generalized theory of the transition time for sequential enzyme reactions" *Biochemical Journal* **199**, 155–161 [Section 3.1.4]

S. Efrat, M. Leiser, Y.-J. Wu, D. Fusco-DeMane, O. A. Emran, M. Surana, T. L. Jetton, M. A. Magnuson, G. Weir and N. Fleischer (1994) "Ribozyme-mediated attenuation of pancreatic β-cell glucokinase expression in transgenic mice results in impaired glucose-induced insulin secretion" *Proceedings of the National Academy of Sciences U.S.A.* **91**, 2051–2055 [Section 2.10.3]

M. Eigen (1954) "Methods for investigation of ionic reactions in aqueous solutions with half times as short as 10^{-9} sec. Application to neutralization and hydrolysis reactions" *Discussions of the Faraday Society* **17**, 194–205 [Section 13.3.7]

R. Eisenthal and A. Cornish-Bowden (1974) "The direct linear plot" *Biochemical Journal* **139**, 715–720 [Section 2.6.5]

R. Eisenthal and A. Cornish-Bowden (1998) "Prospects for antiparasitic drugs: the case of *Trypanosoma brucei*, the causative agent of African sleeping sickness" *Journal of Biological Chemistry* **273**, 5500–5505 [Section 12.10]

K. R. F. Elliott and K. F. Tipton (1974) "A kinetic analysis of enzyme systems involving four substrates" *Biochemical Journal* **141**, 789–805 [Section 7.8]

L. Endrenyi (1981) "Design of experiments for estimating enzyme and pharmacokinetic parameters" pages 137–167 in *Kinetic Data Analysis* (edited by L. Endrenyi), Plenum Press, New York [Section 3.3.1]

O. Exner (1964) "On the enthalpy–entropy relationship" *Collection of Czechoslovakian Chemical Communications* **26**, 1094–1113 [Section 10.4]

H. Eyring (1935) "The activated complex in chemical reactions" *Journal of Physical Chemistry* **3**, 107–115 [Section 1.7.3]

D. A. Fell (1992) "Metabolic control analysis: a survey of its theoretical and experimental development" *Biochemical Journal* **286**, 313–330 [Sections 12.2, 12.6.2]

D. Fell (1997) *Understanding the Control of Metabolism*, Portland Press, London [Section 12.2]

D. A. Fell and H. M. Sauro (1985) "Metabolic control and its analysis. Additional relationships between elasticities and control coefficients" *European Journal of Biochemistry* **148**, 555–561 [Section 12.6.2]

W. Ferdinand (1966) "The interpretation of non-hyperbolic rate curves for two-substrate enzymes. A possible mechanism for phosphofructokinase" *Biochemical Journal* **98**, 278–283 [Section 11.8]

A. Fersht (1977) *Enzyme Structure and Mechanism*, pages 97 and 274–287 [also in pages 377–399 of Fersht (1999)], Freeman, Reading [Section 2.4]

A. Fersht (1999) *Structure and Mechanism in Protein Science* [Section 13.3.4], pages 164–167 [Sections 2.10, 6.1, 13.3], 175–176 [Problem 9.1], Freeman, New York

A. R. Fersht and R. Jakes (1975) "Demonstration of two reaction pathways for the aminoacylation of tRNA. Application of the pulsed quenched flow technique" *Biochemistry* **14**, 3350–3356 [Section 13.3.4]

E. Fischer (1894) "Einfluss der Configuration auf die Wirkung der Enzyme" *Berichte der Deutschen Chemischen Gesellschaft* **27**, 2985–2993 [Section 11.4]

D. D. Fisher and A. R. Schulz (1969) "Connection matrix representation of enzyme reaction sequences" *Mathematical Biosciences* **4**, 189–200 [Section 4.8]

J. R. Fisher and V. D. Hoagland, Jr. (1968) "A systematic approach to kinetic studies of multisubstrate enzyme systems" *Advances in Biological and Medical Physics* **12**, 163–211 [Section 7.2.2]

L. M. Fisher, W. J. Albery and J. R. Knowles (1986) "Energetics of proline racemase: tracer perturbation experiments using [14C]proline that measure the interconversion rate of the two forms of free enzyme" *Biochemistry* **25**, 2538–2542 [Section 8.5.3]

C. Frieden (1959) "Glutamic dehydrogenase. III. The order of substrate addition in the enzymatic reaction" *Journal of Biological Chemistry* **234**, 2891–2896 [Section 7.8]

C. Frieden (1967) "Treatment of enzyme kinetic data. II. The multisite case: comparison of allosteric models and a possible new mechanism" *Journal of Biological Chemistry* **242**, 4045–4052 [Section 11.7]

C. Frieden and R. F. Colman (1967) "Glutamate dehydrogenase concentration as a determinant in the effect of purine nucleotides on enzymatic activity" *Journal of Biological Chemistry* **242**, 1705–1715 [Section 11.7]

H. C. Friedmann (editor, 1981) *Enzymes*, Hutchinson Ross, Stroudsburg, Pennsylvania

W. Friedrich, P. Knipping and M. Laue (1913) "Interferenzerscheinungen bei Röntgenstrahlen" *Annalen der Physik und Chemie*, 4th series **41**, 971–988 [Section 13.3.5]

E. Fulhame (1794) *An Essay on Combustion, with a View to a New Art of Dying and Painting, wherein the phlogistic and antiphlogistic hypotheses are proved erroneous*, published by the author, London [Section 1.6]

E. Garcés and W. W. Cleland (1969) "Kinetic studies of yeast nucleoside diphosphate kinase" *Biochemistry* **8**, 633–640 [Section 7.2.3, Problem 7.4]

D. Garfinkel and B. Hess (1964) "Metabolic control mechanisms. VII. A detailed computer model of the glycolytic pathway in ascites cells" *Journal of Biological Chemistry* **239**, 971–983 [Section 12.10]

Q. H. Gibson and L. Milnes (1964) "Apparatus for rapid and sensitive spectrophotometry" *Biochemical Journal* **91**, 161–171 [Section 13.3.3]

R. Giordani, J. Buc, A. Cornish-Bowden and M. L. Cárdenas (1997) "Kinetics of membrane-bound nitrate reductase A from *Escherichia coli* with analogues of physiological electron donors: Different reaction sites for menadiol and duroquinol" *European Journal of Biochemistry* **250**, 567–577 [Section 5.6.2]

S. Glasstone, K. J. Laidler and H. Eyring (1940) *Theory of Rate Processes*, McGraw-Hill, New York [Section 1.7.3]

A. Goldbeter and D. E. Koshland, Jr. (1981) "An amplified sensitivity arising from covalent modification in biological systems" *Proceedings of the National Academy of Sciences U.S.A.* **78**, 6840–6844 [Section 12.9.2]

A. Goldbeter and D. E. Koshland, Jr. (1982) "Sensitivity amplification in biochemical systems" *Quarterly Reviews of Biophysics* **15**, 555–591 [Section 12.9.2]

A. Goldbeter and D. E. Koshland, Jr. (1984) "Ultrasensitivity in biochemical systems controlled by covalent modification. Interplay between zero-order and multistep effects" *Journal of Biological Chemistry* **259**, 14441–14447 [Section 12.9.2]

Y. E. Goldman, M. G. Hibberd, J. A. McCray and D. R. Trentham (1982) "Relaxation of muscle fibres by photolysis of caged ATP" *Nature* **300**, 701–705 [Section 13.3.5]

A. Goldstein (1944) "The mechanism of enzyme–inhibitor–substrate reactions" *Journal of General Physiology* **27**, 529–580 [Section 6.1]

C. T. Goudar, J. R. Sonnad and R. G. Duggleby (1999) "Parameter estimation using a direct solution of the integrated Michaelis–Menten equation" *Biochimica et Biophysica Acta* **1429**, 377–383 [Section 2.9.4]

M. Gregoriou, I. P. Trayer and A. Cornish-Bowden (1981) "Isotope exchange evidence for an ordered mechanism for rat-liver glucokinase, a monomeric co-operative enzyme" *Biochemistry* **20**, 499–506 [Section 8.5.1]

A. K. Groen, R. J. A. Wanders, H. V. Westerhoff, R. van der Meer and J. M. Tager (1982) "Quantification of the contribution of various steps to the control of mitochondrial respiration" *Journal of Biological Chemistry* **257**, 2754–2757 [Sections 12.4.2, 12.8]

A. K. Groen, R. C. Vervoorn, R. van der Meer and J. M. Tager (1983) "Control of gluconeogenesis in rat liver cells. I. Kinetics of the individual enzymes and the effect of glucagon" *Journal of Biological Chemistry* **258**, 14346–14353 [Section 12.8]

A. K. Groen, C. W. T. van Roermund, R. C. Vervoorn and J. M. Tager (1986) "Control of gluconeogenesis in rat liver cells. Flux control coefficients of the enzymes in the gluconeogenic pathway in the absence and presence of glucagon" *Biochemical Journal* **237**, 379–389 [Section 12.8]

E. A. Guggenheim (1926) "On the determination of the velocity constant of a unimolecular reaction" *Philosophical Magazine, Series VII* **2**, 538–543 [Section 1.5]

J. S. Gulbinsky and W. W. Cleland (1968) "Kinetic studies of *Escherichia coli* galactokinase" *Biochemistry* **7**, 566–575 [Sections 4.6, 7.2.1]

H. Gutfreund (1955) "Steps in the formation and decomposition of some enzyme–substrate complexes" *Discussions of the Faraday Society* **20**, 167–173 [Section 13.2.1]

H. Gutfreund (1965) *An Introduction to the Study of Enzymes*, pages 302–306, Blackwell, Oxford [Section 12.9.1]

H. Gutfreund (1995) *Kinetics for the Life Sciences*, pages 246–248, Cambridge University Press, Cambridge [Sections 10.4, 13.1.2]

J. E. Haber and D. E. Koshland, Jr. (1967) "Relation of protein subunit interactions to the molecular species observed during cooperative binding of ligands" *Proceedings of the National Academy of Sciences U.S.A.* **58**, 2087–2093 [Section 11.6.1]

J. B. S. Haldane (1930) *Enzymes*, Longmans Green, London [Sections 2.7.2, 3.2, 7.2.1]

J. B. S. Haldane (1957) "Graphical methods in enzyme chemistry" *Nature* **179**, 832, reprinted on page 271 of Friedmann (1981) [Section 2.6.4]

G. G. Hammes and P. Fasella (1962) "A kinetic study of glutamic-aspartic transaminase" *Journal of the American Chemical Society* **84**, 4644–4650 [Section 13.4.2]

G. G. Hammes and P. R. Schimmel (1970) "Rapid reactions and transient states" pages 67–114 in *The Enzymes* (edited by P. D. Boyer), 3rd edition, volume 2, Academic Press, New York [Sections 13.1.2, 13.3.7, 13.4.3]

C. S. Hanes (1932) "Studies on plant amylases" *Biochemical Journal* **26**, 1406–1421 [Section 2.6.4]

C. S. Hanes, P. M. Bronskill, P. A. Gurr and J. T. Wong (1972) "Kinetic mechanism for the major isoenzyme of horse liver alcohol dehydrogenase" *Canadian Journal of Biochemistry* **50**, 1385–1413 [Section 7.2.3]

A. V. Harcourt (1867) "On the observation of the course of chemical change" *Journal of the Chemical Society* **20**, 460–492 [Section 1.7.1]

B. S. Hartley and B. A. Kilby (1954) "The reaction of *p*-nitrophenyl esters with chymotrypsin and insulin" *Biochemical Journal* **56**, 288–297 [Sections 13.2.1, 13.4.2]

H. Hartridge and F. J. W. Roughton (1923) "A method for measuring the velocity of very rapid chemical reactions" *Proceedings of the Royal Society, Series A* **104**, 376–394 [Section 13.3.2]

J. W. Hastings and Q. H. Gibson (1963) "Intermediates in the bioluminescent oxidation of reduced flavin mononucleotide" *Journal of Biological Chemistry* **238**, 2537–2554 [Section 13.3.3]

J. Heinisch (1986) "Isolation and characterisation of the two structural genes coding for phosphofructokinase in yeast" *Molecular and General Genetics* **202**, 75–82 [Section 12.4.1]

R. Heinrich and T. A. Rapoport (1973) "Linear theory of enzymatic chains: its application for the analysis of the crossover theorem and of the glycolysis of the human erythrocyte" *Acta Biologica et Medica Germanica* **31**, 479–494 [Section 12.10]

R. Heinrich and T. A. Rapoport (1974) "A linear steady-state theory of enzymatic chains: general properties, control and effector strength" *European Journal of Biochemistry* **42**, 89–95 [Sections 12.2, 12.4.1, 12.6.5]

R. Heinrich and T. A. Rapoport (1975) "Mathematical analysis of multienzyme systems. III. Steady state and transient control" *BioSystems* **7**, 130–136 [Section 12.5]

R. Heinrich and S. Schuster (1996) *The Regulation of Cellular Systems*, Chapman and Hall, New York [Section 12.2]

P. J. F. Henderson (1972) "A linear equation that describes the steady-state kinetics of enzymes and subcellular particles interacting with tightly bound inhibitors" *Biochemical Journal* **127**, 321–333 [Section 6.1]

P. J. F. Henderson (1973) "Steady-state kinetics with high-affinity substrates or inhibitors" *Biochemical Journal* **135**, 101–107 [Section 6.1]

V. Henri (1902) *Comptes Rendus Hebdomadaires des Séances de l'Académie des Sciences, Paris* **135**, 916–919; reprinted in pages 140–143 of Boyde (1980), and in English translation in pages 130–135 [Section 2.2]

V. Henri (1903) *Lois Générales de l'Action des Diastases*, Hermann, Paris; pages 85–93 are reprinted in pages 258–266 of Friedmann (1981) [Sections 2.2, 2.8, 2.9.2, 2.9.4]

J. Higgins (1965) "Dynamics and control in cellular systems" pages 13–46 in *Control of Energy Metabolism* (edited by B. Chance, R. W. Estabrook and J. R. Williamson), Academic Press, New York [Section 12.8]

A. V. Hill (1910) "The possible effects of the aggregation of the molecules of haemoglobin on its dissociation curves" *Journal of Physiology* **40**, iv–vii [Section 11.2.1]

C. M. Hill, R. D. Waight and W. G. Bardsley (1977) "Does any enzyme follow the Michaelis–Menten equation?" *Molecular and Cellular Biochemistry* **15**, 173–178 [Section 3.3.1]

K. Hiromi (1979) *Kinetics of Fast Enzyme Reactions*, Wiley, New York [Sections 13.1.2, 13.3.4]

D. I. Hitchcock (1926) "The formal identity of Langmuir's adsorption equation with the law of mass action" *Journal of the American Chemical Society* **48**, 2870 [Section 2.2]

J.-H. S. Hofmeyr (1989) "Control-pattern analysis of metabolic pathways: flux and concentration control in linear pathways" *European Journal of Biochemistry* **186**, 343–354 [Section 12.10]

J.-H. S. Hofmeyr and A. Cornish-Bowden (1991) "Quantitative assessment of regulation in metabolic systems" *European Journal of Biochemistry* **200**, 223–236 [Section 12.8]

J.-H. S. Hofmeyr and A. Cornish-Bowden (1997) "The reversible Hill equation: how to incorporate cooperative enzymes into metabolic models" *Computer Applications in the Biosciences* **13**, 377–385 [Section 12.10]

J.-H. S. Hofmeyr and A. Cornish-Bowden (2000) "Regulating the cellular economy of supply and demand" *FEBS Letters* **476**, 47–51 [Section 12.8]

J.-H. S. Hofmeyr, H. Kacser and K. J. van der Merwe (1986) "Metabolic analysis of moiety-conserved cycles" *European Journal of Biochemistry* **155**, 631–641 [Section 12.1.2]

B. H. J. Hofstee (1952) "Specificity of esterases" *Journal of Biological Chemistry* **199**, 357–364 [Section 2.6.4]

M. R. Hollaway and H. A. White (1975) "A double-beam rapid-scanning stopped-flow spectrophotometer" *Biochemical Journal* **149**, 221–231 [Section 13.3.4]

M. J. Holroyde, M. B. Allen, A. C. Storer, A. S. Warsy, J. M. E. Chesher, I. P. Trayer, A. Cornish-Bowden and D. G. Walker (1976) "The purification in high yield of rat hepatic glucokinase" *Biochemical Journal* **153**, 363–373 [Section 11.8]

H. T. Huang and C. Niemann (1951) "The kinetics of the α-chymotrypsin catalyzed hydrolysis of acetyl- and nicotinyl-L-tryptophanamide in aqueous solutions at 25° and pH 7.9" *Journal of the American Chemical Society* **73**, 1541–1548 [Section 2.9.4]

C. S. Hudson (1908) "The inversion of cane sugar by invertase" *Journal of the American Chemical Society* **30**, 1564–1583 [Section 3.2]

A. Hunter and C. E. Downs (1945) "The inhibition of arginase by amino acids" *Journal of Biological Chemistry* **157**, 427–446 [Sections 5.2.2, 5.3.2]

R. O. Hurst (1967) "A simplified approach to the use of determinants in the calculation of the rate equation for a complex enzyme system" *Canadian Journal of Biochemistry* **45**, 2015–2039 [Section 4.8]

R. O. Hurst (1969) "A computer program for writing the steady-state rate equation for a multisubstrate enzymic reaction" *Canadian Journal of Biochemistry* **47**, 941–944 [Section 4.8]

D. W. Ingles and J. R. Knowles (1967) "Specificity and stereospecificity of α-chymotrypsin" *Biochemical Journal* **104**, 369–377 [Section 5.9.1]

K. Inouye, I. M. Voynick, G. R. Delpierre and J. S. Fruton (1966) "New synthetic substrates for pepsin" *Biochemistry* **5**, 2473–2483 [Section 9.4.5]

International Union of Biochemistry (1982) "Symbolism and terminology in enzyme kinetics" *European Journal of Biochemistry* **128**, 281–291 [Sections 1.1, 2.3.2, 5.2.1, 7.3.1, 12.3.1]

International Union of Biochemistry and Molecular Biology (1992) *Enzyme Nomenclature*, Academic Press, Orlando [Sections 7.1, 7.2.1]

International Union of Pure and Applied Chemistry (1981) "Symbolism and terminology in chemical kinetics" *Pure and Applied Chemistry* **53**, 753–771 [Sections 5.2.1, 8.5.1, 12.3.1]

International Union of Pure and Applied Chemistry (1988) "Names for hydrogen atoms, ions, and groups, and for reactions involving them" *Pure and Applied Chemistry* **60**, 1115–1116 [Section 8.8 (footnote)]

International Union of Pure and Applied Chemistry (1993) *Quantities, Units and Symbols in Physical Chemistry*, Blackwell, Oxford [Section 2.3.3 (footnote)]

K. Ito, Y. Matsuura and N. Minamiura (1994) "Purification and characterization of fungal nuclease composed of heterogeneous subunits" *Archives of Biochemistry and Biophysics* **309**, 160–167 [Problem 5.4]

J. R. Jacobsen, J. R. Prudent, L. Kochersperger, S. Yonkovich and P. G. Schultz (1992) "An efficient antibody-catalyzed aminoacylation reaction" *Science* **256**, 365–367 [Section 2.10.4]

R. Jarabak and J. Westley (1974) "Enzymic memory: a consequence of conformational mobility" *Biochemistry* **13**, 3237–3239 [Section 7.4.1]

R. L. Jarvest, G. Lowe and B. V. L. Potter (1981) "Analysis of the chirality of $[^{16}O,^{17}O,^{18}O]$phosphate esters by ^{31}P nuclear magnetic resonance spectroscopy" *Journal of the Chemical Society Perkin Transactions 1*, 3186–3195 [Section 8.1]

W. P. Jencks (1969) *Catalysis in Chemistry and Enzymology*, McGraw-Hill, New York [Sections 1.6, 8.6.1, 9.6]

W. P. Jencks (1975) "Binding energy, specificity, and enzymic catalysis: the Circe effect" *Advances in Enzymology* **43**, 219–410 [Section 2.7.3]

R. R. Jennings and C. Niemann (1955) "The evaluation of the kinetic constants of enzyme-catalyzed reactions by procedures based upon integrated rate equations" *Journal of the American Chemical Society* **77**, 5432–5483 [Section 2.9.3]

G. Johansen and R. Lumry (1961) "Statistical analysis of enzymic steady-state rate data" *Comptes Rendus des Travaux du Laboratoire Carlsberg* **32**, 185–214 [Section 14.2.2]

R. A. John (2002) "Photometric assays" pages 49–78 in *Enzyme Assays* (2nd edition, edited by R. Eisenthal and M. J. Danson), Oxford University Press, Oxford [Section 3.1.1]

H. Kacser and J. A. Burns (1973) "The control of flux" *Symposia of the Society for Experimental Biology* **27**, 65–104, revised by H. Kacser, J. A. Burns and D. A. Fell (1995) *Biochemical Society Transactions* **23**, 341–366 [Chapter 12, Section 13.1.3]

W. E. Karsten, C.-C. Hwang and P. F. Cook (1999) "α-Secondary tritium kinetic isotope effects indicate hydrogen tunneling and coupled motion occur in the oxidation of L-malate by NAD-malic enzyme" *Biochemistry* **38**, 4398–4402 [Section 8.6.1]

M. Katz and J. Westley (1979) "Enzymic memory: steady-state kinetic and physical studies with ascorbate oxidase and aspartate aminotransferase" *Journal of Biological Chemistry* **254**, 9142–9147 [Section 7.4.1]

M. Katz and J. Westley (1980) "Enzymic memory studies with nucleoside-5'-diphosphate kinase" *Archives of Biochemistry and Biophysics* **204**, 464–470 [Section 7.4.1]

T. Keleti, R. Leoncini, R. Pagani and E. Marinello (1987) "A kinetic method for distinguishing whether an enzyme has one or two active sites for two different substrates: rat liver L-threonine dehydratase has a single active site for threonine and serine" *European Journal of Biochemistry* **170**, 179–183 [Section 5.6.2]

F. J. Kézdy, J. Jaz and A. Bruylants (1958) "Cinétique de l'action de l'acide nitreux sur les amides. I. Méthode générale" *Bulletin de la Société Chimique de Belgique* **67**, 687–706 [Problem 1.3]

A. R. Khan and M. N. G. James (1998) "Molecular mechanisms for the conversion of zymogens to active proteolytic enzymes" *Protein Science* **7**, 815–836 [Section 5.7.1]

H. C. Kiefer, W. I. Congdon, I. S. Scarpa and I. M. Klotz (1972) "Catalytic accelerations of 10^{12}-fold by an enzyme-like synthetic polymer" *Proceedings of the National Academy of Sciences U.S.A.* **69**, 2155–2159 [Section 2.10.2]

J. Kinderlerer and S. Ainsworth (1976) "A computer program to derive the rate equations of enzyme catalysed reactions with up to ten enzyme-containing intermediates in the reaction mechanism" *International Journal of Biomedical Computing* **7**, 1–20 [Section 4.8]

E. L. King and C. Altman (1956) "A schematic method of deriving the rate laws for enzyme-catalyzed reactions" *Journal of Physical Chemistry* **60**, 1375–1378 [Chapter 4, Sections 7.2.2, 13.4.1]

R. Kitz and I. B. Wilson (1962) "Esters of methanesulfonic acid as irreversible inhibitors of acetylcholinesterase" *Journal of Biological Chemistry* **237**, 3245–3249 [Sections 6.2.2, 6.4.2, Problem 6.2]

I. M. Klotz, F. M. Walker and R. B. Pivan (1946) "The binding of organic ions by proteins" *Journal of the American Chemical Society* **68**, 1486–1490 [Section 11.3.2]

A. A. Klyosov (1996) "Kinetics and specificity of human liver aldehyde dehydrogenases toward aliphatic, aromatic, and fused polycyclic aldehydes" *Biochemistry* **35**, 4457–4467 [Section 2.4 (footnote)]

F. Knoop (1904) *Der Abbau aromatischer Fettsäuren im Tierkörper*, Kuttruff, Freiburg [Section 8.2]

J. R. Knowles (1976) "The intrinsic pK_a-values of functional groups in enzymes: improper deductions from the pH-dependence of steady-state parameters" *Critical Reviews in Biochemistry* **4**, 165–173 [Section 9.6]

J. R. Knowles, R. S. Bayliss, A. J. Cornish-Bowden, P. Greenwell, T. M. Kitson, H. C. Sharp and G. B. Wybrandt (1970) "Towards a mechanism for pepsin" pages 237–250 in *Structure–Function Relationships of Proteolytic Enzymes* (edited by P. Desnuelle, H. Neurath and M. Ottesen), Munksgaard, Copenhagen [Section 9.2]

D. E. Koshland, Jr. (1954) "Group transfer as an enzymatic substitution mechanism" pages 608–641 in *A Symposium on the Mechanism of Enzyme Action* (edited by W. D. McElroy and B. Glass), Johns Hopkins Press, Baltimore [Section 7.2.3]

D. E. Koshland, Jr. (1955) "Isotope exchange criteria for enzyme mechanisms" *Discussions of the Faraday Society* **20**, 142–148 [Section 8.4]

D. E. Koshland, Jr. (1958) "Application of a theory of enzyme specificity to protein synthesis" *Proceedings of the National Academy of Sciences U.S.A.* **44**, 98–104 [Sections 7.2.1, 11.4]

D. E. Koshland, Jr. (1959a) "Mechanisms of transfer enzymes" pages 305–306 in *The Enzymes*, 2nd edition (edited by P. D. Boyer, H. Lardy and K. Myrbäck), volume 1, Academic Press, New York [Sections 7.2.1, 11.4]

D. E. Koshland, Jr. (1959b) "Enzyme flexibility and enzyme action" *Journal of Cellular and Comparative Physiology* **54**, supplement 1, 245–258 [Sections 7.2.1, 11.4]

D. E. Koshland, Jr. (2002) "The application and usefulness of the ratio k_{cat}/K_M" *Bioorganic Chemistry* **30**, 211–213 [Section 2.4]

D. E. Koshland, Jr., G. Némethy and D. Filmer (1966) "Comparison of experimental binding data and theoretical models in proteins containing subunits" *Biochemistry* **5**, 365–385 [Sections 11.6, 12.8]

D. E. Koshland, Jr., D. H. Strumeyer and W. J. Ray, Jr. (1962) "Amino acids involved in the action of chymotrypsin" *Brookhaven Symposia in Biology* **15**, 101–133 [Section 13.2.2]

E. G. Krebs and J. A. Beavo (1979) "Phosphorylation–dephosphorylation of enzymes" *Annual Review of Biochemistry* **48**, 923–959 [Section 12.9.2]

H. A. Krebs (1964) "Gluconeogenesis" *Proceedings of the Royal Society, Series B* **159**, 545–564 [Section 12.9.3]

B. I. Kurganov (1982) *Allosteric Enzymes: Kinetic Behaviour*, pages 151–248, Wiley-Interscience, Chichester [Section 11.7]

P. Kuzmic (1998) "Fixed-point methods for computing the equilibrium composition of complex biochemical mixtures" *Biochemical Journal* **331**, 571–575 [Section 3.4]

J. Kvassman and G. Pettersson (1976) "Kinetic transients in the reduction of aldehydes catalysed by liver alcohol dehydrogenase" *European Journal of Biochemistry* **69**, 279–287 [Section 13.4.2]

K. J. Laidler (1955) "Theory of the transient phase in kinetics, with special reference to enzyme systems" *Canadian Journal of Chemistry* **33**, 1614–1624 [Section 2.5]

K. J. Laidler (1993) *The World of Physical Chemistry*, pages 232–289, Oxford University Press, Oxford [Section 1.2.1]

K. J. Laidler (1998) *To Light such a Candle*, Oxford University Press, Oxford [Preface]

K. J. Laidler and P. S. Bunting (1973) *The Chemical Kinetics of Enzyme Action*, 2nd edition, pages 89–110 [Section 5.2.3], 163–195 [Section 2.9.4], 220–232 [Section 1.7.4], Clarendon Press, Oxford

K. J. Laidler, J. H. Meiser and B. C. Sanctuary (2002) *Physical Chemistry*, 4th edition, pages 819–826, Houghton Mifflin, Boston [Section 1.7.3]

I. Langmuir (1916) "The constitution and fundamental properties of solids and liquids. I. Solids" *Journal of the American Chemical Society* **38**, 2221–2295 [Section 2.2]

I. Langmuir (1918) "The adsorption of gases on plane surfaces of glass, mica and platinum" *Journal of the American Chemical Society* **40**, 1361–1403 [Section 2.2]

S. Lebreton and B. Gontero (1999) "Memory and imprinting in multienzyme complexes: evidence for information transfer from glyceraldehyde-3-phosphate dehydrogenase" *Journal of Biological Chemistry* **274**, 20879–20884 [Section 14.5]

X.-L. Li, X.-D. Lei, H. Cai, J. Li, S.-L. Yang, C.-C. Wang and C.-L. Tsou (1998) "Binding of a burst-phase intermediate formed in the folding of denatured D-glyceraldehyde-3-phosphate dehydrogenase by chaperonin 60 and 8-anilino-1-naphthalene-sulphonic acid" *Biochemical Journal* **331**, 505–511 [Section 6.4.2]

D. M. J. Lilley (2001) "Origins of RNA catalysis in the hairpin ribozyme" *ChemBioChem* **2**, 31–35 [Section 2.10.3]

C. Lindbladh, M. Rault, C. Haggelund, W. C. Small, K. Mosbach and L. Bülow (1994) "Preparation and kinetic characterization of a fusion protein of yeast citrate synthase and malate dehydrogenase" *Biochemistry* **33**, 11692–11698 [Section 12.9.1]

H. Lineweaver and D. Burk (1934) "The determination of enzyme dissociation constants" *Journal of the American Chemical Society* **56**, 658–666 [Sections 2.2, 2.6.2, 2.6.4]

H. Lineweaver, D. Burk and W. E. Deming (1934) "The dissociation constant of nitrogen-nitrogenase in *Azotobacter*" *Journal of the American Chemical Society* **56**, 225–230 [Section 2.6.2]

H. Ma and T. M. Penning (1999) "Conversion of mammalian 3α-hydroxysteroid dehydrogenase to 20α-hydroxysteroid dehydrogenase using loop chimeras: changing specificity from androgens to progestins" *Proceedings of the National Academy of Sciences U.S.A.* **96**, 11161–11166 [Section 2.10.2]

G. C. McBane (1998) "Chemistry from telephone numbers: the false isokinetic relationship" *Journal of Chemical Education* **75**, 919–922 [Section 10.4, Problem 10.2]

N. MacFarlane and S. Ainsworth (1972) "A kinetic study of Baker's-yeast pyruvate kinase activated by fructose 1,6-diphosphate" *Biochemical Journal* **129**, 1035–1047 [Section 5.7.1]

H. A. C. McKay (1938) "Kinetics of exchange reactions" *Nature* **142**, 997–998 [Section 8.2]

A. D. B. Malcolm and G. K. Radda (1970) "The reaction of glutamate dehydrogenase with 4-iodoacetamido salicylic acid" *European Journal of Biochemistry* **15**, 555–561 [Sections 6.2.2, 6.3]

B. Mannervik, I. Jakobson and M. Warholm (1986) "Error structure as a function of substrate and inhibitor concentration in enzyme kinetic experiments" *Biochemical Journal* **235**, 797–804 [Section 14.2.1]

L. A. Marquez, J. T. Huang and H. B. Dunford (1994) "Spectral and kinetic studies on the formation of myeloperoxidase compounds I and II: roles of hydrogen peroxide and superoxide" *Biochemistry* **33**, 1447–1454 [Section 13.3.4]

R. G. Martin (1963) "The first enzyme in histidine biosynthesis: the nature of the feedback inhibition by histidine" *Journal of Biological Chemistry* **238**, 257–268 [Section 11.5.1]

S. F. Martin and P. J. Hergenrother (1999) "Catalytic cycle of the phosphatidylcholine-preferring phospholipase C from *Bacillus cereus*: solvent viscosity, deuterium isotope effects, and proton inventory studies" *Biochemistry* **38**, 4403–4408 [Section 8.8]

J. J. Martínez-Irujo, M. L. Villahermosa, J. Mercapide, J. F. Cabodevilla and E. Santiago (1998) "Effect of two linear inhibitors on a single enzyme" *Biochemical Journal* **329**, 689–698 [Section 5.4]

A. M. Martins, P. Mendes, C. Cordeiro and A. Ponces Freire (2001a) "*In situ* kinetic analysis of glyoxalase I and glyoxalase II in *Saccharomyces cerevisiae*" *European Journal of Biochemistry* **268**, 3930–3936 [Sections 12.1.3, 14.5]

A. M. T. B. S. Martins, C. A. A. Cordeiro and A. M. J. Ponces Freire (2001b) "*In situ* analysis of methylglyoxal metabolism in *Saccharomyces cerevisiae*" *FEBS Letters* **499**, 41–44 [Section 12.1.3]

V. Massey, B. Curti and G. Ganther (1966) "A temperature-dependent conformational change in D-amino acid oxidase and its effect on catalysis" *Journal of Biological Chemistry* **241**, 2347–2357 [Section 10.3]

F. A. Matsen and J. L. Franklin (1950) "A general theory of coupled sets of first order reactions" *Journal of the American Chemical Society* **72**, 3337–3341 [Section 13.4.1]

P. Mendes (1997) "Biochemistry by numbers: simulation of biochemical pathways with Gepasi 3" *Trends in Biochemical Sciences* **22**, 361–363 [Section 12.10]

P. Mendes, D. B. Kell and H. V. Westerhoff (1992) "Channelling can decrease pool size" *European Journal of Biochemistry* **204**, 257–266 [Section 12.10]

S. Merry and H. G. Britton (1985) "The mechanism of rabbit muscle phosphofructokinase at pH 8" *Biochemical Journal* **226**, 13–28 [Section 8.5.1]

J. Messens, J. C. Martins, E. Brosens, K. Van Belle, D. M. Jacobs, R. Willem and L. Wyns (2002) "Kinetics and active site dynamics of *Staphylococcus aureus* arsenate reductase" *Journal of Biological Inorganic Chemistry* **7**, 146–156 [Section 3.2]

L. Michaelis (1926) *Hydrogen Ion Concentration*, translated by W. A. Perlzweig from the 2nd German edition (1921), volume 1, Baillière, Tindall and Cox, London [Section 9.3]

L. Michaelis (1958) "Leonor Michaelis" *Biographical Memoirs of the National Academy of Sciences* **31**, 282–321 [Section 9.1]

L. Michaelis and H. Davidsohn (1911) "Die Wirkung der Wasserstoffionen auf das Invertin" *Biochemisches Zeitschrift* **35**, 386–412; English translation in pages 264–286 of Boyde (1980) [Sections 3.2, 9.1]

L. Michaelis and M. L. Menten (1913) "Kinetik der Invertinwirkung" *Biochemisches Zeitschrift* **49**, 333–369; English translation in pages 289–316 of Boyde (1980) [Chapter 2, Section 6.2.2]

L. Michaelis and H. Pechstein (1914) "Über die verschiedenartige Natur der Hemmungen der Invertasewirkung" *Biochemisches Zeitschrift* **60**, 79–90 [Section 2.8]

L. Michaelis and P. Rona (1914) "Die Wirkungsbedingungen der Maltase aus Bierhefe. III. Über die Natur der verschiedenartigen Hemmungen der Fermentwirkungen" *Biochemisches Zeitschrift* **60**, 62–78 [Section 2.8]

G. A. Millikan (1936) "Photometric methods of measuring the velocity of rapid reactions. III. A portable micro-apparatus applicable to an extended range of reactions" *Proceedings of the Royal Society, Series A* **155**, 455–476 [Section 13.3.3]

O. Monasterio (2001) "Rate constants determined by nuclear magnetic resonance" *Methods, a Companion to Methods in Enzymology* **24**, 97–103 [Section 13.3.6]

O. Monasterio and M. L. Cárdenas (2003) "Kinetic studies of rat-liver hexokinase D ('glucokinase') in non-cooperative conditions show an ordered mechanism with MgADP as the last product to be released" *Biochemical Journal* **371**, 29–38 [Section 7.2.1]

J. Monod, J.-P. Changeux and F. Jacob (1963) "Allosteric proteins and cellular control systems" *Journal of Molecular Biology* **6**, 306–329 [Sections 11.1.4, 11.5.1, 12.8]

J. Monod, J. Wyman and J.-P. Changeux (1965) "On the nature of allosteric transitions: a plausible model" *Journal of Molecular Biology* **12**, 88–118 [Sections 11.5, 12.8]

F. P. Morpeth and V. Massey (1982) "Steady-state kinetic studies on D-lactate dehydrogenase from *Megasphera elsdenii*" *Biochemistry* **21**, 1307–1312 [Section 7.4.1]

J. F. Morrison (1969) "Kinetics of the reversible inhibition of enzyme-catalysed reactions by tight-binding inhibitors" *Biochimica et Biophysica Acta* **185**, 269–286 [Section 6.1]

S. H. Mudd and J. D. Mann (1963) "Activation of methionine for transmethylation. VII. Some energetic and kinetic aspects of the reaction catalyzed by the methionine-activating enzyme of bakers' yeast" *Journal of Biological Chemistry* **238**, 2164–2170 [Section 2.7.3]

P. J. Mulquiney and P. W. Kuchel (2003) *Modelling Metabolism with Mathematica*, CRC Press, Boca Raton [Section 12.10]

D. K. Myers (1952) "Studies on cholinesterase. 7. Determination of the molar concentration of pseudo-cholinesterase in serum" *Biochemical Journal* **51**, 303–311 [Section 6.1]

P. Nash, M. Barry, B. T. Seet, K. Veugelers, S. Hota, J. Heger, C. Hodgkinson, K. Graham, R. J. Jackson and G. McFadden (2000) "Post-translational modification of the myxoma-virus anti-inflammatory serpin SERP-1 by a virally encoded sialyltransferase" *Biochemical Journal* **347**, 375–382 [Section 14.1]

K. E. Neet and D. E. Koshland, Jr. (1966) "The conversion of serine at the active site of subtilisin to cysteine: a 'chemical mutation'" *Proceedings of the National Academy of Sciences U.S.A.* **56**, 1606–1611 [Section 2.10.2]

J. M. Nelson and R. S. Anderson (1926) "Glucose and fructose retardation of invertase action" *Journal of Biological Chemistry* **69**, 443–448 [Section 5.2.2]

N. O. Newell, D. W. Markby and H. K. Schachman (1989) "Cooperative binding of the bisubstrate analog N-(phosphonacetyl)-L-aspartate to aspartate transcarbamoylase and the heterotropic effects of ATP and CTP" *Journal of Biological Chemistry* **264**, 2476–2481 [Section 11.2.3]

L. W. Nichol, W. J. H. Jackson and D. J. Winzor (1967) "A theoretical study of the binding of small molecules to a polymerizing protein system: a model for allosteric effects" *Biochemistry* **6**, 2449–2456 [Section 11.7]

P. Niederberger, R. Prasad, G. Miozzari and H. Kacser (1992) "A strategy for increasing an *in vivo* flux by genetic manipulations: the tryptophan system of yeast" *Biochemical Journal* **287**, 473–479 [Problem 12.5]

H. Niemeyer, M. L. Cárdenas, E. Rabajille, T. Ureta, L. Clark-Turri and J. Peñaranda (1975) "Sigmoidal kinetics of glucokinase" *Enzyme* **20**, 321–333 [Section 11.8]

A. E. Nixon, S. M. Firestine, F. G. Salinas and S. J. Benkovic (1999) "Rational design of a scytalone dehydratase-like enzyme using a structurally homologous protein scaffold" *Proceedings of the National Academy of Sciences U.S.A.* **96**, 3568–3571 [Section 2.10.2]

R. Norris and K. Brocklehurst (1976) "A convenient method of preparation of high-activity urease from *Canavalia ensiformis* by covalent chromatography and an investigation of its thiol groups with 2,2'-dipyridyl disulphide as a thiol titrant and reactivity probe" *Biochemical Journal* **159**, 245–257 [Section 6.4.2, Problem 6.3]

R. G. W. Norrish and G. Porter (1949) "Chemical reactions produced by very high light intensities" *Nature* **164**, 658 [Section 13.3.5]

D. B. Northrop (1977) "Determining the absolute magnitude of hydrogen isotope effects" pages 122–152 in *Isotope Effects on Enzyme-Catalyzed Reactions* (edited by W. W. Cleland, M. O'Leary and D. B. Northrop), University Park Press, Baltimore [Section 8.7]

D. B. Northrop (1981) "Minimal kinetic mechanism and general equation for deuterium isotope effects on enzymic reactions: uncertainty of detecting a rate-limiting step" *Biochemistry* **20**, 4056–4061 [Section 13.1.3]

D. B. Northrop (2001) "Uses of isotope effects in the study of enzymes" *Methods, a Companion to Methods in Enzymology* **24**, 117–124 [Section 13.1.3]

J. W. Nyce and W. J. Metzger (1997) "DNA antisense therapy for asthma in an animal model" *Nature* **385**, 721–725; erratum in *Nature* **390**, 424 (1997) [Section 11.3.2]

J. M. Olavarría (1986) "A systematic method for the compilation of the King and Altman graphs of a steady-state enzyme model suitable for manual or computer use" *Journal of Theoretical Biology* **122**, 269–275 [Section 4.8]

C. O'Sullivan and F. W. Tompson (1890) "Invertase: a contribution to the history of an enzyme or unorganised ferment" *Journal of the Chemical Society (Transactions)* **57**, 834–931; reprinted in part in pages 208–216 of Friedmann (1981) [Sections 2.1, 3.2]

J. Ovádi (1991) "Physiological significance of metabolite channelling" *Journal of Theoretical Biology* **152**, 1–22 [Section 12.9.1]

J. Ovádi, J. Batke, F. Bartha and T. Keleti (1979) "Effect of association–dissociation on the catalytic properties of glyceraldehyde 3-phosphate dehydrogenase" *Archives of Biochemistry and Biophysics* **193**, 28–33 [Section 11.7]

E. Parzen (1980) "Comment" *American Statistician* **34**, 78–79 [Preface]

L. Pasteur (1860) "Mémoire sur la fermentation alcoolique" *Annales de Chimie, troisième Série* **58**, 323–426 [Section 2.1]

A. K. Paterson and J. R. Knowles (1972) "The number of catalytically essential carboxyl groups in pepsin: modification of the enzyme by trimethyloxonium fluoroborate" *European Journal of Biochemistry* **31**, 510–517 [Section 6.4.2]

L. Pauling (1935) "The oxygen equilibrium of hemoglobin and its structural interpretation" *Proceedings of the National Academy of Sciences U.S.A.* **21**, 186–191 [Sections 11.4, 11.6.3]

D. D. Perrin (1965) "Multiple equilibria in assemblages of metal ions and complexing species: a model for biological systems" *Nature* **206**, 170–171 [Section 3.4]

D. D. Perrin and I. G. Sayce (1967) "Computer calculation of equilibrium concentrations in mixtures of metal ions and complexing species" *Talanta* **14**, 833–842 [Section 3.4]

M. F. Perutz, M. G. Rossmann, A. F. Cullis, H. Muirhead, G. Will and A. C. T. North (1960) "Structure of haemoglobin: a three-dimensional Fourier synthesis at 5.5 Å resolution, obtained by X-ray analysis" *Nature* **185**, 416–422 [Section 11.4]

L. C. Petersen and H. Degn (1978) "Steady-state kinetics of laccase from *Rhus vernicifera*" *Biochimica et Biophysica Acta* **526**, 85–92 [Problem 7.2]

G. Pettersson (1976) "The transient-state kinetics of two-substrate enzyme systems operating by an ordered ternary-complex mechanism" *European Journal of Biochemistry* **69**, 273–278 [Section 13.4.2]

G. Pettersson (1978) "A generalized theoretical treatment of the transient-state kinetics of enzymic reaction systems far from equilibrium" *Acta Chemica Scandinavica* **B32**, 437–446 [Section 13.4.1]

H. Pettersson, P. Olsson, L. Bülow and G. Pettersson (2000) "Kinetics of the coupled reaction catalysed by a fusion protein of yeast mitochondrial malate dehydrogenase and citrate synthase" *European Journal of Biochemistry* **267**, 5041–5146 [Section 12.9.1]

V. W. Pike (1987) "Synthetic enzymes" pages 465–485 in *Biotechnology* (edited by H.-J. Rehm and G. Reed) volume 7a, Verlag-Chemie, Weinheim [Section 2.10.2]

K. M. Plowman (1972) *Enzyme Kinetics*, McGraw-Hill, New York [Section 4.7.1]

L. Polgar and M. L. Bender (1966) "A new enzyme containing a synthetically formed active site: thiol-subtilisin" *Journal of the American Chemical Society* **88**, 3153–3154 [Section 2.10.2]

S. J. Pollack, J. R. Atack, M. R. Knowles, G. McAllister, C. I. Ragan, R. Baker, S. R. Fletcher, L. I. Iversen and H. B. Broughton (1994) "Mechanism of inositol monophosphatase, the putative target of lithium therapy" *Proceedings of the National Academy of Sciences U.S.A.* **91**, 5766–5770 [Section 5.2.3]

D. Pollard-Knight and A. Cornish-Bowden (1984) "Solvent isotope effects on the glucokinase reaction: negative co-operativity and a large inverse isotope effect in ^2H_2O" *European Journal of Biochemistry* **141**, 157–163 [Section 8.8]

D. Pollard-Knight, B. V. L. Potter, P. M. Cullis, G. Lowe and A. Cornish-Bowden (1982) "The stereochemical course of phosphoryl transfer catalysed by glucokinase" *Biochemical Journal* **201**, 421–423 [Section 7.2.3]

F. C. V. Portaro, A. B. F. Santos, M. H. S. Cezari, M. A. Juliano, L. Juliano and E. Carmona (2000) "Interactions between subsites of papain and cathepsins B and L" *Biochemical Journal* **347**, 123–129 [Section 14.1]

G. Porter (1967) "Flash photolysis and some of its applications" pages 469–476 in *Proceedings of the 5th Nobel Symposium* (edited by S. Claesson), Interscience, New York [Section 13.3.4]

N. C. Price and L. Stevens (2002) "Techniques for enzyme extraction" pages 209–224 in *Enzyme Assays* (2nd edition, edited by R. Eisenthal and M. J. Danson), Oxford University Press, Oxford [Section 9.4.5]

E. Quéméneur, M. Moutiez, J.-B. Charbonnier and A. Ménez (1998) "Engineering cyclophilin into a proline-specific endopeptidase" *Nature* 391, 301–304 [Section 2.10.2]

B. R. Rabin (1967) "Cooperative effects in enzyme catalysis: a possible kinetic model based on substrate-induced conformation isomerization" *Biochemical Journal* 102, 22C–23C [Section 11.8]

A. Radzicka and R. Wolfenden (1995) "A proficient enzyme" *Science* 267, 90–93 [Section 2.4]

R. T. Raines and J. R. Knowles (1987) "Enzyme relaxation in the reaction catalyzed by triosephosphate isomerase: detection and kinetic characterization of two unliganded forms of the enzyme" *Biochemistry* 26, 7014–7020 [Section 8.5.3]

W. J. Ray, Jr. (1983) "Rate-limiting step: a quantitative definition; application to steady-state enzymic reactions" *Biochemistry* 22, 4625–4637 [Section 13.1.3]

W. J. Ray, Jr. and D. E. Koshland, Jr. (1961) "A method for characterizing the type and numbers of groups involved in enzyme action" *Journal of Biological Chemistry* 236, 1973–1979 [Section 6.4]

C. Reder (1988) "Metabolic control theory; a structural appproach" *Journal of Theoretical Biology* 135, 175–201 [Section 12.5]

R. E. Reeves and A. Sols (1973) "Regulation of *Escherichia coli* phosphofructokinase *in situ*" *Biochemical and Biophysical Research Communications* 50, 459–466 [Section 12.1.3]

J. M. Reiner (1969) *Behavior of Enzyme Systems*, 2nd edition, pages 85–89, Van Nostrand Reinhold, New York [Section 13.4.2]

Z. Ren, B. Perman, V. Srajer, T. Y. Teng, C. Pradervand, D. Bourgeois, F. Schotte, T. Ursby, R. Kort, M. Wulff and K. Moffat (2001) "A molecular movie at 1.8 Å resolution displays the photocycle of photoactive yellow protein, a eubacterial blue-light receptor, from nanoseconds to seconds" *Biochemistry* 40, 13788–13801 [Section 13.3.5]

D. G. Rhoads and M. Pring (1968) "The simulation and analysis by digital computer of biochemical systems in terms of kinetic models. IV. Automatic derivation of enzymic rate laws" *Journal of Theoretical Biology* 20, 297–313 [Section 4.8]

J. Ricard, J.-C. Meunier and J. Buc (1974) "Regulatory behavior of monomeric enzymes. I. The mnemonical enzyme concept" *European Journal of Biochemistry* 49, 195–208 [Section 11.8]

G. V. Richieri, P. J. Low, R. T. Ogata and A. M. Kleinfeld (1997) "Mutants of rat intestinal fatty acid-binding protein illustrate the critical role played by enthalpy–entropy compensation in ligand binding" *Journal of Biological Chemistry* 272, 16737–16740 [Section 10.4]

H. E. Rosenthal (1967) "A graphic method for the determination and presentation of binding parameters in a complex system" *Analytical Biochemistry* 20, 525–532 [Section 11.3.2]

F. B. Rudolph and H. J. Fromm (1971) "Use of isotope competition and alternative substrates for studying the kinetic mechanism of enzyme action. II. Rate equations for three substrate enzyme systems" *Archives of Biochemistry and Biophysics* 147, 515–526 [Section 4.8]

B. Sacktor and E. C. Hurlbut (1966) "Regulation of metabolism in working muscle *in vivo*. II. Concentrations of adenine nucleotides, arginine phosphate, and inorganic phosphate in insect flight muscle during flight" *Journal of Biological Chemistry* 241, 632–635 [Section 12.9.3]

B. Sacktor and E. Wormser-Shavit (1966) "Regulation of metabolism in working muscle *in vivo*. I. Concentrations of some glycolytic, tricarboxylic acid cycle, and amino acid intermediates in insect flight muscle during flight" *Journal of Biological Chemistry* 241, 624–631 [Section 12.9.3]

H. M. Sauro (1990) "Regulatory responses and control analysis: assessment of the relative importance of internal effectors" pages 225–230 in *Control of Metabolic Processes* (edited by A. Cornish-Bowden and M. L. Cárdenas), Plenum Press, New York [Section 12.8]

H. M. Sauro (1993) "SCAMP: a general-purpose simulator and metabolic control analysis program" *Computer Applications in the Biosciences* **9**, 441–450 [Section 12.10]

H. M. Sauro (2000) "JARNAC: a system for interactive metabolic analysis" pages 221–228 in *Animating the Cellular Map* (edited by J.-H. S. Hofmeyr, J. M. Rohwer and J. L. Snoep), Stellenbosch University Press, Stellenbosch [Section 12.10]

M. A. Savageau (1976) *Biochemical Systems Analysis: a Study of Function and Design in Molecular Biology*, Addison-Wesley, Reading, Massachusetts [Section 12.3.1]

G. Scatchard (1949) "The attractions of proteins for small molecules and ions" *Annals of the New York Academy of Sciences* **51**, 660–672 [Section 11.3.2]

R. J. Scheibe and J. A. Wagner (1992) "Retinoic acid regulates both expression of the nerve growth factor receptor and sensitivity to nerve growth factor" *Journal of Biological Chemistry* **267**, 17611–17616 [Section 11.3.2]

M. R. Schiller, L. D. Holmes and E. A. Boeker (1996) "Analysis of wild-type and mutant aspartate aminotransferases using integrated rate equations" *Biochimica et Biophysica Acta* **1297**, 17–27 [Section 2.9.4]

G. R. Schonbaum, B. Zerner and M. L. Bender (1961) "The spectrophotometric determination of the operational normality of an α-chymotrypsin solution" *Journal of Biological Chemistry* **236**, 2930–2935 [Section 13.2.2]

F. Schønheyder (1952) "Kinetics of 'acid' phosphatase action" *Biochemical Journal* **50**, 378–384 [Sections 2.9.3, 2.9.4]

R. L. Schowen (1972) "Mechanistic deductions from solvent isotope effects" *Progress in Physical Organic Chemistry* **9**, 275–332 [Section 8.8]

A. R. Schulz (1994) *Enzyme Kinetics: from Diastase to Multi-enzyme Systems*, Cambridge University Press, Cambridge [Section 4.7.1]

H. L. Segal, J. F. Kachmar and P. D. Boyer (1952) "Kinetic analysis of enzyme reactions. I. Further considerations of enzyme inhibition and analysis of enzyme activation" *Enzymologia* **15**, 187–198 [Section 7.3.1]

I. H. Segel (1975) *Enzyme Kinetics*, Wiley–Interscience, New York [Sections 4.7.1, 7.1]

I. H. Segel and R. L. Martin (1988) "The general modifier ('allosteric') unireactant enzyme mechanism: redundant conditions for reduction of the steady state velocity equation to one that is first degree in substrate and effector" *Journal of Theoretical Biology* **135**, 445–453 [Section 4.6]

M. J. Selwyn (1965) "A simple test for inactivation of an enzyme during assay" *Biochimica et Biophysica Acta* **105**, 193–195 [Section 3.2]

M. J. Selwyn (1993) "Application of the principle of microscopic reversibility to the steady-state rate equation for a general mechanism for an enzyme reaction with substrate and modifier" *Biochemical Journal* **295**, 897–898 [Section 4.6 (footnote)]

R. Serrano, J. M. Gancedo and C. Gancedo (1973) "Assay of yeast enzymes *in situ*: a potential tool in regulation studies" *European Journal of Biochemistry* **34**, 479–482 [Section 12.1.3]

K. Sharp (2001) "Entropy–enthalpy compensation: fact or artifact?" *Protein Science* **10**, 661–667 [Section 10.4]

K. Shatalin, S. Lebreton, M. Rault-Leonardon, C. Vélot and P. A. Srere (1999) "Electrostatic channeling of oxaloacetate in a fusion protein of porcine citrate synthase and porcine mitochondrial malate dehydrogenase" *Biochemistry* **38**, 881–889 [Section 12.9.1]

K. R. Sheu, J. P. Richard and P. A. Frey (1979) "Stereochemical courses of nucleotidyltransferase and phosphotransferase action: uridine diphosphate glucose pyrophosphorylase, galactose-1-phosphate uridylyltransferase, adenylate kinase, and nucleoside diphosphate kinase" *Biochemistry* **18**, 5548–5556 [Section 7.2.3]

J. R. Silvius, B. D. Read and R. N. McElhaney (1978) "Membrane enzymes: artifacts in Arrhenius plots due to temperature dependence of substrate-binding affinity" *Science* **199**, 902–904 [Section 10.3]

M. L. Sinnott and I. J. L. Souchard (1973) "The mechanism of action of β-galactosidase: effect of aglycone nature and α-deuterium substitution on the hydrolysis of aryl galactosides" *Biochemical Journal* **133**, 89–98 [Section 8.6.2]

Z. Songyang, K. L. Carraway III, M. J. Eck, S. C. Harrison, R. A. Feldman, M. Mohammadi, J. Schlessinger, S. R. Hubbard, D. P. Smith, C. Eng, M. L. Lorenzo, B. A. J. Ponder, B. J. Mayer and L. C. Cantley (1995) "Catalytic specificity of protein-tyrosine kinases is critical for selective signalling" *Nature* **373**, 536–539 [Problem 5.8]

S. P. L. Sørensen (1909) "Études enzymatiques. II. Sur la mesure et l'importance de la concentration des ions hydrogène dans les réactions enzymatiques" *Comptes Rendus des Travaux du Laboratoire Carlsberg* **8**, 1–168; partial English translation in pages 272–283 of Friedmann (1981) [Sections 2.2, 9.1]

L. B. Spector (1980) "Acetate kinase: a triple-displacement enzyme" *Proceedings of the National Academy of Sciences U.S.A.* **77**, 2626–2630 [Section 7.2.3]

L. B. Spector (1982) *Covalent Catalysis by Enzymes*, Springer-Verlag, New York [Section 7.2.3]

V. Srajer, Z. Ren, T. Y. Teng, M. Schmidt, T. Ursby, D. Bourgeois, C. Pradervand, W. Schildkamp, M. Wulff and K. Moffat (2001) "Protein conformational relaxation and ligand migration in myoglobin: a nanosecond to millisecond molecular movie from time-resolved Laue X-ray diffraction" *Biochemistry* **40**, 13802–13815 [Section 13.3.5]

H. Stabile, B. Curti and M. A. Vanoni (2000) "Functional properties of recombinant *Azospirillum brasilense* glutamate synthase, a complex iron-sulfur flavoprotein" *European Journal of Biochemistry* **267**, 2720–2730 [Section 14.1]

E. R. Stadtman (1970) "Mechanisms of enzyme regulation in metabolism" pages 397–459 in *The Enzymes* (3rd edition, edited by P. D. Boyer), volume 1, Academic Press, New York [Section 12.8]

E. R. Stadtman and P. B. Chock (1977) "Superiority of interconvertible enzyme cascades in metabolic regulation: analysis of monocyclic systems" *Proceedings of the National Academy of Sciences U.S.A.* **74**, 2761–2766 [Section 12.9.2]

E. R. Stadtman and P. B. Chock (1978) "Interconvertible enzyme cascades in metabolic regulation" *Current Topics in Cellular Regulation* **13**, 53–95 [Section 12.9.2]

J. Steinhardt and J. A. Reynolds (1969) *Multiple Equilibria in Proteins*, pages 176–213, Academic Press, New York [Section 9.2]

B. L. Stoddard (2001) "Trapping reaction intermediates in macromolecular crystals for structural analyses" *Methods, a Companion to Methods in Enzymology* **24**, 125–138 [Section 13.3.5]

B. L. Stoddard, B. E. Cohen, M. Brubaker, A. D. Mesecar and D. E. Koshland, Jr. (1998) "Millisecond Laue structures of an enzyme–product complex using photocaged substrate analogs" *Nature Structural Biology* **5**, 891–897 [Section 13.3.5]

A. C. Storer and A. Cornish-Bowden (1974) "The kinetics of coupled enzyme reactions" *Biochemical Journal* **141**, 205–209 [Section 3.1.4]

A. C. Storer and A. Cornish-Bowden (1976a) "Concentration of MgATP^{2-} and other ions in solution" *Biochemical Journal* **159**, 1–5 [Section 3.4]

A. C. Storer and A. Cornish-Bowden (1976b) "Kinetics of rat-liver glucokinase: co-operative interactions with glucose at physiologically significant concentrations" *Biochemical Journal* **159**, 7–14 [Section 11.8]

A. C. Storer and A. Cornish-Bowden (1977) "Kinetic evidence for a 'mnemonical' mechanism for rat liver glucokinase" *Biochemical Journal* **165**, 61–69 [Sections 5.3.1, 5.7.1, 7.2.1, 11.8]

A. C. Storer, M. G. Darlison and A. Cornish-Bowden (1975) "The nature of experimental error in enzyme kinetic measurements" *Biochemical Journal* **151**, 361–367 [Section 14.2.1]

"Student" (W. S. Gosset) (1908) "The probable error of a mean" *Biometrika* **6**, 1–25 [Section 14.4]

D. Su and J. F. Robyt (1994) "Determination of the number of sucrose and acceptor binding sites for *Leuconostoc mesenteroides* B-512FM dextransucrase, and the confirmation of the two-site mechanism for dextran synthesis" *Archives of Biochemistry and Biophysics* **308**, 471–476 [Section 11.3.2]

C. G. Swain, E. C. Stivers, J. F. Reuwer, Jr. and L. J. Schaad (1958) "Use of hydrogen isotope effects to identify the attacking nucleophile in the enolization of ketones catalyzed by acetic acid" *Journal of the American Chemical Society* **80**, 5885–5893 [Section 8.6.1]

J. R. Sweeny and J. R. Fisher (1968) "An alternative to allosterism and cooperativity in the interpretation of enzyme kinetic data" *Biochemistry* **7**, 561–565 [Section 7.2.2]

E. S. Swinbourne (1960) "Method for obtaining the rate coefficient and final concentration of a first-order reaction" *Journal of the Chemical Society* 2371–2372 [Problem 1.3]

E. Szathmáry and J. Maynard-Smith (1993) "The evolution of chromosomes. II. Molecular mechanisms" *Journal of Theoretical Biology* **164**, 447–454 [Section 2.10.1]

K. Taketa and B. M. Pogell (1965) "Allosteric inhibition of rat liver fructose 1,6-diphosphatase by adenosine 5'-monophosphate" *Journal of Biological Chemistry* **240**, 651–662 [Section 11.2.2]

J. W. Teipel, G. M. Hass and R. L. Hill (1968) "The substrate specificity of fumarase" *Journal of Biological Chemistry* **243**, 5684–5694 [Section 2.4]

T. M. Thomas and R. K. Scopes (1998) "The effects of temperature on the kinetics and stability of mesophilic and thermophilic 3-phosphoglycerate kinases" *Biochemical Journal* **330**, 1087–1095 [Section 10.2]

D. Thusius (1973) "Simultaneous determination of equilibrium constants and thermodynamic functions by means of relaxation amplitude measurements" *Biochimie* **55**, 277–282 [Section 13.4.1]

K. F. Tipton and H. B. F. Dixon (1979) "Effects of pH on enzymes" *Methods in Enzymology* **63**, 183–234 [Section 9.6]

C. M. Topham and K. Brocklehurst (1992) "In defence of the general validity of the Cha method of deriving rate equations: the importance of explicit recognition of the thermodynamic box in enzyme kinetics" *Biochemical Journal* **282**, 261–265 [Section 4.6]

C. M. Topham, S. Gul, M. Resmini, S. Sonkaria, G. Gallacher and K. Brocklehurst (2000) "The kinetic basis of a general method for the investigation of active site content of enzymes and catalytic antibodies: first-order behaviour under single-turnover and cycling conditions" *Journal of Theoretical Biology* **204**, 239–256 [Section 13.4.2]

C. Tsou (1962) "Relation between modification of functional groups of proteins and their biological activity. I. A graphical method for the determination of the number and type of essential groups" *Scientia Sinica* **11**, 1536–1538, first published in Chinese in *Acta Biochimica et Biophysica Sinica* **2**, 203–217 (1962) [Section 6.4]

E. R. Tufte (1983) *The Visual Display of Quantitative Information*, Graphics Press, Cheshire, Connecticut [Section 14.5]

E. R. Tufte (1990) *Envisioning Information*, Graphics Press, Cheshire, Connecticut [Section 14.5]

E. R. Tufte (1997) *Visual Explanations: Images and Quantities, Evidence and Narrative*, Graphics Press, Cheshire, Connecticut [Section 14.5]

P. M. Turner, K. M. Lerea and F. J. Kull (1983) "The ribonuclease inhibitors from porcine thyroid and liver are slow, tight-binding inhibitors of bovine pancreatic ribonuclease A" *Biochemical and Biophysical Research Communications* **114**, 1154–1160 [Section 6.1]

H. E. Umbarger (1956) "Evidence for a negative-feedback mechanism in the biosynthesis of isoleucine" *Science* **123**, 848 [Section 12.8]

A. Vandercammen and E. Van Schaftingen (1991) "Competitive inhibition of liver glucokinase by its regulatory protein" *European Journal of Biochemistry* **200**, 545–551 [Problem 5.7]

D. D. Van Slyke and G. E. Cullen (1914) "The mode of action of urease and of enzymes in general" *Journal of Biological Chemistry* **19**, 141–180 [Sections 2.2, 13.1.3]

J. H. van't Hoff (1884) *Études de Dynamique Chimique*, pages 114–118, Muller, Amsterdam [Section 1.7.1]

R. Varón, M. García-Moreno, C. Garrido and F. García-Carmona (1992) "The steady-state rate equation for the general modifier mechanism of Botts and Morales when the quasi-equilibrium assumption for the binding of modifier is made" *Biochemical Journal* **288**, 1072–1073 [Section 4.6 (footnote)]

E. O. Voit and A. E. Ferreira (2000) *Computational Analysis of Biochemical Systems*, Cambridge University Press, Cambridge [Section 12.10]

M. V. Volkenstein and B. N. Goldstein (1966) "A new method for solving the problems of the stationary kinetics of enzymological reactions" *Biochimica et Biophysica Acta* **115**, 471–477 [Section 4.5]

G. B. Warren and K. F. Tipton (1974a) "Pig liver pyruvate carboxylase: the reaction pathway for the carboxylation of pyruvate" *Biochemical Journal* **139**, 311–320 [Section 7.2.1]

G. B. Warren and K. F. Tipton (1974b) "Pig liver pyruvate carboxylase: the reaction pathway for the decarboxylation of oxaloacetate" *Biochemical Journal* **139**, 321–329 [Section 7.2.1]

H. Watari and Y. Isogai (1976) "A new plot for allosteric phenomena" *Biochemical and Biophysical Research Communications* **69**, 15–18 [Problem 11.1]

G. Weber and S. R. Anderson (1965) "Multiplicity of binding: range of validity and practical test of Adair's equation" *Biochemistry* **4**, 1942–1947 [Sections 4.6, 11.6.3]

J. A. Wells, W. J. Fairbrother, J. Otlewski, M. Laskowski, Jr. and J. Burnier (1994) "A reinvestigation of a synthetic peptide (TrPepz) designed to mimic trypsin" *Proceedings of the National Academy of Sciences U.S.A.* **91**, 4110–4114 [Problem 2.8]

P. Wentworth, Jr. (2002) "Antibody design by man and nature" *Science* **296**, 2247–2249 [Sections 2.10.1, 2.10.4]

H. V. Westerhoff and Y.-D. Chen (1984) "How do enzyme activities control metabolite concentrations? An additional theorem in the theory of metabolic control" *European Journal of Biochemistry* **142**, 425–430 [Section 12.6.1]

J. Westley (1969) *Enzymic Catalysis*, page 21, Harper and Row, New York [Problem 2.3]

E. Whitehead (1970) "The regulation of enzyme activity and allosteric transition" *Progress in Biophysics and Molecular Biology* **21**, 321–397 [Section 11.8]

E. P. Whitehead (1978) "Co-operativity and the methods of plotting binding and steady-state kinetic data" *Biochemical Journal* **171**, 501–504 [Section 11.2.5]

L. F. Wilhelmy (1850) "Über das Gesetz, nach welchem die Einwirkung der Säuren auf Rohrzucker stattfinden" *Poggendorff's Annalen der Physik und Chemie* **81**, 413–433, 499–526 [Section 1.2.1]

G. N. Wilkinson (1961) "Statistical estimations in enzyme kinetics" *Biochemical Journal* **80**, 324–332 [Section 14.2.3, 14.4]

J. T. Wong (1975) *Kinetics of Enzyme Mechanisms*, pages 10–13 [Section 2.5]; 19–21 [Section 4.7.1], Academic Press, London

J. T. Wong and C. S. Hanes (1962) "Kinetic formulations for enzymic reactions involving two substrates" *Canadian Journal of Biochemistry and Physiology* **40**, 763–804 [Sections 4.4, 7.2.1]

J. T. Wong and C. S. Hanes (1969) "A novel criterion for enzymic mechanisms involving three substrates" *Archives of Biochemistry and Biophysics* **135**, 50–59 [Section 7.8]

B. Woolf (1929) "Some enzymes of *B. coli communis* which act on fumaric acid" *Biochemical Journal* **23**, 472–482 [Section 7.2.1]

B. Woolf (1931) "The addition compound theory of enzyme action" *Biochemical Journal* **25**, 342–348 [Section 7.2.1]

B. Woolf (1932) cited by J. B. S. Haldane and K. G. Stern in pages 119–120 of *Allgemeine Chemie der Enzyme*, Steinkopff, Dresden and Leipzig; reprinted in pages 269–270 of Friedmann (1981) [Section 2.6.4]

A. Wurtz (1880) "Sur la papaïne: nouvelle contribution à l'histoire des ferments solubles" *Comptes Rendus Hebdomadaires des Séances de l'Académie des Sciences, Paris* **91**, 787–791 [Section 2.1]

K. Yagi and T. Ozawa (1960) "Complex formation of apo-enzyme, coenzyme and substrate of D-amino acid oxidase. I. Kinetic analysis using inhibitors" *Biochimica et Biophysica Acta* **42**, 381–387 [Section 5.4]

C. Yanofsky (1989) "A second reaction catalyzed by the tryptophan synthetase of *Escherichia coli*" *Biochimica et Biophysica Acta* **1000**, 133–137 [Section 12.9.1]

R. A. Yates and A. B. Pardee (1956) "Control of pyrimidine biosynthesis in *Escherichia coli* by a feed-back mechanism" *Journal of Biological Chemistry* **221**, 757–770 [Section 12.8]

T. Yonetani and H. Theorell (1964) "Studies on liver alcohol dehydrogenase complexes. III. Multiple inhibition kinetics in the presence of two competitive inhibitors" *Archives of Biochemistry and Biophysics* **106**, 243–251 [Section 5.4]

A. J. Zaug and T. R. Cech (1986) "The intervening sequence RNA of *Tetrahymena* is an enzyme" *Science* **231**, 470–471 [Section 2.10.3]

B. Zerner, R. P. M. Bond and M. L. Bender (1964) "Kinetic evidence for the formation of acyl-enzyme intermediates in the α-chymotrypsin-catalyzed hydrolysis of specific substrates" *Journal of the American Chemical Society* **86**, 3674–3679 [Section 7.4.1]

Solutions and Notes to Problems

1.1   Order 1/2 with respect to A, order 1 with respect to B. The value of 1/2 can be rationalized by supposing that A exists predominantly as a dimeric molecule but the actual reactant is the corresponding monomer $A_{0.5}$. In this case $[A_{0.5}]^2 = K[A]$, where K is an equilibrium constant, and so a rate that is proportional to $[A_{0.5}]$ will be proportional to $[A]^{0.5}$, and the total A concentration, strictly $[A] + 0.5[A_{0.5}]$ or $[A] + 0.5(K[A])^{0.5}$, will be approximately the same as $[A]$ if K is small.

1.2   (a) The slope and intercept are both inconsistent; (b) consistent; (c) slope consistent, intercept inconsistent. In all these examples, and, more important, in all similar cases, remember that although "inconsistent" certainly implies "false", "consistent" does not necessarily mean "true".

1.3   Slope $= \exp[-(k_1 + k_{-1})\tau]$; point of intersection $= (p_\infty, p_\infty)$.

1.4   It implies that they are typically about $50\,kJ\,mol^{-1}$.

1.5 $0.0033\,K^{-1}$.

1.6   If the actual reacting species in these reactions is a dimer N_2O_2 of NO that exists in equilibrium with NO at low concentrations, then its concentration is proportional to $[NO]^2$, so in a reaction with O_2, for example, the rate may be expected to be proportional to $[NO]^2[O_2]$. If the equilibrium between the two forms of NO varies with temperature such that the proportion of dimer decreases steeply with temperature, then this decrease in the concentration of the reacting species can be more than sufficient to compensate for the expected increase with temperature of the rate constant.

2.1   (a) $K_m/9$; (b) $9K_m$; (c) 81.

2.2   The rate equation is $v = kKe_0a/(K + a)$; hence $V = kKe_0$, $K_m = K$.

2.3   If the derivation is unclear, see Westley (1969) or Cornish-Bowden (1976c). The expression for A is $-VK_m$. The value $x = 0$ implies a vertical asymptote given by $a + K_m = 0$, or $a = -K_m$, and $y = 0$ similarly implies a horizontal asymptote given by $v = V$. These are illustrated in Figure 2.2.

2.4   The relative rate for a positive error of 10% is $1.1(1 + a/K_m)/(1 + 1.1a/K_m)$, which is 1.01 for $a/K_m = 8.18$. For a negative error the relative rate is $0.9(1 + a/K_m)/(1 + 0.9a/K_m)$, which is 0.99 for $a/K_m = 10.0$. So to give satisfactory results regardless of the sign of the error a/K_m must be at least 10.

2.5   Equilibrium constant $= 0.25 \times 3.8/(0.11 \times 1.7) = 5.1$. The second experiment gives 1.5 for the same equilibrium constant. A change of enzyme could not by itself account for this difference. Therefore, either the reported values are unreliable or the experimental conditions were different in some unspecified way (for example, the experiment was done at a different temperature or pH).

2.6   The more accurate methods should give values close to $K_m = 10.6\,mM$, $V = 1.24\,mM\,min^{-1}$.

2.7   (b) $V^{app} = k_{cat}e_0/(1 - K_{mA}/K_{sP})$, $K_m^{app} = K_{mA}(1 + a_0/K_{sP})/(1 - K_{mA}/K_{sP})$; (c) When K_{sP} is less than K_{mA}.

2.8   None of the k_{cat} or k_{cat}/K_m values are large compared with those for many enzyme-catalysed reactions with specific substrates, and both of the two substrates used are unnatural substances not found in nature, so there is no possibility that either enzyme is optimized to act on them. The results do not therefore allow any meaningful comparison between the catalytic power of the "pepzymes" and that of real enzymes. Moreover, as hydrolysis of these compounds has no obvious industrial or medical use there is no basis for drawing conclusions about possible applications.

2.9   The plot of v against $\log a$ is the only one of those mentioned in Chapter 2 that would allow all three lines to be easily distinguished. In a plot of a/v against a, for example, the line for the isoenzyme with largest K_m could only be accommodated by choosing a scale that would make the line for the isoenzyme with smallest K_m almost coincident with the horizontal axis.

3.1   For [glucose 6-phosphate] = 0.1 mM and $v_1 = 0.1\,\text{mM min}^{-1}$, the rate condition is $0.1 \leq 0.1 V_2/(0.11 + 0.1)$, hence $V_2 \geq 0.21\,\text{mM min}^{-1}$.

3.2   Although some of the details will vary on account of formation of species such as Mg_2ATP, the general principles for handling binary complexes do not depend on which component is held in constant excess over the other, so maintaining the total ATP concentration 5 mM in excess over the total $MgCl_2$ concentration would be about as effective as the inverse design for controlling the $MgATP^{2-}$ concentration, but it would allow effects due to $MgATP^{2-}$ and to ATP^{4-} to be distinguished.

3.3   Selwyn plots of the two time courses are not even approximately superimposable, the initial rate of increase for p_b being about half that of p_a, instead of about double, as one would expect for double the amount of enzyme. The results could be explained by a polymerizing enzyme for which the more highly associated species was less active.

4.1   The right-hand side of equation 4.2 is a constant, so its time derivative is necessarily zero. If the time derivatives of all except one of the concentrations on the left-hand side are defined as zero, the exception must be zero as well, as otherwise the sum could not be zero. So defining the last one as zero adds no information to what is already known from the others.

4.2   The full equation in coefficient form is given as equation 7.7 (Section 7.3.3) and the abbreviated form ignoring terms in product concentrations is given as equation 7.18 (Section 7.4.5). Unlike equation 4.12, these equations contain no constant term in the denominator.

4.3   The second and third denominator terms in equation 7.7, the terms in b and p respectively, must be multiplied by $(1 + b/K_{siB})$.

4.4   If the rate is defined as $-da/dt$ it includes molecules of A transformed into P′ as well as ones transformed into P; if it is defined as dp/dt it includes only molecules transformed into P.

4.5   The conclusion follows because every valid King–Altman pattern contains an arrow out of every enzyme form except one. If two enzyme forms that have reactant concentrations associated with all of the arrows that lead out of them, at least one of these enzyme forms must appear in any King–Altman pattern, so at least one concentration must be associated with every pattern.

5.1   Under the original hypothesis the rate could be written as $v = Va_L/(2 + a_L)$, but this must be revised to read $v = V'a_L/[K_m(1 + a_D/K_{ic}) + a_L)]$. For $a_D = a_L$ and $K_{ic} = K_m$ this is $v = V'a_L/(K_m + 2a_L)$. To give the correct limit of V when a_L is very large, V' must be $2V$, so $Va_L/(2 + a_L) = 2Va_L/(K_m + 2a_L)$, hence $K_m = 4\,\text{mM}$.

5.2 $K_{ic} = 4.9\,\text{mM}$, $K_{iu} \gg 5\,\text{mM}$. The data do not allow a definite distinction between pure competitive and mixed inhibition, because the highest substrate concentration is less than K_m, and thus too small to provide much information about K_{iu}. In addition, the inhibitor concentrations do not extend to a high enough value to define either inhibition constant accurately.

5.3   (a) The double-reciprocal plot; (b) ordinate; (c) as negative intercepts on the abscissa.

5.4   A–C (because it gives the largest value of V/K_m).

5.5   (a) L-Ala–L-Ala–L-Ala (because it has the largest value of k_{cat}/K_m); (b) yes (compare Table 2.1 and the discussion of it in Section 5.6.1).

5.6   If $a < K_m$, then $K_m(1 + i/K_{ic})$ is larger than $a(1 + i/K_{iu})$ for any ratio of inhibitor concentration to inhibition constant. Hence in these conditions a competitive inhibitor increases the denominator of the rate expression more than an uncompetitive inhibitor at the same concentration, that is to say it decreases the rate more. The reverse applies when $a > K_m$. A competitive inhibitor exerts its effect by binding to the enzyme form that predominates at low substrate concentrations, an uncompetitive inhibitor by binding to the form that predominates at high substrate concentrations.

5.7   The logic of the analysis is essentially the same as for the Dixon plot (Section 5.3), but the result is the opposite, as the lines intersect where $i = -K_{iu}$, not $i = -K_{ic}$. The results for hexokinase D imply that N-acetylglucosamine and glucosamine bind at the same site and that only one can be bound at a time, so K'_j is infinite. By contrast, the value of K'_j for the regulatory protein is the same as or similar to the inhibition constant K_j, implying that it binds at a different site from that for N-acetylglucosamine and that neither inhibitor affects the binding of the other. In the terminology of Chou and Talalay, N-acetylglucosamine and glucosamine bind exclusively, and N-acetylglucosamine and the regulatory protein inhibit synergistically.

5.8   (a) 17 possibilities at each of 8 positions gives $17^8 \approx 7 \times 10^9$ sequences. (b) $1\,\text{mg} = 10^{-3}/1600 = 6 \times 10^{-7}\,\text{mol}$ without regard to sequence. Thus each of 7×10^9 species is represented on average by $8.5 \times 10^{-17}\,\text{mol}$. Multiplying by the Avogadro constant 6×10^{23} gives about 5×10^7 molecules — ample for a representative sampling even if the 17 possibilities at each locus are not evenly and independently distributed. (c) $(8.5 \times 10^{-17}\,\text{mol})/(300 \times 10^{-6}\,\text{l}) = 2.8 \times 10^{-13}\,\text{M}$. The extreme smallness of this concentration in relation to a typical K_m is irrelevant as specificity is not determined by K_m; as the experiment required the enzymes to choose between a (large) variety of substrates simultaneously available it offered an excellent test of specificity.

6.1   (a) 0.667 (from equation 5.1); (b) 0.732 (from equation 6.4). Equation 5.1 does not apply in the conditions of problem (b), and will not give the right result. Equation 6.4 is valid in both cases, but needs to be applied with care if it is to give the correct result in problem (a): it is easy to show that $i_{0.5}^{app} = 1\,\text{nM}$, but less easy to obtain the correct result after making the substitutions in equation 6.4, because one is tempted to neglect small values prematurely. To avoid this error, while at the same time avoiding the need to do the arithmetic with 12 significant figures, use the approximate expression $(1 - x)^{0.5} \approx 1 - 0.5x$, which is valid for very small x.

6.2   If $K_i = (k_{-1} + k_2)/k_1$, and if this expression is very different from the equilibrium constant k_{-1}/k_1, as claimed by Childs and Bardsley (1975), then k_{-1} must be small compared with k_2. So k_1 is about $5 \times 10^{-3}/10^{-4}$, or $50\,\text{M}^{-1}\,\text{s}^{-1}$. This would make k_1 smaller by a factor of about 10^4 than typical values for *on* rate constants observed with numerous enzymes: not impossible, but not likely either, so it suggests that the original analysis of Kitz and Wilson (1962) was valid. For more detailed discussion, see Cornish-Bowden (1979).

6.3 (a) 1; (b) $24/6 = 4$.

7.1 Simple bimolecular nucleophilic substitution reactions (called S_N2 reactions in textbooks on organic reaction mechanisms) commonly proceed with inversion of configuration at the substituted atom. Retention, as with α-amylase, can occur as a consequence of two successive substitutions, as in a substituted-enzyme or double-displacement mechanism. Net inversion, as with β-amylase, suggests an odd number of inversions, as in a ternary-complex or single-displacement mechanism. (The stereochemical data do not exclude a triple-displacement mechanism: see Section 7.2.3.)

7.2 The mechanism as described contains no first-order steps, so the net rate can in principle be increased without limit by increasing the substrate concentrations. (In most mechanisms there is at least one first-order step, and saturation arises because the chemiflux through such a step cannot be raised indefinitely by increasing the substrate concentrations.)

7.3 $v = Vab/(K_{sA}K_{sB} + K_{sA}a + ab)$ (no denominator term in b); plots of b/v against b would be parallel lines with slope $1/V$.

7.4 (a) Hyperbolic; (b) parallel straight lines with slope $1/V$. This approach has been little used, but it provides a quicker and simpler way of distinguishing between ternary-complex and substituted-enzyme mechanisms than some of the better known methods. For an application to nucleoside diphosphate kinase, see Garcés and Cleland (1969).

7.5 Competitive inhibition for all substrate–product pairs.

7.6 Uncompetitive with respect to A; competitive with respect to B.

7.7 $\phi_0 = e_0/V$; $\phi_1 = e_0K_{mA}/V$; $\phi_2 = e_0K_{mB}/V$; $\phi_{12} = e_0K_{iA}K_{mB}/V$; the lines intersect where $S_1 = -\phi_{12}/\phi_2$, $S_1/v = (\phi_1 - \phi_0\phi_{12}/\phi_2)/e_0$.

7.8 (a) There is no term in b, because as long as the three product-release steps are irreversible it is not possible to draw a King–Altman pattern that contains b but not either a or c; (b) none, because no single product can alter the need for b to be associated with a or c; (c) Q, because in the absence of P and R the concentration q can only occur in patterns that terminate at EQR, and all such patterns contain abc; (d) R, because no pattern can be drawn that contains ar.

8.1 The capacity to catalyse a half-reaction is characteristic of an enzyme that follows a substituted-enzyme mechanism. In this case the substituted enzyme is likely to be a glucosylenzyme. Competitive inhibition of exchange by glucose indicates the occurrence of a *distinct* non-covalent enzyme–glucose complex: if this were the same as the glucosylenzyme the enzyme would be a good catalyst for hydrolysis of glucose 1-phosphate.

8.2 The equation as presented may be written as $k_{1H}/k_{3H} = (k_{1H}/k_{3H})^{1.442}/(k_{2H}/k_{3H})^{1.442}$, hence $(k_{1H}/k_{3H})^{-0.442} = (k_{2H}/k_{3H})^{-1.442}$, or $k_{1H}/k_{3H} = (k_{2H}/k_{3H})^{1.442/0.442}$, thus $k_{1H}/k_{3H} = (k_{2H}/k_{3H})^{3.26}$.

8.3 Use of kinetic isotope effects depends on assuming that the isotopic substitution not only affects the rate constants for the chemical step, but also that it has no *other* effects. Chemical substitution would satisfy the first but not the second requirement.

9.1   (a) Let $y = K_m/\tilde{K}_m = (K_E + h)/(K_{EA} + h)$. Then $dy/d\ln h = h\,dy/dh = h(K_{EA} - K_E)/(K_{EA} + h)^2$, and $d^2y/d(\ln h)^2 = h\,dy/d\ln h = h(K_{EA} - K_E)(K_{EA} - h)/(K_{EA} + h)^3$, which is zero if $h = K_{EA}$, i.e. if $pH = pK_{EA}$; (b) $pH = pK_E$ (by a similar argument).

9.2   An expression such as $\log K_m$ is really a shorthand for $\log(K_m/K_m^0)$, where K_m^0 (most likely $= 1\,mM$) is an implied standard state, and $\log V$ and $\log(V/K_m)$ are to be interpreted similarly. The definitions of the implied standard states (the choice of units of measurement) affect the heights of the curves above the abscissa axis, but they have no effect on their shapes and in particular they have no effect on the pH values at which there are changes in slope. Thus the standard states can be left undefined without invalidating the determination of pK_a values.

9.3   The width at the half-height is $7.5 - 5.7 = 1.8$, and from Table 9.1 this corresponds to a difference of 1.22 between the pK values, which are thus $pK_1 = 6.6 - 0.61 = 5.99$; and $pK_2 = 6.6 + 0.61 = 7.21$. If the value of 6.1 refers to pK_{11}, then $pK_{12} = 6.64$, $pK_{22} = 7.10$, $pK_{21} = 6.56$; if it refers to pK_{22}, then $pK_{11} = 7.10$, $pK_{12} = 6.03$, $pK_{21} = 7.17$. In both cases $pK_1 + pK_2 = pK_{11} + pK_{22} = pK_{21} + pK_{12}$.

9.4   First note that the rate constants for binding and release of substrate and product can only be independent of the state of protonation if H_2E, H_2EA and H_2EP have the *same* pair of dissociation constants K_1 and K_2. It is thus unnecessary to define $f(h)$ separately for H_2E, H_2EA and H_2EP. Then $K_m = (k_{-1}/k_1) + k_2(k_3 - k_{-1})f(h)/[(k_2 + k_{-2})f(h) + k_3]k_1$. This can be independent of pH either if k_2 is very small or if $k_3 = k_{-1}$. In both cases it simplifies to $K_m = k_{-1}/k_1$.

10.1   No: an Arrhenius plot of the data shows pronounced curvature, characteristic of combined effects on two or more rate constants, rather than the straight line that would be expected for the temperature dependence of an elementary rate constant.

10.2   The details of the plot of entropy of activation against enthalpy of activation will vary with the particular numbers used, but it is likely to show an almost perfect correlation between the two parameters. However, as the numbers had no biological significance whatsoever, the supposed "correlation" cannot have one either. The problem was adapted from one suggested by McBane (1998), and his article should be consulted for more information.

11.1   Slope $= h - 1$. This plot expresses the cooperativity more directly than the Hill plot, because its slope always has the same sign as the sign of the cooperativity.

11.2 $v = [V_1 a/(K_{m1} + a)] + [V_2 a/(K_{m2} + a)]$; $dv/da = [K_{m1}V_1/(K_{m1} + a)^2] + [K_{m2}V_2/(K_{m2} + a)^2]$; $d^2v/da^2 = -[2K_{m1}V_1/(K_{m1} + a)^3] - [2K_{m2}V_2/(K_{m2} + a)^3]$. As the kinetic parameters and the substrate concentration must all be positive, this expression for the second derivative can only be negative. A plot of v against a therefore cannot be sigmoid.

11.3 $h = 1 + [a/(K_2 + a)] - [a/(K_1 + a)]$. The extreme value of $2/[1 + (K_2/K_1)^{0.5}]$ occurs when $a = (K_1 K_2)^{0.5}$.

11.4 $K_1 = K_4$. To show this, derive an expression for $y/(1 - y)$ from equation 11.15, and then find the condition for identity of the two expressions when a is negligible and when a is very large.

11.5   At low concentrations there are abundant sites for both substrate and inhibitor, and there is very little competition between them; thus the predominant effect of inhibitor is to promote the conformational change that favours substrate binding, so it appears as an activator. At high concentrations the ordinary competition predominates and so the result is the expected inhibition.

11.6   The general expression for h can be written $\{[A]/[y(1-y)]\}\,dy/d[A]$. For the specific case of equation 11.42, and writing $a = [A]/\bar{K}$, $dy/da = (c + 2a + ca^2)/(1 + 2ca + a^2)^2$. Substituting this into the expression for h and taking the case of $a = 1$, or $[A] = \bar{K}$, gives a limiting value of $h = 2/(c+1)$.

12.1 $dv/da = VhK_{0.5}^h a^{h-1}/(K_{0.5}^h + a^h)^2$. The elasticity ε_a^v is obtained by multiplying by a/v, or $\varepsilon_a^v = hK_{0.5}^h/(K_{0.5}^h + a^h)$, which has values $h/2$ at half-saturation, h at very low a, and 0 at very large a.

12.2 $C_1^J = 0.5$, $C_2^J = 0.333$, $C_3^J = 0.167$ (from equations 12.30–32).

12.3 $C_j^J = -0.15 \times (-0.25)/0.2 = 0.1875$ (from equation 12.38).

12.4   Consider a block of three enzymes with control coefficients C_1^J, C_2^J and C_3^J for some system variable J, which can be, but does not need to be, a flux. If the activity of E_1 is increased by a factor $1 + \alpha$ (with $\alpha \ll 1$) then J increases to $J(1 + C_1^J \alpha)$, and similarly for increases of the activities of E_2 and E_3. If all three are increased one after another by the same factor $1 + \alpha$ then J increases to $J(1 + C_1^J \alpha + C_2^J \alpha + C_3^J \alpha) = J[1 + \alpha(C_1^J + C_2^J + C_3^J)]$. If, however, the same three enzymes are treated as a block with control coefficient C_{123}^J for the same variable, then the same modulation would be considered to cause an increase of J to $J(1 + \alpha C_{123}^J)$. Comparing the two expressions for the same modulation it is clear that $C_{123}^J = C_1^J + C_2^J + C_3^J$, and more generally, any control coefficient for a block of enzymes is the sum of the control coefficients for the same enzymes considered separately.

12.5   A method based on overexpressing rate-limiting enzymes can only work if rate-limiting enzymes exist and remain rate-limiting when their activities are increased. Metabolic control analysis suggests not only that rate-limiting enzymes do not exist, but also that the flux control coefficient of any enzyme tends to decrease when its activity in the system is increased. Thus the strategy outlined cannot work. For more discussion of this point, see Cornish-Bowden (1995c), and for a practical example, see Niederberger and co-workers (1992).

13.1   The enzyme concentration of $1\,\text{mg/ml}$ is $1\,\text{M}/50000 = 20\,\mu\text{M}$, and that of $5\,\text{mg/ml}$ is $100\,\mu\text{M}$, both large compared with the substrate concentration of $1\,\mu\text{M}$. The two relaxations suggest a mechanism as shown in equation 13.35, and the simplest assumption to make is that $k_{-2} = 0$. Applying equations 13.28–29 (with $k_{-2} = 0$) the sum of relaxation frequencies at $20\,\mu\text{M}$ enzyme is $220 + 11 = 20k_1 + k_{-1} + k_2$ and the product is $220 \times 11 = 20k_1 k_2$; the corresponding equations at $100\,\mu\text{M}$ enzyme are $360 + 34 = 100k_1 + k_{-1} + k_2$ and $360 \times 34 = 100k_1 k_2$. Solving any three of these four equations for the three rate constants yields $k_1 \approx 2 \times 10^6\,\text{M}^{-1}\,\text{s}^{-1}$, $k_2 \approx 60\,\text{s}^{-1}$, $k_{-1} \approx 130\,\text{s}^{-1}$. As one equation is unused it is not necessary to assume $k_{-2} = 0$, but the solution is more complicated if this assumption is not made.

13.2   The data were calculated from $y = 16.37 + 46.3\exp(-0.269t) + 29.7\exp(-0.0876t)$, but the following values would be typical of the accuracy one could expect to obtain with the peeling procedure described in Section 13.4.2: $A = 17$, $B = 42.5$, $C = 33.4$, $\lambda_1 = 294\,\text{s}^{-1}$, $\lambda_2 = 97.1\,\text{s}^{-1}$.

13.3   Note that the final concentrations after mixing were the same in both experiments, and the substrate concentration was high enough to compete effectively with the inhibitor. The different time constants indicate that the slower process ($\lambda = 67\,\text{s}^{-1}$)

approximates to the release of inhibitor from the enzyme–inhibitor complex, that is to say that $k_{off} \approx 67 \, s^{-1}$. Hence $k_{on} \approx 3.3 \times 10^6 \, M^{-1} \, s^{-1}$. The observation that inhibitor release is rate-limiting in experiment (b) has no bearing on the interpretation of K_i as an equilibrium constant, which is true because it refers to a dead-end reaction (see Section 4.7.3).

14.1   Least squares: $\hat{K}_m = 1.925 \, mM$, $\hat{V} = 0.666 \, mM \, min^{-1}$; distribution-free: $K_m^* = 1.718 \, mM$, $V^* = 0.636 \, mM \, min^{-1}$. The trend in the residuals suggests that the enzyme is subject to substrate inhibition, so that it would be invalid to draw any conclusions about appropriate weights until the data have been fitted to an equation more suitable than the Michaelis–Menten equation. It would also be desirable to have considerably more than ten observations.

14.2 $\hat{K}_m = 2.747 \, mM$, $\hat{V} = 0.737 \, mM \, min^{-1}$; $K_m^* = 1.718 \, mM$, $V^* = 0.638 \, mM \, min^{-1}$ (if you obtained $K_m^* = 1.677 \, mM$, $V^* = 0.632 \, mM \, min^{-1}$ you should read the discussion in Section 14.3.5). The results, when compared with those for Problem 14.1, illustrate the general point that least-squares estimates are much more sensitive than distribution-free estimates to the presence of exceptionally poor observations.

14.3 $\hat{V} = \{\sum(v_i^3/a_i^2)\sum v_i^3 - [\sum(v_i^3/a_i)]^2\}/[\sum(v_i^3/a_i^2)\sum v_i^2 - \sum(v_i^3/a_i)\sum(v_i^2/a_i)]$;

$\widehat{V/K_m} = \{\sum(v_i^3/a_i^2)\sum v_i^3 - [\sum(v_i^3/a_i)]^2\}/[\sum v_i^3 \sum(v_i^2/a_i) - \sum(v_i^3/a_i)\sum v_i^2]$;

$\hat{K}_m = [\sum v_i^3 \sum(v_i^2/a_i) - \sum(v_i^3/a_i)\sum v_i^2]/[\sum(v_i^3/a_i^2)\sum v_i^2 - \sum(v_i^3/a_i)\sum(v_i^2/a_i)]$.

In the numerator of the expression for $\hat{V}$ both terms have the dimensions of v^6/a^2, whereas those in the denominator have dimensions of v^5/a^2, and dividing one by the other produces a quantity with dimensions of v, which is correct. Similar analysis shows the expression for $\widehat{V/K_m}$ to have dimensions of v/a and the expression for $\hat{K}_m$ to have dimensions of a.

14.4   No, it implies the opposite. Note first that weights v^3 for $1/v$ correspond approximately to weights $1/v$ for v, with a variance of about v for v, or a standard deviation of $v^{0.5}$: far from being approximately constant at small values but increasing steeply at high values, it increases steeply from the origin and tails off as v increases.

Index